PLANT CHEMICAL BIOLOGY

PLANT CHEMICAL BIOLOGY

Edited by

DOMINIQUE AUDENAERT
PAUL OVERVOORDE

WILEY

Published by John Wiley & Sons, Inc., Hoboken, New Jersey
Published simultaneously in Canada

Library of Congress Cataloging-in-Publication Data:

Plant chemical biology / edited by Dominique Audenaert, Paul Overvoorde.
p. cm.
Includes bibliographical references and index.
ISBN 978-0-470-94669-5 (hardback)
1. Botanical chemistry. I. Audenaert, Dominique, 1976– editor of compilation. II. Overvoorde, Paul, 1968– editor of compilation.
[DNLM: 1. Plants–chemistry. 2. Biochemical Phenomena. 3. Plant Physiological Phenomena. QK 861]
QK861.P525 2014
572′.2–dc23

2013035521

Printed in the United States of America

ISBN: 9780470946695

10 9 8 7 6 5 4 3 2 1

CONTENTS

PREFACE

Plant biologists have a long history of using small molecules (e.g., MW < 500 Da) to explore developmental, physiological, and metabolic pathways. Compared to molecular genetic approaches used to study biological processes in plants, the roots of plant chemical biology became well established in the first decades of the twentieth century with the discovery of plant growth regulators or plant hormones. Since this time, the capacity of organic synthetic chemists to create and characterize the effect of small molecules provided the basis for understanding processes such as cell elongation through the action of auxins, the stimulation of cell division via the activity of cytokinins, the control of fruit senescence by gibberellins, and the influence of abscisic acid on seed dormancy. In addition, naturally occurring molecules as well as synthetic analogues formed the basis for genetic screens that looked for resistant or hypersensitive mutant plants. During the past 30 years, this approach led to the discovery of cellular components and pathways that integrate environmental and developmental information into regulated phenotypic outcomes.

Plant Chemical Biology provides tools that overcome the pleiotropic effects of plant hormones. A single hormone makes defining the role of specific components or signaling pathways a challenge. The capacity to identify novel compounds that stimulate or inhibit a subset of a signaling cascade holds tremendous promise for refining our understanding of these pathways. The coupling of this potential with advances in combinatorial chemistry offers exciting opportunities to develop new tools for basic and applied applications. In recent years, the querying of large collections of diverse compounds through the use of robust assays has led to an impressive array of new discoveries in the field of plant biology.

This book intends to provide advanced students and professionals in plant biology—molecular biologists, physiologists, biochemists, or those active in agriculture, horticulture, or agronomy—a summary of key aspects of plant chemical biology. Even as the pace of progress in this subfield accelerates, we hope that you will discover in the following chapters a summary of key findings to date, as well as an overview of the available compound collections, principles of assay design, and the translatability of these new tools to applied settings. In this day and age, no volume provides an encyclopedic review. We hope that our selection of topics reveals one of the ways that the quest for understanding biological processes drives the innovation of new technologies, which in turn propel our understanding of the biology.

We thank the authors of this collection. Their expertise and experience allows each topic to be handled with insight and depth. Without their contributions and patience, this book would not have been possible. We thank our institutions, Macalester College and VIB, for their support during the preparation of this book. Finally, we acknowledge the steadfast encouragement of our spouses, Lynn Overvoorde and Eva Vermeulen, and our children, Mia, Anika, Vigdis, Sigun, and Eldrid, during this process.

Dominique Audenaert
Paul Overvoorde

CONTRIBUTORS

Frank W. An Center for the Science of Therapeutics, The Broad Institute of Harvard and MIT, Cambridge, MA, USA

Tadao Asami Graduate School of Agricultural and Life Sciences, The University of Tokyo, Bunkyo-ku, Tokyo, Japan, and CREST, Japan Science and Technology, Chiyoda-ku, Tokyo, Japan

Dominique Audenaert Compound Screening Facility, VIB, Ghent University, Ghent, Belgium

Michelle Q. Brown Department of Botany and Plant Sciences, Center for Plant Cell Biology, University of California, Riverside, CA, USA

Seth DeBolt Department of Horticulture, Agricultural Science Center North, University of Kentucky, Lexington, KY, USA

Danny Geelen Department of Plant Production, Faculty of Bioscience Engineering, Ghent University, Ghent, Belgium

Ash Haeger Department of Forest Genetics and Plant Physiology, SLU/Umeå Plant Science Centre, Umeå, Sweden

Darby Harris Department of Horticulture, Agricultural Science Center North, University of Kentucky, Lexington, KY, USA

Alan L. Harvey Strathclyde Institute of Pharmacy and Biomedical Science, University of Strathclyde, Glasgow, UK

Ken-ichiro Hayashi Department of Biochemistry, Okayama University of Science, Ridai-cho, Okayama, Japan

Yuming Hu Department of Physiology, Laboratory for Functional Plant Biology, Faculty of Sciences, Ghent University, Ghent, Belgium

Tae-Houn Kim Department of Prepharm-Med/Health Functional Biomaterials, Duksung Women's University, Seoul, Korea

Robert Klein Bayer CropScience AG, Frankfurt am Main, Germany

Erich Kombrink Department of Plant Microbe Interactions, Max Planck Institute for Plant Breeding Research, Köln, Germany

Malgorzata Łangowska Department of Forest Genetics and Plant Physiology, SLU/Umeå Plant Science Centre, Umeå, Sweden

Stephen D. Lindell Bayer CropScience AG, Frankfurt am Main, Germany

Christian Meesters Department of Plant Microbe Interactions, Max Planck Institute for Plant Breeding Research, Köln, Germany

Hans Motte Department of Plant Production, Faculty of Applied Bioscience Engineering, University College Ghent, Ghent, Belgium (Member of the Ghent University Association), and Department of Plant Production, Faculty of Bioscience Engineering, Ghent University, Ghent, Belgium

Takeshi Nakano RIKEN Advanced Science Institute, Hirosawa, Wako, Japan, and CREST, Japan Science and Technology, Chiyoda-ku, Tokyo, Japan

Paul Overvoorde Department of Biology, Macalester College, St. Paul, MN, USA

Eunjoo Park Department of Prepharm-Med/Health Functional Biomaterials, Duksung Women's University, Seoul, Korea

Jose R. Perez Center for the Science of Therapeutics, The Broad Institute of Harvard and MIT, Cambridge, MA, USA

Natasha V. Raikhel Department of Botany and Plant Sciences, Center for Plant Cell Biology, University of California, Riverside, CA, USA

Reg Richardson ChemBridge Corporation, San Diego, CA, USA

Stéphanie Robert Department of Forest Genetics and Plant Physiology, SLU/Umeå Plant Science Centre, Umeå, Sweden

Abel Rosado Department of Botany and Plant Sciences, Center for Plant Cell Biology, University of California, Riverside, CA, USA

Dominique Van Der Straeten Department of Physiology, Laboratory for Functional Plant Biology, Faculty of Sciences, Ghent University, Ghent, Belgium

Filip Vandenbussche Department of Physiology, Laboratory for Functional Plant Biology, Faculty of Sciences, Ghent University, Ghent, Belgium

Terence A. Walsh Dow AgroSciences LLC, Indianapolis, IN, USA

Stefaan Werbrouck Department of Plant Production, Faculty of Applied Bioscience Engineering, University College Ghent, Belgium, and Department of Plant Production, Faculty of Bioscience Engineering, Ghent University, Ghent, Belgium

PART ONE

INTRODUCTION

1.1

FROM HERBAL REMEDIES TO CUTTING-EDGE SCIENCE: A HISTORICAL PERSPECTIVE OF PLANT CHEMICAL BIOLOGY

Michelle Q. Brown, Abel Rosado, and Natasha V. Raikhel

1.1.1 HERBAL REMEDIES AND PHARMACOLOGY IN THE ANCIENT WORLD

Herbal medicine is the oldest form of pharmacology known to mankind, and the evolution of its applications has depended upon the availability of plants to emerging cultures over time. Many of the medicinal uses for plants have developed through observations of wild animals and by trial-and-error approaches. As cultures evolved, the acquired knowledge of the medicinal herbs became systematized in order to increase the health and productivity of these civilizations [1]. As early as 2735 B.C., the Chinese emperor Shennong wrote a herbal treatise recommending the use of Ma Huang (*Ephedra sinica*, which contains the decongestant ephedrine) against respiratory distress. The records of King Hammurabi of Babylon (1800 B.C.) prescribed the use of mint (genus *Mentha*) for digestive disorders, and multiple texts surviving from the ancient cultures of India, Mesopotamia, and Egypt illustrate the varied application of medicinal plant products, including white poppies (*Papaver somniferum*), linseed (*Linum usitatissimum*), and castor (*Ricinus communis*) oils, to treat diseases [2].

Deriving knowledge from the medical treatises and methods of those ancient civilizations and other conquered cultures such as Greeks, Etruscans, and Persians

Plant Chemical Biology, First Edition. Edited by Dominique Audenaert and Paul Overvoorde.

allowed the Roman civilization to devise one of the best and most sophisticated medical systems of the ancient world. Roman contributions to the development of chemical biology include the cultivation of medicinal herbs for commercial purposes along with the introduction of a variety of painkillers and sedatives including extracts of opium poppies (morphine) and of henbane seeds (*Hyoscyamus niger*, scopolamine), which Roman physicians tested on soldiers wounded in battle [3]. The Roman physician Galen of Pergamon (130–200 A.D.) outlined the basic principles of pharmacology by developing a systematic approach to drugs based on reasoning "logos." For each extract, a defined system of Galenic degrees of potency was used to treat varying levels of dysfunction. This approach enabled physicians and pharmacists to more precisely deliver and monitor the effects of a medicinal substance in a dosage-dependent manner [4]. For almost a thousand years, the drugs and other methods of treatment popularized by Galen were used as standard protocol in European medicine [5].

With the advent and flourishing of the Islamic culture in Europe, the Persian text Qanun (Ibn Sina, also known as Avicenna, 1025), translated into Latin as *Canon medicinae*, displaced the works of Galen and became the textbook for medical education in European universities up to 1650 [6]. The *Canon* established the basis of modern clinical pharmacology by introducing concepts such as clinical trials, randomized controlled trials, and efficacy tests laying out rules and principles for testing the effectiveness of new drugs and medications. The *Canon* was also used for medieval medical practitioners as a drug encyclopedia. It listed, with commentary, more than 700 preparations, including plant and mineral substances [7–11].

1.1.2 ALCHEMY, CHEMISTRY, AND THE ISOLATION OF THE BIOACTIVE METABOLITES

By the seventeenth century, the knowledge of herbal medicine was widely disseminated throughout Europe, and many natural remedies were dependent on plant-based substances. In that period, the study of alchemy and the quest for a "panacea," a treatment to cure all diseases as well as to prolong life indefinitely, was sought by such renowned scientists as Sir Isaac Newton, Roger Bacon, and Tycho Brahe [12]. Although the final objective was never achieved, alchemy established the experimental basis of modern chemistry and was a very important milestone for the development of extractive techniques to isolating plant compounds for medicines, dyes, and perfumes preparation [13]. Thus, as the science of chemistry matured, chemists began extracting and purifying the biologically active molecules from the natural substances that had been used as remedies for millennia. Included in this list of purifications were the isolations of the cardiac glycoside, digoxin from foxglove (*Digitalis lanata*), which was commonly used as a treatment for edema or dropsy [14], the addictive analgesic morphine from the seed capsules of poppy flowers [15], and the anti-inflammatory salicin from bark of willow (genus *Salix*) trees [16] (Fig. 1.1.1 Left Panel).

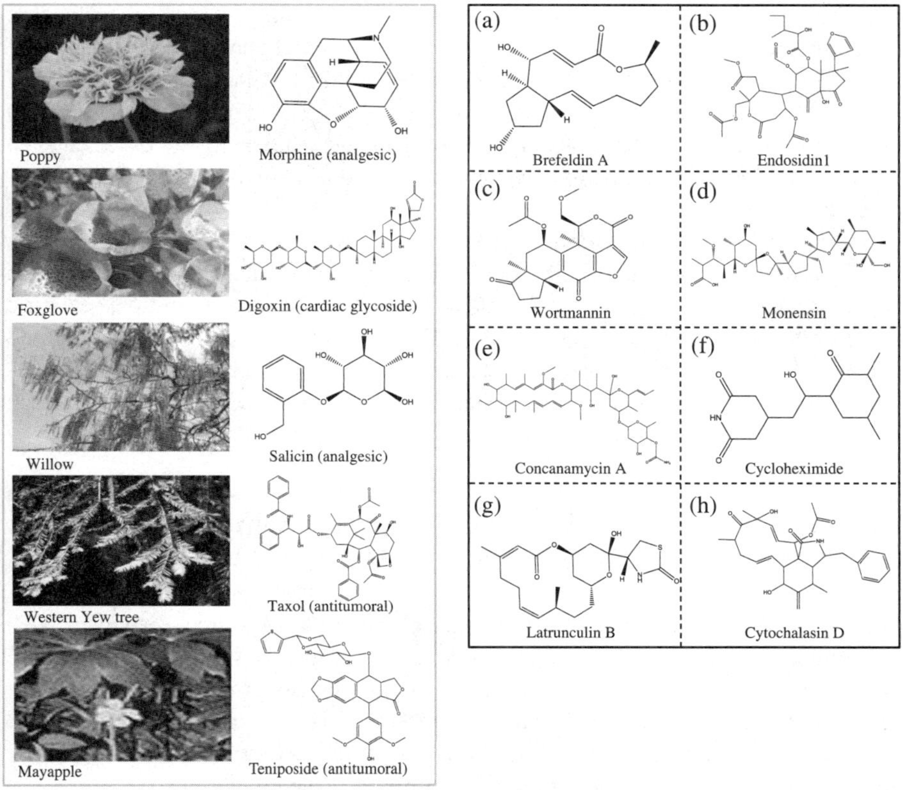

FIGURE 1.1.1 Many plant metabolites have been utilized as pharmaceuticals, such as morphine, digitalis, and taxol. Several compounds have been utilized in cell biology research. These include compounds that inhibit protein trafficking (**a**)–(**e**), protein synthesis inhibitor (**f**), and cytoskeleton-inhibiting drugs (**g**)–(**h**). For color detail, please see color plate section.

With increased purity and knowledge of the active plant extracts and the advances in pharmacology, the first purely synthetic drugs based on natural products were elaborated in the middle of the nineteenth century. For example, in 1839, studies of the willow bark extract showed that salicylic acid was the active ingredient responsible for its analgesic properties [17]. In 1853, chemist Charles Frédéric Gerhardt mixed acetyl chloride with sodium salicylate to produce synthetic acetylsalicylic acid for the first time [18]. Academic chemists established the compound's chemical structure and devised more efficient methods of synthesis that finally led to Felix Hoffman's formulation of acetylsalicylic acid named Aspirin®, which helped to found the pharmaceutical industry in 1899 and remains the most widely used synthetic drug today [19].

In parallel with the isolation and synthesis of bioactive metabolites derived from plants, the search for antibiotics began in the late 1800s, with the growing acceptance of the germ theory of disease, a theory that linked microbes to the causation of a variety of illnesses [20]. Initially, antibiotics were identified as natural substances released by moss and fungi to inhibit growth of other organisms. Many extracts with antibiotic properties were reported in the literature during the late 1800s. Perhaps the most significant milestone in antibiotic research was the identification and isolation of the antibiotic penicillin from the *Penicillium* mold [21]. The subsequent development by Sir Howard Florey in 1942 of a process for manufacturing penicillin G has enabled the use of this small molecule to have more positive impact on human health than any other natural substances [22].

The technological innovations that led to the isolation of bioactive compounds toward the end of the nineteenth century impacted more than the area of human health. In fact, these successes set the stage for the development of multiple scientific disciplines, including plant biology. Out of the capacity to identify hormonal and nutritional requirements for plant growth and new ways of applying small molecules to manipulate and interrogate plant systems, the field of plant physiology emerged. Aspects of these advances are highlighted in following sections.

1.1.3 THE DISCOVERY OF PHYTOHORMONES AND THE FOUNDATION OF MODERN PLANT CHEMICAL BIOLOGY

Since antiquity, people have sought to explain plant physiological responses such as gravitropism and touch sensitivity via mechanical or passive physiochemical forces. In this context, Aristotle's conclusion that plants were insensitive and lack active motion against external stimuli pronounced in the third century B.C. was still prevalent and widely used by eighteenth-century taxonomists to classify and differentiate plants from other organisms within the graduated scale of nature [23]. The 1880 publication of *The Power of Movement in Plants* by Charles Darwin with assistance from his son Francis was a landmark attempt to integrate plant movements into a biological perspective of behavior by introducing the concepts of plant sensitivity and active perception of external stimuli [24]. The recognition that plants might respond to the environment in much the same way as animals was a paradigm shift for taxonomists and plant physiologists and became the seminal idea that led to the discovery of the first plant growth factor, the auxin phytohormone indole-3-acetic acid (IAA).

The Darwins used experiments on gravi- and phototropism to show that some internal signal or "influence" moved from coleoptile tips of the annual canary grass (*Phalaris canariensis*) to control cell elongation after stimulation [24]. The suggestion of such signaling molecules led Hans Fitting (1909) to introduce the term "hormone" when referring to substances that stimulate specific plant developmental processes [25]. In 1913, Danish botanist P. Boysen Jensen suggested that the phototropic signal described by the Darwins was indeed a chemical messenger moving down from the tip of the coleoptile [26]. During the early 1920s, several botanists provided circumstantial evidence of the presence of such a chemical in different plants, and finally in 1928, the Dutch physiologist Frits Went defined an experimental procedure for isolating this chemical and quantifying its physiological activity [27]. A next important step concerned the purification of sufficient amounts of this chemical for analytical purposes.

Five years of attempts made by K. Thimann, and J. Haagen-Smit among others, finally succeeded, and the Darwinian "influence" was identified as IAA [28, 29].

The identification of IAA as the growth factor responsible for the gravitropic and phototropic responses crowned a half century of research that revealed the use of chemical signals in plants and opened the door to the development of compounds that could contribute to modern agriculture by stimulating growth or eliminating weeds.

1.1.4 THE DAWN OF PLANT SYNTHETIC CHEMISTRY

The implementation of synthetic and combinatorial chemistry in the agrochemical industry coincided with the development of the modern pharmaceutical industry. Both of these fields were fueled by the diversity and increased availability of organic compounds derived from coal tar, plants, and microorganisms. In this context, pharmacological studies by P. Ehrlich (1913) led to the receptor theory of small-molecule action, and receptors were defined as specific body constituents where the active drugs selectively bind [30]. After the identification of IAA as a plant growth factor, the 1940s witnessed the creation of multiple academic and industrial laboratories in which detailed structure–activity analysis along with combinatorial chemistry led to the development of novel compounds that mimic the activity of the endogenous hormone or that could function as herbicides. As a result, a wide array of IAA analogs and derivatives including 1-naphthaleneacetic acid (1-NAA), 2-methyl-4-chlorophenoxyacetic acid (MCPA), and 2,4-dichlorophenoxyacetic acid (2,4-D) were synthesized. Further research demonstrated that these derivatives elicited the same type of plant responses as IAA such as elongation in aerial tissues and stimulated lateral root formation but with activity that lasted longer and was more potent than the endogenous molecule because of their relatively slower metabolism by plants [31–33].

With increased concentration, stability, and activity in plant tissues, synthetic auxins gained considerable importance for probing auxin function in basic research as well as for applied civil and military uses. Examples of civil uses range from the commercialization of the synthetic auxin 1-NAA marketed as "Hortomone A" to stimulate lateral rooting in crops [34] to the commercial elaboration in 1945 of the first systemic herbicide based on synthetic auxins marketed as "Weedone," a product that started a new era of weed control in modern agriculture [35]. In parallel, the development of synthetic auxins for military purposes was fast-tracked during World War II, and early investigations regarding the use of synthetic auxins were kept secret because these novel herbicides were deemed putative bioweapons [36]. It was not until the Vietnam Conflict nearly twenty years later, when hormone-based weed killers would become infamous due to the extensive use of the defoliant *Agent Orange*, an equal mixture of 2,4-D and 2,4,5-trichlorophenoxyacetic acid, as part of the U.S. herbicidal warfare program. The use of *Agent Orange* in the Vietnam War caused the loss of millions of acres of upland and mangrove forests and crops, and its side effects on humans and the environment awakened interest in the effects of synthetic chemicals on the natural environment [37].

In the post-World War II era, the excitement caused by the synthesis of the novel antibiotic penicillin and its impact for human health, the increased crop production after the use of synthetic herbicides, and the development of affordable synthetic chemicals stimulated the search for more natural and synthetic substances derived from plants and microorganisms. These advances had profound implications in the field of plant chemical biology because multitudes of small molecular probes became available to explore biological processes without the need for stringent pharmacokinetic or safety profiles required for medicinal applications. Many compounds such as cycloheximide, Brefeldin A, cytochalasins, and latrunculins remain powerful and much used probes that enable answers to basic and applied questions to be found.

1.1.5 SERENDIPITY VERSUS RATIONAL DESIGN OF CHEMICAL LIBRARIES

By the end of the 1940s, the use of natural and synthetic compounds as probes to modulate biological pathways in plants was extended, and synthetic organic chemistry had progressed to the point where the large-scale preparation of synthetic analogs of bioactive molecules such as antibiotics and plant hormones was economically feasible. At the same time, the introduction of powerful commercial spectrometers (i.e., nuclear magnetic resonance, 1952) and separation techniques (i.e., high-performance liquid chromatography, 1961) for determining the structures of minute quantities of biologically active natural products boosted the identification and refinement of natural drugs through structure–activity relation analysis. Thus, the resulting "methyl, ethyl, isopropyl" analog model for drug discovery, although empirical, was quite successful for several decades. These collections of natural compound and their synthetic derivatives were, however, limited by two factors. First, the basis of drug discovery relied heavily on the serendipity of finding biologically active products from a relatively small number of plant species, and second, these collections contained a limited number of compounds representing only a minute fraction of the structural diversity possibilities in the chemical space [38].

For the well-resourced biomedical community, the development of computer tools, and concurrent advances in structural biology, chemistry, and protein crystallography in the 1980s, a shift to a "rational" computational archetype for drug discovery emerged, and computational drug discovery was born [39]. The assumption underpinning this approach was that a continuous stream of new protein structures and associated active sites and binding pockets would provide the basis for the computational-assisted rational drug design. Limitations in the crystallographic techniques such as the difficulty in generating protein crystals, the slow pace of protein structure determination, and the time-consuming experimental validation of the candidate molecules have limited the number of successes obtained by this approach. In addition, for plant biologists, limitations to bioavailability due to the protective nature of the cell wall and rapid xenobiotic metabolism further tempered enthusiasm for this approach and led to a return of largely empirical methods of small-molecule library synthesis and high-throughput screening (HTS). Technological advances in robotics and biological techniques for simultaneously assaying thousands of compounds in concert during the 1990s fueled this paradigm shift.

1.1.6 THE DEVELOPMENT OF COMBINATORIAL CHEMISTRY

After the partial drawback of the rational drug design, chemists and biologists realized that large collections of structurally distinct molecules should be generated to perform efficient pharmacological studies and therefore, the traditional synthesis and purification of one compound at a time for screening was no longer a feasible approach. An important step toward increasing the diversity and the speed in which structurally related compounds were generated was the application of combinatorial chemistry techniques involving systematic mixing and matching of various chemical building blocks to generate libraries of small molecules. The development of these techniques, running in parallel with advances in automation and robotics, enabled organic chemists Furka and Lam to apply Merrifield's solid-phase synthesis of peptides to the combinatorial synthesis of diverse sets of peptides on solid phase using mix and split reactions pioneered by them [40–42]. The success of combinatorial chemistry to generate libraries of peptides and peptidemimetics positively impacted pharmacological studies because this approach involved iterative reactions, was amenable to automation, and allowed detritus from the products to be removed by washing the resins.

Early combinatorial chemistry techniques successfully generated enormous numbers of compounds; still their "diversity" was limited to variations in appendages attached to a small number of common skeletons and the compounds frequently lacked chirality. Since most of the small molecules known to disrupt biological systems were structurally complex natural products, the organic chemist Stuart Schreiber reasoned that maximizing the number of structures and scaffolds by increasing the size and number of rigidifying elements (macrocycles, polycycles, olefins, etc.) in small-molecule libraries would be the most efficient way to populate the largest amount of chemical space. Hence, the creation of smaller diversity-oriented library of compounds containing wider spectra of molecular shapes would provide both more hits and a wider range of biological targets than the traditional combinatorial libraries [43].

As the number and diversity of characterized bioactive synthetic compounds increased, the biochemist Chris Lipinski and his colleagues identified the general properties and structural features of molecules that had strong biological activity [44]. What they observed, often referred to as the "Lipinski Rule of Five," is that molecules that are likely to be orally active drugs in humans meet four criteria—they contain not more than 5 hydrogen bond donors, they contain not more than 10 hydrogen bond acceptors, they are less than 500 daltons, and they have $\log P_{oct/wat}$ less than 5. These guidelines were quickly adopted by the pharmaceutical industry as an early filter for the development of orally active drugs. For plant biologist, these same rules apply for agrochemicals [45, 46].

By the end of the millennia, an important limitation to expanding and defining the possibilities of combinatorial chemistry for basic plant research was the closed environment of the biotechnology and pharmaceutical industries in which most of the research projects were carried out. In this context, a milestone for the development of plant chemical biology as a subject for basic science was the creation of publicly available HTS facilities such as the Harvard's Institute of Chemistry and Cell Biology (1998), whose highly successful Investigator-Initiated Screening Program facilitated the development of small-molecule screening projects for basic research and oriented the pursuit of chemical biology as an academic discipline. These efforts were also supported by the availability of libraries from both academic (e.g., the LATCA library) and commercial libraries that continue to grow in size and complexity.

1.1.7 PLANT CHEMICAL BIOLOGY IN THE -OMICS ERA

As chemical biology was gaining prominence as an academic discipline, plant biologists were also gaining a new host of molecular and genetic tools allowing plant genetics to undergo its own development into the age of genomic research. The release of the complete sequence of the model plant *Arabidopsis thaliana* (2000), together with many unique genetic/genomic tools developed in this plant system such as whole-genome microarrays, a large collection of knockout and activation-tagged mutations, a rich array of mutants, and the ease of cloning a gene by map-based cloning aided by DNA microarrays, allowed the integration of chemical screens at various biological levels ranging from subcellular to the whole organism.

The development of sophisticated imaging techniques such as confocal laser scanning microscopy (CLSM) that allows for greater resolution than traditional fluorescence microscopy and that can generate a 3 D representation of the specimen, coupled with the development of plant-specific probes such as engineered green fluorescent protein (GFP), originally isolated from the jellyfish *Aequorea victoria*, fused to the now sequenced *Arabidopsis* genes, allowed researchers to view the dynamic, *in vivo* subcellular localization of proteins instead of the static state of fixed specimens [47]. Thus, rather than have to wait for growth or gross morphological plant phenotypes, plant researchers could now observe subcellular phenotypes caused by small molecules based on the aberrant localization of proteins in real time. The successes of this approach are documented in other chapters in this book. Collectively, the benefits of obtaining novel insights by using chemical tools are reflected in the growing number of publications during the past two decades that use chemical biology approaches for plant research (Fig. 1.1.2).

1.1.8 INTRODUCTION OF BIOINFORMATICS AND CHEMINFORMATICS

The deluge of information generated by the -omics analyses could only be managed with the help of computers and their programmers, and thus, the field of bioinformatics was born. Facilitating the growth of bioinformatics were the Internet-accessible public

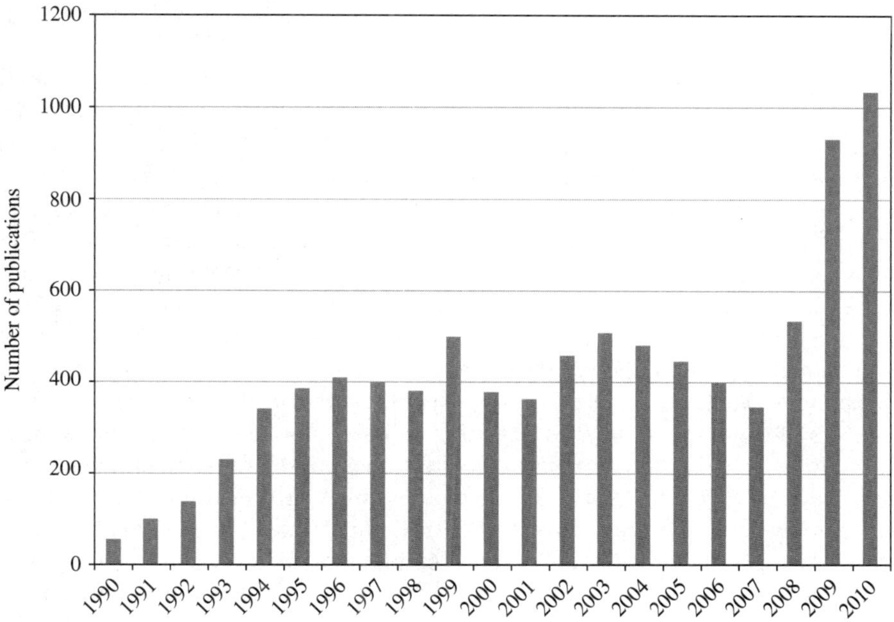

FIGURE 1.1.2 Evolution of the number of publications using chemical genomic approaches in the plant model system *Arabidopsis thaliana*. *Source*: National Center for Biotechnology Information (NCBI).

repositories, such as the National Center for Biotechnology Information (NCBI). At first, bioinformatics merely managed the wealth of sequences and structures. Algorithms to analyze homology of gene sequences, such as Basic Local Alignment Search Tool (BLAST), were necessary to navigate the over 100 million sequence records contained within GenBank at NCBI. This helped to identify the members of large gene families and genes similar in sequence to ones with known function in other sequenced organisms, such as yeast. Bioinformatics has helped identify common promoter motifs among genes that promote both brassinosteroid and auxin pathways [48]. Bioinformatics also provides a critical tool in the field of comparative genomics. Now, with a wealth of analysis and modeling algorithms and tools available, bioinformatics represents just a portion of the field of computational biology.

As high-throughput chemical biology screens continue to be performed, the need for new tools and data repositories resulted in the formation of cheminformatics. The formation of the NCBI's PubChem open-access database in 2004 was initially a cause of concern for the American Chemical Society Chemical Abstract Service (CAS) Registry, which maintained a subscription-based comprehensive database of compounds dating back to the early 1900s. Since the inception of PubChem, however, several publicly accessible compound databases have been established (e.g., ChemDB [49], DrugBank [50], ChemmineR [51]). In these databases, 3 D compounds are described using the simplified molecular input line entry specification (SMILES).

These descriptors along with physiochemical descriptors can be mined for structural and activity relationships among compounds.

Future chemical biology approaches will likely involve better and more efficient computational starting points. While a little more than 10 million pure compounds are described in chemical databases, the potential chemical diversity (defined as the number of unique chemical structures) of compounds composed of carbon, hydrogen, nitrogen, oxygen, sulfur, phosphorous, and the halogens (the organic chemist's periodic table) of molecular weight less than 1000 likely exceeds 10^{60} [52]. However, in a time when Moore's Law (1963), which states that computer processing speed doubles every two years, still prevails, it is not utopist to think that in the near future the virtual vastness of chemical spaces could be overcome with enough computing power. The virtual screening of compounds is extremely appealing because it saves the time and expense of physically screening extensive numbers of compounds. *In silico* screening requires either the identification of the ligand to be bound or the specifications of the docking site for competitive inhibition. Docking programs that generate the possible conformations of interaction between ligand and docking site are of critical importance. Maintaining the open-access example of the databases, many docking programs such as AutoDock [53] and ZINC [54] are publicly accessible. *In silico* screening allows an enrichment of putatively active compounds based on the ligand or docking site. These compounds then need to be actively assayed to determine actual activity. To demonstrate the utility of this approach, Doman et al. [55] in a screen to identify novel inhibitors of tyrosine phosphatase 1B utilized both *in silico* chemical genomics and HTS. The *in silico* program determined 365 putative hits and 127 of these were confirmed, for a hit rate of 34.8%. The HTS screen of over 400,000 compounds produced only 85 hits, for a hit rate of 0.021%. With *in silico* applications, however, results need to be verified with wet lab experiments when the targets are known. Without a directed chemical screen, compound targets must be identified through mutational analysis and mapping or biochemical means.

1.1.9 UNIQUE CHALLENGES IN THE FIELD OF CHEMICAL GENOMIC RESEARCH

Plant chemical genomics, defined as the use of small molecules to study the functions of the plant cell at the genome level, is a developing field that provides valuable tools and novel perspectives for plant research. However, the use of small molecules in plant systems is associated with a unique set of challenges. Still unresolved is the synthesis of enough quantities of compound required for larger-scale studies and structure–activity relation analyses. In some cases, obtaining sufficient supply of a particular bioactive compound can be difficult when dealing with natural metabolite compounds too difficult to synthesize in a laboratory and found in minimal concentrations in nature [56]. Also, the compound isolation from natural sources can provoke intellectual property and environmental conservation arguments for regionally endogenous and endangered species. To cope with these challenges, the future chemical genomics will undoubtedly have to incorporate and streamline novel technologies for increased synthesis, utility, and understanding in a sustainable and environmentally friendly setup, as well as to implement efficient intellectual property protocols.

A challenge of a different nature is derived from the plant system itself. Being sessile organisms, plants have adapted to deal with a multitude of environmental and toxicological stresses. These adaptations can help a plant to habituate to a compound over time at the cellular level, as in the case of Brefeldin A [57], and at the gross morphology level, as evidenced by the regeneration of shoots in non-hormone-containing media from *calli* of *Nicotiana bigelovii* habituated to the herbicide 2,4-D [58]. However, sometimes it is not always the plant metabolic pathways that are altered, but the structure of the compound. Chemical compounds can act as precursor molecules, metabolizing into the smaller bioactive molecule, or be modified by the plant's metabolism *in vivo*. Two compounds that are now known to be modified *in planta* are sirtinol and hypostatin. Sirtinol was originally identified in a yeast screen as an inhibitor of the Sirtuin family of nicotinamide adenine dinucleotide (NAD)-dependent deacetylases [59], was later characterized as an activator of auxin-inducible genes through the degradation of negative auxin regulators, and finally determined to be converted *in vivo* to an active auxin that is transported in cells through a polar-independent pathway [60, 61]. Hypostatin was originally identified as a cell expansion inhibitor and found to be glycosylated by the UDP-glycosyltransferase HYPOSTATIN RESISTANCE 1 (HYR1) *in vivo*. These examples underscore the need to ascertain and verify *in vivo* the compound responsible for the biological phenotypes observed.

1.1.10 CONCLUDING REMARKS

Despite the unique challenges in the field of chemical biology, the use of small compounds to circumvent the problems of gene redundancy and lethality encountered by classical genetics has proven to be a useful approach in interrogating biological systems. The discipline has always been an interface between chemistry and biology, maturing as advances in the separate fields occurred and eventually including computer science through cheminformatics, *in silico* screening, and docking programs. As the future of chemical genomics is pondered, it is easy to imagine the continued incorporation of other scientific disciplines as well.

In contrast to classical biologists, taking a reductionist view of the plant systems and reducing them to basic parts and pathways, the new breed of plant system biologists look for the integration of the whole organism into a complex network of pathways, with the belief that unique properties can only be observed through the study of the whole system. The next natural step would be to incorporate chemical genomics into the systems biology paradigm as it has been done in study with yeast [62]. Thus, chemical genomics will continue its interdisciplinary journey, including various biological fields such as genomics, epigenomics, metabolomics, and phenomics, with chemistry and computer sciences.

ACKNOWLEDGMENTS

We would like to thank Dr. Sean Cutler for critical reading of the manuscript. This work was supported by the National Science Foundation (DGE-0504249 to M.Q.B.) and by the DOE/BES Physical Biosciences program (DE–FG02–02ER15295 to A.R. and N.V.R.)

REFERENCES

1. Saad B and Said O (2011) *Greco-Arab and Islamic Herbal Medicine: Traditional System, Ethics, Safety, Efficacy, and Regulatory Issues*. John Wiley and Sons, Hoboken, NJ.
2. Alam A et al. (2008) Green Pharmacy: An Innovative Pharmaceutical Emergence. Medvarsity Online Limited. Available at http://www.medvarsity.com/ejournals/18_5_08_article.html. Accessed July 29, 2013.
3. Medicine and Surgery in the Ancient Rome. Available at http://www.crystalinks.com/romemedicine.html. Accessed April 10, 2011.
4. Debru A. (1997) *Galen on Pharmacology: Philosophy, History, and Medicine: Proceedings of the Vth International Galen Colloquium, Lille, 16–18 March 1995*. Brill Academic Publishers, Boston, MA.
5. Porter R et al. (1995) *The Western Medical Tradition: 800 BC to AD 1800*. Cambridge University Press, Cambridge, UK.
6. Ashoor S. (1984) Islamic Medicine in the European Universities. The Bulletin of the 3rd International Conference on Islamic Medicine. Islamic Organization for Medical Sciences, Kuwait City, Kuwait, 77–104.
7. Jacquart D (2008) Islamic pharmacology in the Middle Ages: Theories and substances. *European Review*, *16* (2): 219–227.
8. Craig BD and Daly WJ (2000) Clinical pharmacology in the Middle Ages: Principles that presage the 21st century. *Clinical Pharmacology & Therapeutics*, *67* (5): 447–450.
9. Eldredge JD (2003) The randomised controlled trial design: Unrecognized opportunities for health sciences librarianship. *Health Information and Libraries Journal*, *20*: 34–44.
10. Craig BD and Daly W (2000) Medieval contributions to the search for truth in clinical medicine. *Perspectives in Biology and Medicine*, *43* (4): 530–540.
11. Langermann YT (ed.) (2010) *Avicenna and his Legacy. A Golden Age of Science and Philosophy*. Brepols Publishers, Turnhout, Belgium.
12. Williams HS and Williams EH (2009) *A History of Science (Volume 2) The Beginnings of Modern Science*. General Books LLC Publishers, Memphis, TN.
13. Newman WR and Principe LM (1998) Alchemy vs. chemistry: The etymological origins of a historiographic mistake. *Early Science and Medicine*, *3*: 32–65.
14. Withering W (1785) *An Account of the Foxglove and Some of its Medical Uses: With Practical Remarks on Dropsy and Other Diseases*. J and J Robinson, London.

15. Sertürner FW (1806) Darstellung der reinen Mohnsäure (Opiumsäure) nebst einer Chemischen Untersuchung des Opiums mit vorzüglicher Hinsicht auf einen darin neu entdeckten Stoff und die dahin gehörigen Bemerkungen. *Journal der Pharmacie fuer Aerzte und Apotheker, 14*: 47–93.

16. Leroux H (1830) Découvert de la salicine. *Journal de Chimie Médicale, 6*: 341.

17. Piria R (1839) Untersuchungen über das Salicin und die daraus entstehenden Producte. *Annalen der Pharmacie, 29*: 300–306.

18. Gerhardt C (1853) Untersuchungen über die wasserfreien organischen Säuren. *Annalen der Chemie und Pharmacie, 87*: 149–179.

19. Andermann AAJ (1996) Physicians, fads and pharmaceuticals: A history of aspirin. *McGill Journal of Medicine, 2*: 115–120.

20. Beale LS (1870) *Disease Germs: Their Supposed Nature, an Original Investigation, with Critical Remarks*. J. Churchill, London.

21. Fleming A (1929) On the antibacterial action of cultures of a penicillium with special reference to their use in isolation of *B. influenzae*. *British Journal of Experimental Pathology, 10*: 226–236.

22. Mailer JS and Mason B. Penicillin. Medicine's Wartime Wonder Drug and Its Production at Peoria, Illinois. Available at http://www.lib.niu.edu/2001/iht810139.html. Accessed April 10, 2011.

23. Stafleau FA (1971) *Linnaeus and the Linnaeans: The Spreading of Their Ideas in Systematic Botany, 1753–1789*. Oosthoek, Utrecht.

24. Darwin C and Darwin F (1880) *The Power of Movements in Plants*. John Murray Publishing, London.

25. Fitting H (1909) Die Beeinflussung der Orchideenbluten durch die Bestaübing und durch andere Umstände. Ein entwicklunsphysiologische Studie aus den Tropen. *Zeitschrift für Botanik, 1*: 1–86.

26. Boysen Jensen P (1913) Über die Leitung des phototropischen Reizes in der Avenakoleoptile. *Berichte des Deutschen Botanischen Gesellschaft, 31*: 559–566.

27. Went FW (1928) Wuchstoff and Wachstum. *Recueil des travaux botaniques neerlandais, 25*: 1–116.

28. Thimann KV and Kipefli JB (1935) Identity of the growth-promoting and root-forming substances of plants. *Nature, 135*: 101–102.

29. Haagen Smit AJ and Went FW (1935) A physiological analysis of the growth substance. *Proceedings Koninklijke Akademie van Wetenschappen the Amsterdam, 38*: 852–857.

30. Ehrlich P (1913) Address in Pathology, ON CHEMIOTHERAPY: Delivered before the Seventeenth International Congress of Medicine. *British Medical Journal, 2* (2746): 353–359.

31. Cobb AH (1992) Auxin-type herbicides. In *Herbicides and Plant Physiology*. Chapman and Hall, London, 82–106.

32. Sterling TM and Hall JC (1997) Mechanism of action of natural auxins and the auxinic herbicides. In *Herbicide Activity: Toxicology, Biochemistry and Molecular Biology*. Eds. Roe RM, Burton JD, and Kuhr RJ. IOS Press, the Netherlands, 111–141.

33. Fedtke C and Duke SO (2005) Herbicides. In *Plant Toxicology*. Eds. Hock B and Elstner EF. Marcel Dekker, New York, 247–330.

34. Taylor FS (1940) *Science Front 1939*. Ayers Publishing, Trenton, FL.
35. Troyer JR (2001) In the beginning: The multiple discovery of the first hormone herbicides. *Weed Science*, *49*: 290–297.
36. Martin JW, Christopher GW, and Eitzen EM (2007) History of biological weapons: From poisoned darts to intentional epidemics. In *Medical Aspects of Biological Warfare*. Ed. Dembek ZF. The Borden Institute, Washington, DC, 1–20.
37. Kolko G (1994) *Anatomy of a War: Vietnam, the United States, and the Modern Historical Experience*. Panteon Publishing, New York.
38. A Short Story of Drug Discovery. Available at www.pharmsci.uci.edu/history.php. Accessed April 10, 2011.
39. Craig PN (1980) Guidelines for drug and analog design. In *Burger's Medicinal Chemistry; Part 1, The Basis of Medicinal Chemistry*, 4th ed. Ed. Wolff ME. Wiley, New York, 331–348.
40. Merrifield RB (1963) Solid phase peptide synthesis. 1. Synthesis of a tetrapeptide. *Journal of the American Chemical Society*, *85* (14): 2149–2154.
41. Lam KS et al. (1991) A new type of synthetic peptide library for identifying ligand-binding activity. *Nature*, *354*: 82–84.
42. Furka A (1995) History of combinatorial chemistry. *Drug Development Research*, *36*: 1–12.
43. Schreiber SL (2000). Target-oriented and diversity-oriented organic synthesis in drug discovery. *Science*, *287* (5460): 1964–1969.
44. Lipinski CA et al. (2001) Experimental and computational approaches to estimate solubility and permeability in drug discovery and development settings. *Advanced Drug Delivery Reviews*, *46* (1–3): 3–26.
45. Tice CM (2001) Selecting the right compounds for screening: does Lipinski's Rule of 5 for pharmaceuticals apply to agrochemicals? *Pest Management Science*, *57*: 3–16.
46. Lindell SD, Pattenden LC, and Shannon J (2009) Combinatorial chemistry in the agrosciences. *Bioorganic Medicinal Chemistry*, *17* (12): 4035–4046.
47. Cutler SR et al. (2000) Random GFP: cDNA fusions enable visualization of subcellular structures in cells of Arabidopsis at a high frequency. *Proceedings of the National Academy of Sciences*, *97* (7): 3718–3723.
48. Nemhauser JL, Mockler TC, and Chory J (2004) Interdependency of brassinosteroid and auxin signaling in Arabidopsis. *Public Library of Science Biology*, *2* (9): E258.
49. Chen J et al. (2005) ChemDB: A public database of small molecules and related chemoinformatics resources. *Bioinformatics*, *21* (22): 4133–4139.
50. Wishart DS et al. (2006) DrugBank: A comprehensive resource in silico drug discovery and exploration. *Nucleic Acids Research*, *34*: D668–D672.
51. Cao Y et al. (2008) ChemmineR: A compound mining framework for R. *Bioinformatics*, *24* (15): 1733–1734.
52. Bohacek RS, McMartin C, and Guida WC (1996) The art and practice of structure-based drug design: A molecular modeling perspective. *Medicinal Research Reviews*, *16*: 3–50.
53. Goodsell DS, Morris GM, and Olson AJ (1996) Automated docking of flexible ligands: Applications of AutoDock. *Journal of Molecular Recognition*, *9* (1): 1–5.
54. Irwin JJ and Shoichet BK (2005) ZINC – A free database of commercially available compounds for virtual screening. *Journal of Chemical Information Modelling*, *45*: 177–182.

55. Doman TN et al. (2002) Molecular docking and high-throughput screening for novel inhibitors of protein tyrosine phosphatase-1B. *Journal of Medicinal Chemistry*, *45* (11): 2213–2221.

56. Ortholand JY and Ganesan A (2004) Natural products and combinatorial chemistry: Back to the future. *Current Opinion in Chemical Biology*, *8* (3): 271–280.

57. Grebe M et al. (2003) *Arabidopsis* sterol endocytosis involves actin-mediated trafficking via ARA6-positive early endosomes. *Current Biology*, *13*: 1378–1387.

58. Bennici A and Bruschi P (1999) Stability of habituation in *Nicotiana bigelovii* var. *bigelovii* shoots regenerated from habituated callus. *Plant Science*, *143* (2): 195–199.

59. Grozinger CM et al. (2001) Identification of a class of small molecule inhibitors of the sirtuin family of NAD-dependent deacetylases by phenotypic screening. *Journal of Biological Chemistry*, *276*: 38837–38843.

60. Zhao Y et al. (2007) Chemical genetic interrogation of natural variation uncovers a molecule that is glycoactivated. *Nature Chemical Biology*, *3* (11): 716–721.

61. Dai X et al (2005) Genetic and chemical analyses of the action mechanisms of sirtinol in Arabidopsis. *Proceedings of the National Academy of Sciences of the United States of America*, *102*: 3129–3134.

62. Hillenmeyer ME et al. (2011) Systematic analysis of genome-wide fitness data in yeast reveals novel gene function and drug action. *Genome Biology*, *11* (3): R30.

PART TWO

SOURCES OF SMALL MOLECULES

2.1

COMPOUND COLLECTIONS

REG RICHARDSON

2.1.1 INTRODUCTION

Discovery science imposes constantly changing targets and responsibilities on those trying to create future agrochemical materials. This chapter reviews the information and chemical libraries that are available for use in this environment and for the sourcing and selection of such compounds destined to be screened as those potential new agents.

While the cost of getting new chemical entities (NCEs) to the pharma market continues to grow, it is now estimated to cost \$1 billion (Fig. 2.1.1) to achieve this; that within the crop science arena is modest at approximately \$250 million. That said, the marketplace for crop science materials is not close to that of human pharmaceuticals, and while financially comparable "blockbusters" do exist within the crop science space, the definition of a blockbuster here is generally accepted as a substance attracting sales of \$200 million p.a. or more.

Glyphosate, *N*-(phosphonomethyl)glycine, commercialized in the United States in 1976 [1], is somewhat atypical as it remains the only crop science agent attracting sales in excess of \$1 billion p.a. (last known figure was \$4.95 billion p.a.) followed by the insecticide imidacloprid (Bayer) at \$950 million, the fungicide azoxystrobin (Syngenta) at \$910 million, and "old school" products such as paraquat (*N*,*N*′-dimethyl-4,4′-bipyridinium dichloride), discovered in 1882 but only developed as a herbicide in the 1950s, showing respectable revenues around \$650 million p.a. With revenues generally considered to be modest at best, the costs applied to crop science research are set in accordance with this; funding is only a fraction of research and development expenditures in pharma and discoveries need to be inexpensive to develop and inexpensive to manufacture.

Plant Chemical Biology, First Edition. Edited by Dominique Audenaert and Paul Overvoorde.

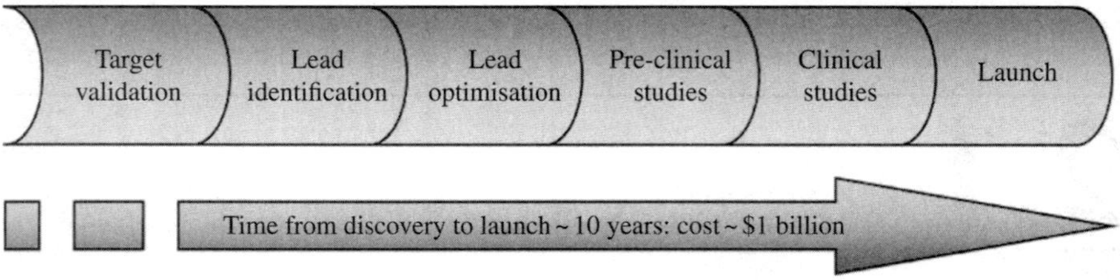

FIGURE 2.1.1 The cost of advancing an NCE to the pharma marketplace.

The remit of research and development in the agro-industry, as with pharmaceutical research and development, is that in order to discover new agents, NCEs are required. These compounds need to be designed and synthesized or sourced and acquired and then screened and evaluated. External compound suppliers offer a significant number of potential new screening candidates that can supplement the work of company scientists. With the number of theoretically possible small molecules conservatively estimated to be 10^{60} compounds and with a reported 60,000,000–120,000,000 [2] being available on the shelves of screening facilities worldwide, it is easy to see why research organizations are attracted to external sources for novel chemical samples. With funding for the sourcing of external materials on the increase, the use of these external sources of NCEs will grow in line with demand.

Commercial screening chemical suppliers provide a source of constantly changing stock acquired from a multitude of sources including academic institutions or developed through an internal chemistry resource. A growing number of providers have made available an interesting pool of generally nonexclusive small molecules; others ensure exclusivity by in-house design and synthesis or by isolation of previously unreported products from natural sources. Some larger providers are able to approach the business from multiple directions.

Internal compound collections within the crop science industry very often reflect the research history of that company's agrochemical focus. As a result, legacy collections may be narrow in diversity and even narrower in substructure. The current providers of commercial libraries are able to supply millions of previously unreported and, more importantly, untested compounds that have novel structures and that explore new regions of chemical and pharmacophore space. The number of commercially available compounds is small compared to the number of small molecules theoretically possible; nonetheless, this resource is valuable and must not be underestimated in its potential.

2.1.2 COMMERCIAL SOURCES

There are many sources of commercially available collections of small molecules; those sources would acknowledge that not all of the molecules they offer are suitable as potential agrochemical materials and that many more could only be considered as reagents or building blocks/intermediates to other chemical targets. However, there are still a very large number of providers giving access to high-quality small molecules, which may be considered good candidates for further development whenever they would be identified as a hit in a biological screen.

The commercial suppliers fall into four broad categories:

1. Shared-pool, nonexclusive, providers. Typically, these companies source compounds from the former Russian states, other Eastern Bloc countries, India, and China. In these countries, compounds are available in relatively large numbers at a modest cost and each compound is available from one or more providers.
2. Nontargeted, nonexclusive library providers. Typically, these companies design and synthesize libraries of compounds for commercial, off-the-shelf sale. These compounds are usually available at a higher cost.
3. Targeted library providers. These providers design and synthesize libraries against particular biological targets or target families. These generally have a higher cost than nontargeted collections, reflecting the design input costs.

TABLE 2.1.1 Providers of stock small-molecule libraries

No.	Company name	Web
1	Albany Molecular Research Inc.	www.albmolecular.com
2	ART-CHEM GmbH	www.art-chem-berlin.de/
3	Asinex	www.asinex.com
4	AsisChem	www.asischem.com
5	ChemBridge Corporation	www.chembridge.com
6	ChemDiv	www.chemdiv.com
7	Enamine	www.enamine.net/
8	Fluorochem	www.fluorochem.co.uk/
9	InterBioScreen	www.ibscreen.com
10	IVK Laboratories	www.ivklabs.com
11	Key Organics	www.keyorganics.ltd.uk
12	Life Chemicals	www.lifechemicals.com
13	Maybridge	www.maybridge.com
14	Nanosyn	www.nanosyn.com
15	Oakwood Chemicals	www.oakwoodchemical.com
16	Otava	www.otava.com.ua
17	Peakdale	www.peakdale.co.uk
18	Pharmeks	www.pharmeks.com
19	Princeton BioMolecular Research	www.princetonbio.com
20	SALOR	www.sigmaaldrich.com
21	Specs	www.specs.net
22	Spectrum	www.spectrum.kiev.ua
23	TimTec	www.timtec.com
24	TOSLab	www.toslab.com
26	UORSY	www.uorsy.com
27	Vitas-M	www.vitasmlab.com
28	Open NCI database	Cactus.nci.nih.gov

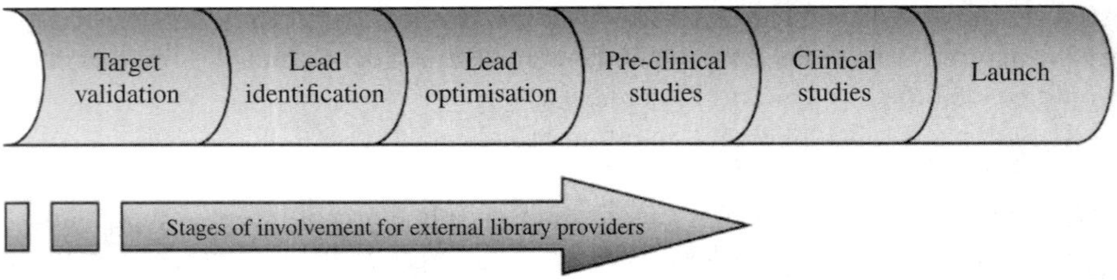

FIGURE 2.1.2 The involvement of library providers4 in the discovery pipeline.

4. Proprietary, exclusive providers. These companies perform contract research, by which they offer compound design and/or synthesis services on a fee-for-service or FTE basis. Such arrangements tend to come with the highest cost and in a shared risk model, which may be associated with downstream milestones and/or royalties payable on a success-related basis in order to make the initial outlay more reasonable.

Table 2.1.1 lists a representative (nonexhaustive) selection of companies recognized as providers of high-quality collections of small molecules that are available from stock. The involvement of library providers, regardless of their business model, is generally restricted to the early (preclinical) stages of the discovery process (Fig. 2.1.2).

2.1.3 COMPANIES PROVIDING NONPROPRIETARY, NONPARALLEL SYNTHESIZED LIBRARIES SOURCED EXTERNALLY TO THE COMPANY

These providers number well over 50, although the better recognized companies are limited to less than 30. Combined, these companies are capable of supplying several millions of compounds; some are unique to individual providers, while others are offered by multiple providers. These so-called "historical" compound collections have been viewed with more than a little suspicion with respect to the structures and the purity of the samples offered. Much of this suspicion has been offset by companies offering detailed analytical data to support the proposed structure and its purity. That analytical support usually comes in the form of NMR or LC/MS data. Feedback from the companies to their (often academic) sources has resulted in a significant improvement in structural quality and increased complexity. Due to this improved quality, valuable leads are now being found in the inexpensive collections from almost all providers.

The "collected collections" often originate from academic institutes throughout the former Russian states, other Eastern Bloc countries, India, China, and the Far East in general. The compounds are generally classically synthesized, perhaps as part of academic training, and are acquired by vendors who now put the samples through rigorous quality control (QC) routines to prove structure and purity. In most cases, all molecules have available QC data.

As part of the selection procedure during the acquisition process, many providers apply multiple structural and cheminformatic filters to remove molecules that are unsuitable for discovery purposes. The profile of these compound collections has, in most instances, improved dramatically as the pharmaceutical and agroscience industries have made it clear which types of compounds are considered useful. The sheer number in these collections, however, ensures that a large number of diverse and interesting compounds are available, from which researchers are able to "cherry-pick" according to their selection protocols and criteria.

The shared-pool suppliers are capable of providing from 10,000 to 750,000+ compounds each from stock, generally as dry powder/dry film or as solutions in DMSO (from cherry-picked selections or as pre-plated sets). The cost of these libraries from a per-compound viewpoint is relatively low, allowing scientists the opportunity to screen a large number of structurally and pharmacophorically diverse compounds easily and cheaply. End users can utilize the hits as a basis for further search queries in sourcing analogs for rapid structure–activity relationship (SAR) generation without the need to undertake any synthetic chemistry at all. One potential drawback of these "collected collections" is that they may contain singletons (i.e., compounds for which no close analogs are present); while many assay managers try to avoid screening singletons, many others see them as valid members of a diverse collection of compounds. The inherent problem with singletons is that analogs may be difficult or impossible to obtain, thus requiring synthesis before an SAR can be generated. It takes time and effort to develop a synthetic route to such analogs, although it often does clear the way to claim valuable, additional IP at a very early stage.

Next to collections of synthetic molecules, there has been a mercurial interest in collections of natural products, although at present the use of such libraries has declined for a number of reasons. The continued preference for screening individual compounds means that isolation of pure components from natural extracts is a requirement. The purification and identification of individual components is often complex, especially if the number of components in a mixture is high. The structural complexity of many natural products may also prove unattractive for chemists trying to convert a complicated hit structure into a useful small molecule that is easy and cheap to produce in bulk quantities. Finally, the situation with respect to intellectual property is fast becoming unclear as territories try to protect so-called biodiversity rights, covering all natural products found within their boundaries. While purified (although not necessarily to the point of being a single agent) natural extracts may be acceptable in use, there is concern that a strong IP position on such extracts may become more difficult to obtain.

2.1.4 COMPANIES PROVIDING IN-HOUSE DESIGNED, PARALLEL SYNTHESIZED LIBRARIES

While there are a significant number of providers positioning themselves as providers of both externally sourced collections and in-house designed libraries, the number of companies specializing only in the latter is growing.

In the late 1980s and early 1990s, the advent and rapid development of combinatorial chemistry and parallel synthesis methods, along with the associated but necessary high-throughput screening techniques, promised to revolutionize discovery research. The vast numbers of small molecules that could be made quickly and inexpensively were expected to provide a huge increase in the number of hits and useful leads. Although the results fell short of these promises, many of the concepts of combinatorial chemistry, in particular strategies for parallel synthesis driven by cheminformatic concepts, are being used by many discovery chemistry groups.

The failure of many of the initial combinatorial chemistry approaches can be traced to two problems. First, the early production of complex compound mixtures by the so-called split-and-mix method [3] led to considerable difficulties in the screening step, necessitating resynthesis or other "deconvolution" methods to identify the active components. Second, in many cases the design methods employed resulted in libraries that were very similar in structure, with little diversity in substructures or pharmacophores. Thus, while the number of hits may have increased, all that was ultimately being gained was a faster SAR, not necessarily more good lead molecules. Current library providers and drug discovery research teams use advanced cheminformatic methodology to assist in library design for almost all applications. Some of these basic cheminformatic strategies will be discussed later in describing approaches to compound selection.

Companies that provide these libraries can be split roughly into two overlapping groups: those that provide diverse libraries and those that provide targeted/focused libraries; either type of library may be "off-the-shelf" or offered through custom synthesis. A reasonably comprehensive compendium of molecular diversity and chemistry service providers (covering more than 90 suppliers) has been published [4], although details regarding the libraries that are actually stocked are somewhat limited. In addition to the databases made available by individual library providers (either freely or under protection of non-disclosure agreements), a number of other, accumulated databases can be accessed. These datasets include the Available Chemicals Directory (ACD) from Accelrys, the Screening Compounds Directory (SCD) also from Accelrys, and DiscoveryGate (www.discoverygate.com), the online extension to the SCD. The SCD claims to have over 9.7 million structures from numerous suppliers of screening compounds. ChemNavigator's iResearch Library (www.chemnavigator.com) is another source of multiple databases as well as procurement services; the listing claims to have access to almost 60 million unique structures from more than 300 compound providers. DiscoveryGate and ChemNavigator are subscription services that require users to pay for access to the databases. Of course, these costs have to be offset against the time and expense that would be required to assemble, format, and update individual databases in-house. Sources such as eMolecules (www.emolecules.com) and MolPort (www.molport.com), for example, also offer both search and purchase services.

At face value, it would appear that the current small-molecule providers listed in Table 2.1.1 could supply a total number of compounds in excess of 8,000,000 (Table 2.1.2); however, in reality there is a significant amount of overlap between these simple, and essentially design-free, collections. In fact, the total number of

TABLE 2.1.2 Analysis of 7.9 million compounds supplied by 27 vendors

Supplier name	Date of library	Library size	Exclusive compounds	% exclusive	% of all exclusive
Albany Molecular Research	2008.10	196,064	194,109	99.0%	4.9%
ART-CHEM GmbH	2008.06	110,873	25,124	22.7%	0.6%
Asinex	2008.09	457,842	238,236	52.0%	6.0%
AsisChem	2008.10	32,749	8,976	27.4%	0.2%
ChemBridge	2009.03	741,176	321,033	43.3%	8.0%
ChemDiv	2008.10	785,740	407,920	51.9%	10.2%
Enamine	2009.03	1,221,957	894,588	73.2%	22.4%
Fluorochem	2008.10	23,498	5,760	24.5%	0.1%
InterBioScreen	2008.10	466,671	162509	34.8%	4.1%
IVK Labs	2008.06	46,515	9,544	20.5%	0.2%
Key Organics	2008.10	47,656	38,880	81.6%	1.0%
Life Chemicals	2008.10	426,135	274,590	64.4%	6.9%
Maybridge	2008.10	69,862	50,314	72.0%	1.3%
Nanosyn	2008.10	62,597	10,984	17.5%	0.3%
Oakwood Chemicals	2008.10	12,621	928	7.4%	0.0%
Otava Chemicals	2008.10	173,941	66,266	38.1%	1.7%
Peakdale	2008.10	14,576	14,398	98.8%	0.4%
Pharmeks	2008.10	155,800	5,494	3.5%	0.1%
Princeton Chemicals	2008.10	380,424	40,191	10.6%	1.0%
SALOR	2008.06	48,693	22,407	46.0%	0.6%
Specs	2008.10	223,630	60,710	27.1%	1.5%
Spectrum	2008.10	8497	132	1.6%	0.0%
TimTec	2008.06	674,773	408,778	60.6%	10.2%
TOSLab	2008.10	26,713	10,399	38.9%	0.3%
Tripos	2008.10	154,604	135,216	87.5%	3.4%
UORSY	2009.03	794,997	421,091	53.0%	10.5%
Vitas-M Lab	2008.10	442,971	138,345	31.2%	3.5%
Total		7,859,402	3,995,521	50.8%	
Total unique		5,183,506	3,995,521	77.1%	
Open NCI database	2008.10	231,458	200,990	86.8%	

unique, structurally different compounds is something less than half that once the overlap is taken into account, and, of these, just 800,000 might be considered "lead-like" once a set of suitable filters have been applied.

The suppliers listed in Table 2.1.1 are represented in the databases used in the study of Chuprina and coworkers [5]. The objective of this study was to identify compounds available in commercial databases that may be considered lead-like or drug-like, according to their criteria. The principle of their analysis was quite general; it was only the detailed metrics that are likely to change from operator to operator, not the comparative operations themselves. Thus, different parameters and metrics may be employed by different research groups and in different disciplines (e.g., pharma, crop science, animal health), but the overall principle is the same. Indeed, the operations described in the following are applied, with many variants, by most of the major research organizations, in many cases using their own preferred or proprietary metrics.

The first question addressed by Chuprina and coworkers was, "how many unique compounds can be identified from all of the databases?" Based on data provided by 27 suppliers, the total number of compound entries was close to eight million. Removing duplicates from this list, mainly from the "shared-pool" providers, left over five million "unique" structures.

The next consideration was to select the appropriate filters to apply to determine lead- or drug-likeness, with the caveat that all of these filters would likely be subject to different interpretations and implementations by different research groups [6]. Nevertheless, there is a fairly general consensus within the industry regarding functional groups that render a compound unacceptable as a candidate agrochemical. Figure 2.1.3 lists functional groups/substructures that may lead to rejection of a compound for screening; this list is not necessarily complete, and some structures in the list may not be rejected by all researchers. However, since these functional groups/substructures are often associated with generation of false-positives because of generalized cytotoxicity or unwanted ligand–receptor interactions such as nonspecific (frequent-hitter) or irreversible binding [7], it is usually desirable to remove compounds that contain them from screening collections.

There are, in addition to these simple functional group filters, a number of property-based filters that may be applied. These filters take the form of calculated or even experimental metrics, such as molecular weight, solubility, hydrogen bond donors/acceptors, total polar surface area (tPSA), cLog*P*, metabolic filters (cADMET, absorption–distribution–metabolism–excretion–toxicity), and bioavailability.

2.1.5 RESULTS OF DATABASE ANALYSIS

It is instructive to show real examples of how database management and analysis techniques have been used to interrogate chemical databases in order to retrieve compounds useful for screening. The publications by Chuprina [5], Baurin [8], and Webb [9] illustrate how commercial databases can be analyzed to identify drug-like or lead-like compounds and how assembly and screening of a diverse "internal" collection

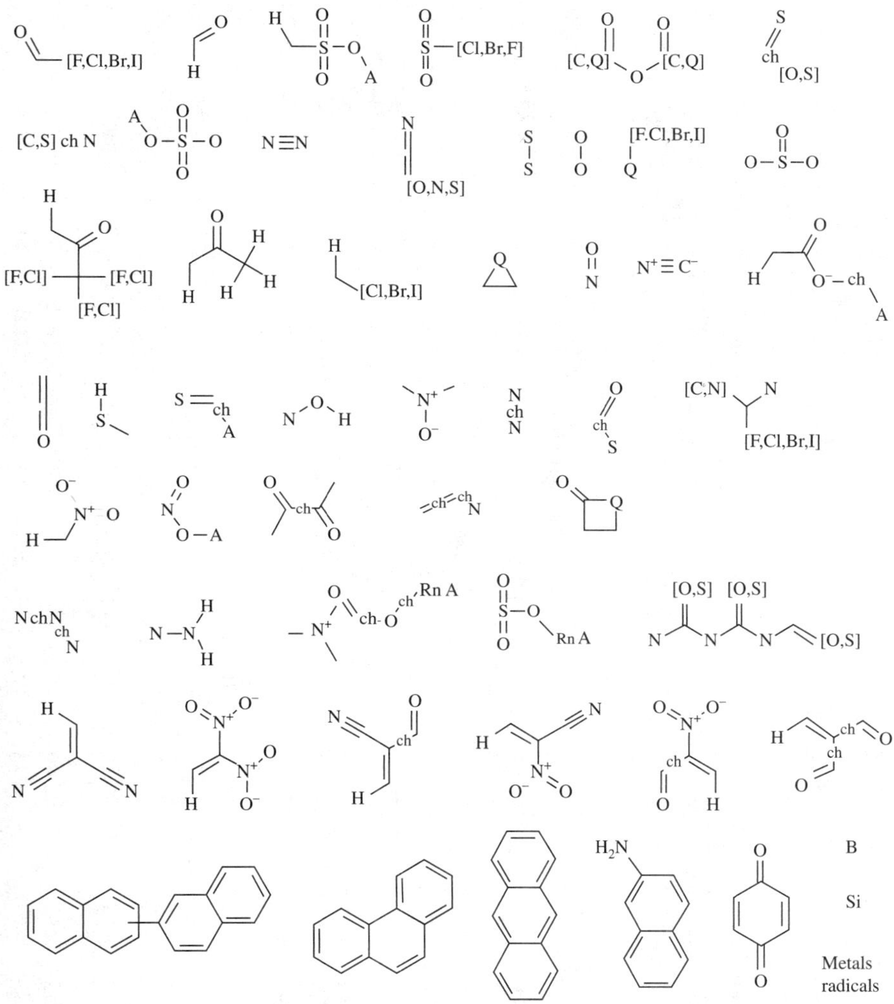

FIGURE 2.1.3 Functional groups that may lead to rejection of compounds as hits or candidates.

can generate potential new lead-like or drug-like compounds for a target based on pharmacophore information.

As described earlier, Chuprina and coworkers [5] analyzed commercial databases provided by a number of compound library vendors. Table 2.1.3 and Table 2.1.4 list the results of their analysis. The tables illustrate how millions of compounds can be reduced to a more manageable number of molecules useful as possible leads. These tables are not included as an endorsement for either the methods or the algorithms used in the calculations; the data are reproduced with permission to illustrate the functions that can be applied to a master database prior to the application of further selection criteria.

TABLE 2.1.3 Analysis of drug-like properties computed for compounds from 27 suppliers[a]

Supplier	Library size	3 of 4 Lipinski		4 of 4 Lipinski		Veber		Absence of toxic/ reactive fragments		Total filtered drug-like	
		No. of compounds passed	%	No. of compounds passed	%	No. of compounds passed	%	No. of compounds passed	%	No. of compounds passed	%
Albany Molecular Research	196,064	168,512	86	104,334	53	171,617	88	186,901	95	66,780	34
ART-CHEM GmbH	110,873	101,959	92	73,540	66	94,653	85	69,628	63	35,040	32
Asinex	457,842	443,550	97	350,843	77	399,447	87	378,697	83	212,472	46
AsisChem	32,749	31,834	97	24,827	76	30,118	92	24,491	75	14,031	43
ChemBridge	741,176	728,794	98	600,477	81	678,668	92	607,613	82	386,566	52
ChemDiv	785,740	747,052	95	540,565	69	720,162	92	650,781	83	330,144	42
Enamine	1,221,957	1,160,859	95	1,002,004	82	1,063,102	87	1,026,443	84	672,076	55
Fluorochem	23,498	23,040	98	14,150	60	22,027	94	15,744	67	7435	32
InterBioScreen	466,671	444,365	95	317,429	68	405,300	87	324,547	70	144,850	31
IVK Labs	46,515	45,144	97	34,059	73	41,926	90	28,966	62	13,688	29
Key Organics	47,656	46,301	97	33,383	70	44,393	93	33,065	69	17,984	38
Life Chemicals	426,135	401,386	94	269,897	63	372,876	88	350,943	82	153,850	36
Maybridge	69,862	68,088	97	48,988	70	63,683	91	46,238	66	25,225	36
Nanosyn	62,597	59,158	95	42,241	67	54,502	87	37,063	59	17,924	29
Oakwood Chemicals	12,621	12,308	98	6,715	53	12,046	95	8500	67	3796	30
Otava Chemicals	173,941	159,692	92	103,736	60	150,082	86	122,534	70	49,281	28
Peakdale	14,576	14,072	97	10,246	70	14,009	96	13,903	95	7209	49
Pharmeks	155,800	138,626	89	94,023	60	127,982	82	95,591	61	34,811	22

Princeton Chemicals	380,424	352,728	93	240,877	63	335,530	88	246,944	65	107,968	28
SALOR	48,693	44,053	90	24,386	50	40,715	84	21,846	45	8300	17
Specs	223,630	207,769	93	140,776	63	201,236	90	159,595	71	71,882	32
Spectrum	8,497	7,997	94	5,607	66	7,769	91	7,083	83	3,590	42
TimTec	674,773	609,034	90	431,965	64	579,690	86	390,235	58	151,164	22
TOSLab	26,713	22,748	85	13,505	51	23,045	86	17,236	65	5,212	20
Tripos	154,604	150,949	98	115,517	75	141,993	92	129,010	83	72,350	47
UORSY	794,997	683,318	86	596,784	75	591,882	74	532,054	67	277,088	35
Vitas-M Lab	442,971	419,064	95	302,404	68	382,768	86	327,505	74	143,146	32
Unique	5,183,506	4,909,717	95	3,716,537	72	4,545,593	88	4,010,470	77	2,020,027	39
Open NCI database compounds	231,458	210,856	91	161,543	70	182,841	79	140,826	61	95,588	41

[a]From Reference [5].

TABLE 2.1.4 Analysis of lead-like properties computed for compounds from 27 suppliers[a]

		Total filtered lead-like:		Total filtered lead-like, exclusive:		
Supplier	Library size	Compounds	%	Compounds	%	% of total lead-like exclusives
Albany Molecular Research	196,064	31,512	16	30,773	16	3.6
ART-CHEM GmbH	110,873	21,882	20	3215	3	0.4
Asinex	457,842	131,093	29	66,006	14	7.7
AsisChem	32,749	7420	23	1876	6	0.2
ChemBridge	741,176	250,740	34	141,438	19	16.6
ChemDiv	785,740	188,588	24	107,468	14	12.6
Enamine	122,1957	291,751	24	223,668	18	26.3
Fluorochem	23,498	6435	27	1472	6	0.2
InterBioScreen	466,671	81,515	17	29,490	6	3.5
IVK Labs	46,515	8877	19	1195	3	0.1
Key Organics	47,656	12,094	25	9458	20	1.1
Life Chemicals	426,135	76,726	18	43,187	10	5.1
Maybridge	69,862	17,533	25	11,432	16	1.3
Nanosyn	62,597	11,354	18	1852	3	0.2
Oakwood Chemicals	12,621	3305	26	197	2	0.0
Otava Chemicals	173,941	27,086	16	6325	4	0.7

Peakdale	14,576	4373	30	4291	29	0.5
Pharmeks	155,800	18,949	12	367	0	0.0
Princeton Chemicals	380,424	62,285	16	4289	1	0.5
SALOR	48,693	5638	12	2925	6	0.3
Specs	223,630	43,042	19	9837	4	1.2
Spectrum	8497	2051	24	33	0	0.0
TimTec	674,773	86,482	13	38,256	6	4.5
TOSLab	26,713	2934	11	835	3	0.1
Tripos	154,604	44,159	29	38,448	25	4.5
UORSY	794,997	118,159	15	53,939	7	6.3
Vitas-M Lab	442,971	80,961	18	15,147	3	1.8
Unique	5,183,506	1,645,602	32	851,954	16	
Open NCI database compounds	231,458	38,613	17	31,745	14	

[a]From Reference [5].

From the total of 7.9 million compounds available from the 27 suppliers, 5.18 million unique structures were identified, of which just over 2 million (39%) pass all the applied filters for drug-likeness. The databases for all 27 suppliers would need to be analyzed to gain access to all 2 million structures. The question thus arises: could that operation be made more manageable if only a subset of suppliers were to be considered? In other words, how would few suppliers be able to provide a workable proportion of those unique compounds? Using the 10 largest suppliers, 90% of the unique, drug-like compounds would still be available for selection.

This analysis demonstrates that it is a relatively simple matter to reduce an otherwise overpowering number of compounds to a manageable number of higher-quality structures primed for more individualized selection processes.

2.1.6 COMPOUND SELECTION AND DATABASE FILTERING

If potentially millions of small molecules are available, then how should the compounds be selected from a seething pool of diversity? The first question could be, "do we need to select compounds at all?," and the answer is not always "yes, we do!" Sometimes a random selection can perform as well as a preselected diversity set, particularly when assaying orphan targets or those for which little or no information is available regarding target protein structures.

Companies planning to assemble a new, diversity-based corporate screening collection have several options available to them. The most straightforward might be to take a diverse collection of compounds as offered by the provider as a starting point and build upon that basic collection. Alternatively, given that the selection software is available to the end user, scientists can collect individual vendors' databases and analyze the entire collection for diversity, which allows to keep the entire selection process internal. Such selections can be based, for example, on substructure similarity/dissimilarity calculations, pharmacophore analyses, or clustering; each approach will generally lead to different selections, even within a single set of compounds.

2.1.7 SUBSTRUCTURE SIMILARITY/DISSIMILARITY

Substructure diversity may be investigated using a method such as the Tanimoto similarity index. This method is based on one-dimensional "fingerprints" from two-dimensional compounds in question (Fig. 2.1.4). One of the companies that extensively developed this approach is Daylight Chemical Information Systems [10]. For each predetermined substructure fragment that is present in the molecule, one

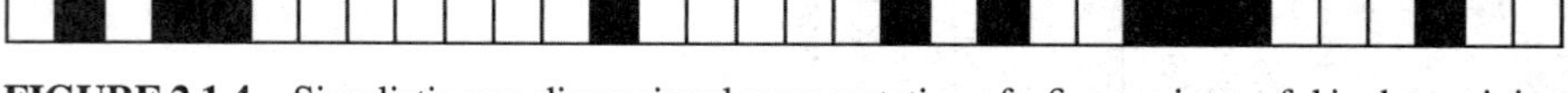

FIGURE 2.1.4 Simplistic one-dimensional representation of a fingerprint useful in determining similarity.

"bit" is set within the linear fingerprint. As such, a compound can be represented as 01011000000010000010100111000100 with each bit assigned a value of 1 if that predetermined substructure is present or 0 if it is not. This analysis is carried out for all compounds in the database to generate a complete set of molecular "fingerprints" for the selected compound collection. Comparison of these fingerprints to a random or preselected reference compound is useful in searching for either similarity or dissimilarity. The comparison involves the calculation of Tanimoto coefficients (Eq. 2.1.1) and Tanimoto dissimilarity index (Eq. 2.1.2):

$$\text{Tanimoto coefficient} = T(x,y) = N(x,y)/(N(x) + N(y) - N(x,y)) \\ = \text{a measure of similarity} \tag{2.1.1}$$

$$\text{Tanimoto dissimilarity index} = 1 - T(x,y) \quad \text{a measure of dissimilarity} \tag{2.1.2}$$

The basis of the calculation is defined as the number of bits equal to 1 that are common to both structures, defined as $N(x,y)$, versus the number of bits equal to 1 in either set, defined as $N(x)$, $N(y)$. The selection process can then proceed as compounds are compared and levels of similarity (or dissimilarity) are set.

As a simple example, a reference compound is set as the starting point. Pair-wise comparison of all compounds to the reference structure identifies the most dissimilar compound, which is added to the collection. The software calculates the mean Tanimoto coefficient from the growing diverse collection ($\overline{x} = \Sigma x / n$: the sum of all coefficients divided by the number of compounds in the growing collection), and the compound with the coefficient furthest from that mean is added. Repetition of this process continues until no more compounds need to be added or until a coefficient threshold is reached. Many other approaches have been developed for measuring similarity and dissimilarity, and a number of detailed papers and reviews are available; particularly recommended for advanced theory are these of Raymond [11] and Agrafiotis [12].

2.1.8 CLUSTER ANALYSIS

One of the currently preferred tools for interrogating databases and making a (diverse) selection is clustering [13]. This method consigns individual compounds into "bins" or clusters based on a common property. Clustering is common in everyday life; you just need to walk through a supermarket to see it in action! Walking up the aisles of the supermarket, you see that items are located in sections containing similar things; these may be cleaning items, canned food, cereals, drinks, and so on. Each of these may be further refined so that foods can be broken down into fruit, vegetables, bread, meat, and fish. These are all "clusters" of household items; the contents of each cluster have something in common.

The clustering of small molecules in a database follows the same principles; characteristics or descriptors need to be identified which can be applied to clusters, and mathematical tools need to be developed to determine the cluster in which a compound fits best. There are several methods of cluster analysis including hierarchical clustering and *k*-means.

The next example will illustrate the employment of pair-wise comparisons of Tanimoto fingerprints for cluster analysis. From a database of randomly organized compounds, compound 1 is taken and assigned as the centroid of the first cluster. Subsequently, the fingerprint of compound 1 is compared with that of compound 2. If the similarity is above a set minimum value, it is added to the cluster and taken out of the database list; if it is not, then it remains and the next compound is analyzed. This process is repeated for the entire database thus creating the first cluster. After the first cluster is completely generated, the first compound not assigned to that first cluster will be assigned as the centroid of the second cluster. The Tanimoto comparison will be repeated to create the second cluster and eventually until all compounds have been assigned to clusters. Certain clusters may only contain the centroid molecule; these molecules are referred to as singletons.

2.1.9 PHARMACOPHORE ANALYSIS

The generation of a diverse set of compounds based on the pharmacophores they contain is more complex, requiring the generation of a pharmacophore fingerprint for all three-dimensional conformations of every molecule in the database. The pharmacophore concept discards the traditional perceptions of chemical structure, focusing instead on a series of descriptors that represent functional groups and that could interact with a receptor. The obvious assumption is that a molecule must have valid pharmacophores to offer to a receptor in order to be considered useful. As a consequence, compounds that have no valid pharmacophores should be removed from a database [14]. However, the emergence of fragment-based screening has shown that this is not as straightforward as it may seem since not all fragments possess what might be considered a valid three-point pharmacophore motif commonly considered necessary for a ligand to bind to a biological target.

The number of descriptors commonly used in a pharmacophore analysis is refreshingly small. On the other hand, there are a huge number of conformations that flexible molecules could potentially adopt. The commonly used descriptors include:

- Hydrogen bond donor (H_{don})
- Hydrogen bond acceptor (H_{acc})
- Charged center
- Aromatic ring (Arom)
- Hydrophobic group
- Acid (including Lewis acid)
- Base (including Lewis base)

The generation of every three-center pharmacophore is the method most commonly used. Figure 2.1.5 shows how a single small molecule can display multiple pharmacophores

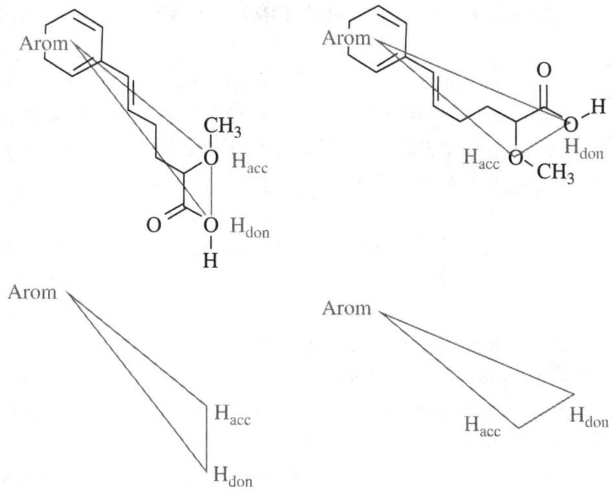

FIGURE 2.1.5 The multiple pharmacophore concept from a single molecule. For color detail, please see color plate section.

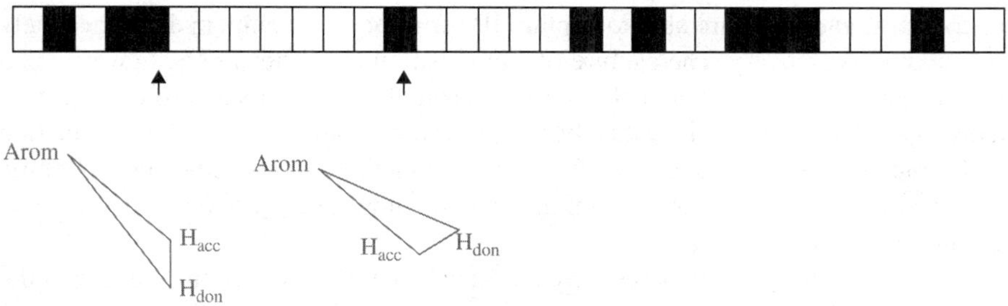

FIGURE 2.1.6 Simplistic representation of pharmacophore fingerprints.

from the same structural elements. Using all available structural conformations, the pharmacophores represented in a single molecule can also be encoded in much the same way as the fingerprints described earlier for dissimilarity searching (Fig. 2.1.6) using distance metrics [10]. Pair-wise comparison of these fingerprints, similar to the cluster analysis described earlier, is one way of using this information to generate a pharmacophorically diverse set of compounds. Since the Tanimoto and pharmacophore analyses measure different structural characteristics, a library that is "diverse" by one criterion may not be diverse by the other.

Substructure-, similarity-, and pharmacophore-based search queries can also be used to generate targeted or focused compound sets. Pharmacophore-based methods also enable "lead-hopping" [15], a technique that can identify molecules with diverse core substructures that have the potential to match the ligand–target interactions of a hit compound [9]. The use of such structural and pharmacophore-based analysis is currently gaining support in the form of library annotation, a process being adopted for the generation of collections of compounds having improved lead-like or drug-like properties.

2.1.10 COMPOUND ACQUISITION FORMAT AND STORAGE

Most companies listed in Table 2.1.1 have compounds available in the form of dry powders, while compounds pre-dissolved in DMSO at fixed concentrations are becoming more readily available in either pre-plated or cherry-pickable formats. Due to the extended stability/shelf life and simpler storage of dry solids, dry samples are generally preferable.

DMSO is frequently used as a solvent to dissolve compounds. DMSO-dissolved compounds are stored frozen between +4 °C and −80°C. Automated plate duplication and cherry-picking of individual wells are becoming commonplace procedures making the creation of assay plates or "use-once" daughter plates simple. Industry standard 96-well plates can be combined to create higher-density plates with 384-well and 1536-well operations emerging. The use of solution-phase samples also allows a significant reduction in deliverable sample sizes, often well below the limits of accurate dry powder weighing. Evidently, it is faster to duplicate a 96-well plate than to freshly weigh 96 compounds. Higher-density screening plates such as 384- and 1536-well formats come with a concomitant reduction in the amount of sample used, although this does not necessarily result in an overall reduction in cost!

IP remains a delicate subject but with most compounds being provided without IP restrictions, end users are able to capture IP for use of the compounds immediately on making a discovery. The capture of Composition of Matter rights, however, can only be gained through chemical modifications to the initial hits identified as part of a lead optimization effort. Fortunately, there is a vast amount of molecular information and structural diversity available from the suppliers of small-molecule screening libraries. These resources are becoming more and more a key part of the research and development process.

So, the commercial databases have been analyzed, your selections made, and the screening compounds delivered—what happens next? Go discover!

ACKNOWLEDGMENTS

Thanks to Dr. Oleg Lukin (ChemBioCenter, Kiev) and Dr. Robert Scoffin (Cresset) for useful discussions and information, some of which has been reproduced or adapted here.

REFERENCES

1. http://www.monsanto.com/whoweare/Pages/monsanto-history.aspx. Accessed on July 29, 2013.
2. Hann, M.M., Oprea, T.I. (2004). Pursuing the Leadlikeness Concept in Pharmaceutical Research. *Curr. Opin. Chem. Biol.*, **8**, 255–263.
3. Lam, K.S., et al. (1991). A New Type of Synthetic Peptide Library for Identifying Ligand-Binding Activity. *Nature* **354** (6348), 82–84.

4. Warr, W.A., *The Directory of Molecular Diversity Suppliers*, Select Biosciences Ltd., Sudbury, 2004.
5. Chuprina, A., et al. (2010). Drug- and Lead-likeness, Target Class, and Molecular Diversity Analysis of 7.9 Million Commercially Available Organic Compounds Provided by 29 Suppliers. *J Chem Inf.*, **50** (4), 470–479.
6. Zernov, V.V., et al. (2003). Drug Discovery Using Support Vector Machines. The Case Studies of Drug-likeness, Agrochemical-Likeness and Enzyme Inhibition Predictions. *J. Chem. Inf. Comput. Sci.*, **43**, 2048–2056.
7. Rishton, G.M. (1997). Reactive Compounds and In Vitro False Positives in HTS. *Drug Discov. Today*, **2**, 382–384.
8. Baurin, N., et al. (2004). Drug-like Annotation and Duplicate Analysis of a 23-Supplier Chemical Database Totalling 2.7 Million Compounds. *J. Chem. Inf. Comput. Sci.*, **44**, 643–651.
9. Webb, T.R., et al. (2000). The Utilization of a Unified Pharmacophore Query in the Discovery of New Antagonists of the Adenosine Receptor Family. *Bioorg. Med. Chem. Lett.*, **10**, 31–34.
10. http://www.daylight.com/meetings/summerschool98/course/basics/fp.html. Accessed on July 1, 2013.
11. Raymond, J.W., Blankley, C.J., Willett, P. (2003). Comparison of Chemical Clustering Methods Using Graph-Based and Fingerprint-Based Similarity Measures. *J. Mol. Graph. Model.*, **21** (5), 421–433.
12. Agrafiotis, D.K. (1998). Diversity of chemical libraries. *The Encyclopedia of Computational Chemistry*; Schleyer, P.v.R. et al., Eds., John Wiley and Sons, Chichester, **1**, pp. 742–761.
13. Brown, R.D., Martin, Y. C. (1996). Use of Structure–Activity Data to Compare Structure-Based Clustering Methods and Descriptors for Use in Compound Selection. *J. Chem. Inf. Comput. Sci.*, **36**, 572–584.
14. Muegge, I. (2003). Selection Criteria for Drug-like Compounds. *Med. Res. Rev.*, **23**, 302–321.
15. Cramer, R. D., et al. (1999). Prospective Identification of Biologically Active Structures by Topomer Shape Similarity Searching. *J. Med. Chem.*, **42**, 3919–3933.

The University of Orleans has also published data on a selection of chemical library providers. This information was part of the 9th and 10th Electronic Computational Chemistry Conferences (ECCC9 and ECCC10) and are available only online: Mozziconacci, J-C., Arnoult, E., Baurin, N., Marot, C. and Morin-Allory, L., "Preparation of a Molecular Database From a Set of 2 Million Compounds for Virtual Screening Applications: Gathering, Structural Analysis and Filtering," http://www.univ-orleans.fr/icoa/eposter/eccc9/ECCC9i.htm and Monge, A., Arrault, A., Marot, C. and Morin-Allory, L., "Analysis of a set of 2.6 million unique compounds gathered from the libraries of 32 chemical providers," http://www.univ-orleans.fr/icoa/eposter/eccc10/monge/

2.2

COMBINATORIAL CHEMISTRY LIBRARY DESIGN

ROBERT KLEIN AND STEPHEN D. LINDELL

2.2.1 INTRODUCTION

Combinatorial chemistry and high-throughput screening (HTS) techniques have seen widespread application within the agrochemical and pharmaceutical industries over the past two decades. The utilization of these technologies has had a dramatic effect on the number of compounds screened for biological activity each year. For example, in the agroscience area, the number of compounds passing through a larger agrochemical company's *in vivo* biological screens has increased from around 10,000 in the 1980s to literally hundreds of thousands in the 2000s [1, 2]. However, despite many impressive technological achievements, there seems to be a widespread concurrence that early combinatorial chemistry failed to deliver the numbers of hits and leads that had been expected at the time the technique was implemented [3]. The reasons for this are many and complex, but the manner in which early libraries were designed was certainly an important contributing factor. In this chapter, we will seek to explain the reasons behind this early disappointment and show how modern library design is improving the outlook and changing the way combinatorial chemistry is used in the discovery process. The term "library design" is usually taken to mean the design process that occurs prior to chemical synthesis in which a set of structures are selected from a large virtual library for preparation. However, it is important to keep in mind that many library design methodologies can equally well be used to select out compounds from an actual collection of commercially available or proprietary compounds. Both strategies are in effect designing screening libraries, and both help conserve synthesis and testing resources. This chapter will concentrate on the design of

Plant Chemical Biology, First Edition. Edited by Dominique Audenaert and Paul Overvoorde.

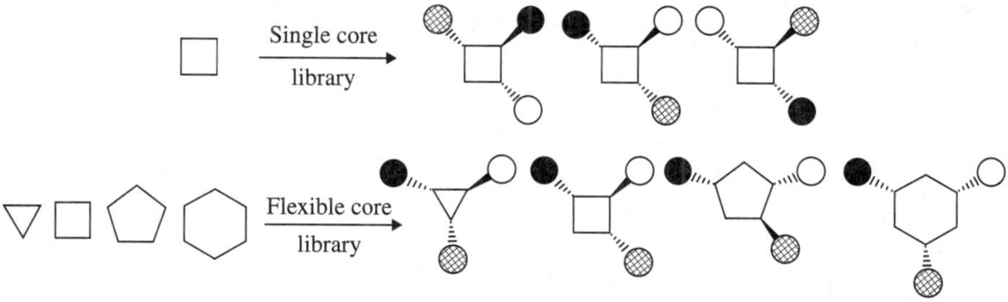

FIGURE 2.2.1 Diversity-oriented synthesis: a variable core varies not only the functionalities but also the orientation of the display. *Source*: Adapted from Reference [5a].

general screening or discovery libraries and will not cover the use of combinatorial chemistry in the optimization of an already discovered lead structure. Important design concepts will be illustrated wherever possible using agrochemically relevant examples. We will not attempt to comprehensively cover the use of combinatorial chemistry in the agrosciences, since this has been the subject of several recent reviews [4].

When the first combinatorial libraries were produced in the early 1990s, the technological platform used for production was incompletely developed and somewhat difficult to use. As a consequence, the target structure of an initial biologically inspired library idea was often quite substantially altered in order to fit the chemistry better with what could realistically be synthesized in the laboratory. Since those earlier days, tremendous advances have been made and many of the technological problems have been solved. The act of producing a library has become more routine, meaning that undesired structural compromises made for technical reasons are less common than was previously the case. Diversity was an important design factor in early libraries. Typically a library was constructed around a central core or scaffold by adding on between two and four differently functionalized diversity groups. The diversity of the proposed library was measured, for example, by using structural fingerprints and the Tanimoto similarity index. Based upon such criteria, the ensuring libraries were indeed diverse, but the diversity was confined to the periphery of the molecule with the central core remaining constant. This peripheral diversity was rather different to that found in traditional collections of individually crafted molecules, where variations in the central core were more commonly found. Modern combinatorial library design strategies falling under the heading of "diversity-oriented synthesis" seek to address this problem by counting the central core as one of the variables [5]. Libraries designed in this way change not just the substituent functionalities but also their orientation and are able to explore a much greater area of real or virtual chemical space (Fig. 2.2.1).

2.2.2 BIOAVAILABILITY

The predominance of the aforementioned peripheral diversity in the design of earlier combinatorial libraries tended to lead to rather large high-molecular-weight compounds with different physical chemical properties to those shown by the majority of commercial

TABLE 2.2.1 Physical chemistry-based bioavailability guidelines

	Lipinski "Rule of Five" for oral absorption (1997)	Briggs "Rule of Three" for agrochemical *in vivo* HTS (1997)
MW	<500	300 ± 100
log *P*octanol/water	<5	3 ± 3
H-bond donors (sum OH + NH)	<5	≤3*
H-bond acceptors (sum O + N)	<10	—
Δlog *P* ($\log P_{octanol/water} - \log P_{hexane/water}$)	—	≤3*
mp (°C)	—	≤300
EH	—	30 ± 5

*Unless water solubility > 3 g/l.

agrochemicals. While this is not necessarily bad, it is important that library members have a certain bioavailability in order to stand a chance of being detected in an agrochemical *in vivo* HTS test [2]. Commercial agrochemicals and drugs clearly fulfill general bioavailability criteria, and based upon an analysis of their physical properties, several researchers have proposed sets of rules for ensuring good uptake and distribution within an organism. The first and most famous of these was the "Rule of Five" for oral absorption of drugs published by Lipinski and coworkers in 1997 [6]. Poor absorption was predicted to be more likely for compounds with physical properties falling outside the limits set by Lipinski's rules (Table 2.2.1). Subsequently, Briggs conducted an analysis of commercial agrochemicals and proposed a "Rule of Three" for selection of compounds for agrochemical *in vivo* HTS testing [7]. A compound was predicted to have a higher probability of being active in an *in vivo* screen if its physical properties fell within the guidelines set by Briggs (Table 2.2.1). Many of the compounds synthesized in early combinatorial libraries had physical chemical properties well outside the limits shown in Table 2.2.1, and this is an important contributing factor explaining the often disappointing biological results seen with these libraries.

Three easily calculated criteria are common to the Lipinski and Briggs rules, namely, MWt, log *P*, and the number of H-bond donors. Briggs introduced three new properties: Δlog *P*, melting point (mp), and equivalent hydrocarbon (EH) number. Whereas log *P* is a good measure of lipophilicity, Δlog *P*, which is ($\log P_{octanol/water} - \log P_{hexane/water}$), is a measure of polarity and the likelihood of cell membrane penetration [8]. As a general rule of thumb, polar compounds with $\Delta \log P > 3$ do not efficiently penetrate cell membranes. Because Δlog *P* cannot be easily calculated, it has been proposed that it should be replaced by polar surface area (PSA) [9]. Both Δlog *P* and PSA attempt to quantify the polarity of a molecule, but PSA has the important advantage that it can be readily calculated [10].

The mp of a substance was included in Briggs rules due to its correlation with water solubility [11]. Library compounds with an mp > 300 °C are predicted to have very low water solubility and will, therefore, be less bioavailable. Since the mp of compounds in combinatorial libraries is almost never determined, Briggs proposed that a very approximate mp, good enough to use for estimating water solubility, could be calculated using Equation 2.2.1.

$$\mathrm{mp}(^{\circ}\mathrm{C}) \approx 30\Delta\log P + 60(\text{No.rigid rings}) - 20\ (\text{No.rotatable bonds}) \qquad (2.2.1)$$

The equivalent hydrocarbon (EH) number is used to estimate the volatility of a compound [12]. The vapor pressures of the simple hydrocarbons have been determined, so converting other molecules containing heteroatoms to the "equivalent hydrocarbon" allows their vapor pressures to be estimated (i.e., a compound with an EH value = n is predicted to have a similar vapor pressure to the hydrocarbon C_nH_{2n+2}). This is important because compounds with a low EH ($\leq$ 24) may be lost during evaporation when preparing plates for HTS and may be too volatile for field testing (potential loss of $\geq$ 500 g/ha/day at 25 °C). An EH value of 35 is the upper limit for vapor activity that in some cases is a very desirable property (e.g., soil insecticides).

Comparisons of the physical chemical properties of commercial agrochemicals [13] and pharmaceuticals [14] with those of lead compounds have shown that their properties are not radically different. This indicates that despite being derived using commercial products, the Briggs and Lipinski rules can be usefully applied for the selection of potential lead structures. However, lead optimization is most commonly achieved by adding on additional substitutes or groups in a process that increases the complexity of a molecule. To allow for this effect, Clark and Delany [15] have proposed a "Guide of 2" in which the probability of successful lead progression is increased when MWt = 200 – 400, $\Delta\log P \leq 4$, $\Delta\log P \leq 2$, H-bond donors $\leq$ 2, mp $\leq$ 200 °C.

It is important to keep in mind that all of these physical chemistry-based "rules" are meant to act more as guidelines that can and should be adjusted to meet specific requirements. In this regard, Genix and coworkers [9] have proposed modified guidelines for fungicide selection, and Tice has proposed different guidelines for the selection of post- and preemergence herbicides and insecticides [16]. It should also be noted that not each and every rule needs to be fulfilled when selecting a compound for synthesis. For example, compliance with three of four criteria based upon M.Wt., log P, PSA, and EH, has been proposed for preselecting compounds for fungicide screening [9]. The computer-aided screening of large virtual compound libraries in order to select molecules with physical properties falling within a certain predetermined bioavailability range is quick and easy to do. Consequently, it is widely used as a selection criterion to help select the molecules from a library that are most likely to be bioavailable. Of course, just because a molecule is predicted to be bioavailable does not necessarily mean that it will also be biologically active.

2.2.3 CHEMICAL SPACE

It has been estimated that around 10^{60} molecules containing up to 30 carbon, nitrogen, oxygen, or sulfur atoms are theoretically accessible [17]. The total global number of all organic substances synthesized to date is statistically insignificant in comparison to such an enormous number. Even within a single library, the potential numbers can be huge. For example, Bleicher et al. [18] considered a library of benzimidazoles in which phenylenediamines were elaborated with carboxylic acids, alkyl halides, amines, and boronic acids. If 100 examples of each building block were used, then a library of 10^8 benzimidazoles would be produced. A large proportion of these molecules could fulfill bioavailability guidelines, so these would not help to significantly reduce the numbers, but clearly the number must be reduced in order to conserve valuable research resources. In the past this was often achieved using a shotgun-type strategy aimed at achieving a maximally diverse library. The idea was to analyze the chemical space occupied by the whole virtual library (e.g., using the Tanimoto index as similarity measure) and then to select reagents that would lead to a real library of 500–1000 compounds exhibiting maximum coverage of the total diversity space. The trouble with the approach is that the "spaces" between the compounds (or shotgun pellets) are so large that they only stand a small chance of finding (hitting) the "target" activity within a given structural type. Put another way, it is well known that a small seemingly trivial structural change in a molecule can make the difference between inactivity and an excellent biological activity. In order to stand a realistic chance of detecting such small but crucially important structural requirements, it is necessary to concentrate synthesis around small targeted areas with high biological potential (i.e., increase the density of shotgun pellets in these areas).

One way to increase the focus of a combinatorial library is to synthesize incomplete combinatorials [19]. Traditionally, in a combinatorial library, a number of starting materials A_1–A_n, are reacted with reagents B_1–B_n to yield a matrix of products A_nB_n. These products can then be reacted further with reagents C_1–C_n to yield a larger product matrix $A_nB_nC_n$, etc. While this process enables a very high synthetic efficiency, it inevitably leads to the synthesis of less or undesirable molecules (e.g., nondrug-like products). In order to avoid making these substances, the combinatorial matrix is divided into sub-matrices in which properties can be more easily tailored. Since these sub-matrix libraries no longer need to be made using the same reaction conditions, this is essentially the same as making a series of small targeted libraries. In cases where an active molecule already exists and where the synthetic goal is to efficiently make a library of close analogs, then the question of how to select and target a library is relatively easily answered. However, in cases where the objective is to discover a new lead structure, then the answer is more difficult and represents a very significant challenge.

2.2.4 PRIVILEGED STRUCTURES

In the agrochemical sciences, there are many examples where new active substances have been discovered by systematically modifying the structure of a known molecule exhibiting biological activity in a completely different area. For example, the insecticides fenazaquin [20] and flubendiamide [21] were discovered through chemical modification of structurally related fungicidal and herbicidal lead structures, respectively. Although neither of these compounds was discovered with the help of combinatorial chemistry, it is common to use existing active molecules as a basis around which to design small targeted exploratory libraries. This approach is closely related to the idea of using a privileged structure as the basis around which to design a new library [22]. The term privileged structure, first coined by Evans et al., was defined as a single molecular framework able to provide ligands for diverse receptors [23].

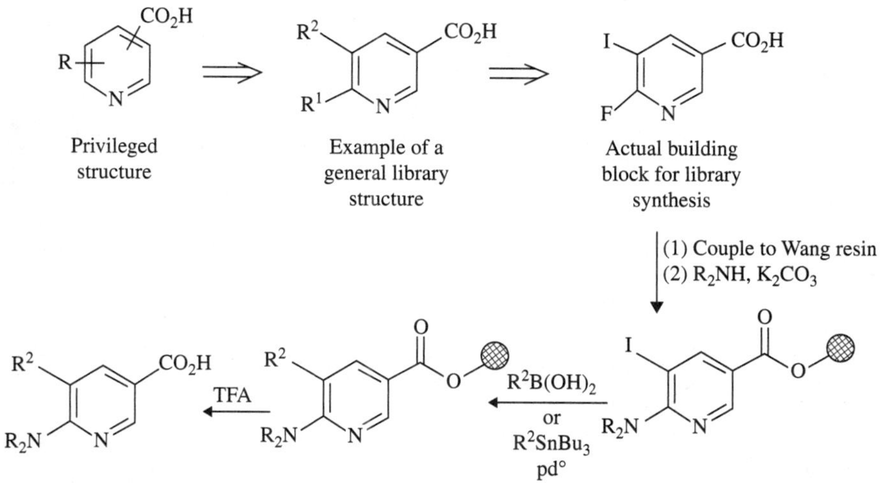

FIGURE 2.2.2 Design and synthesis of a library of pyridine carboxylic acids.

A diverse collection of scaffolds have been identified as privileged structures including benzamides, benzodiazepines, benzopyrans, benzylpiperidines, biphenyltetrazoles, indoles, β-glucose, monosaccharides, and spiropiperidines [24]. Natural products are sometimes classified as privileged structures because during their biosynthesis they interact with multiple proteins and quite often exhibit more than one type of biological activity due to binding with different proteins [25]. The term biology-oriented synthesis (BIOS) has been introduced to describe libraries based upon natural product scaffolds [26]. Basing a library upon a privileged, natural product-like or drug-like structure means that the compounds should have an increased probability of binding to potential target proteins. Regarding library design, this usually means that the library core or scaffold is based upon such a structure.

A good agrochemical example of this approach has been reported by Turner et al. who synthesized a series of 10 combinatorial libraries based upon pyridine carboxylic acids using solid- and liquid-phase technologies [27]. Substituted pyridylcarboxylic acids are found in a number of herbicides possessing different modes of action (e.g., acetohydroxyacid synthase (AHAS) inhibitors (imazapyr, imazaquin), auxins (clopyralid, picloram), auxin transport inhibitors (diflufenzopyr), and microtubulin inhibitors (thiazopyr)) [28] and as such can be considered to be privileged structures. The basic library design principle, together with an example of an actual library synthesis, is illustrated in Figure 2.2.2. The synthesized libraries ranged in size from 176 to 2112 compounds, and biological hit rates, after testing in a herbicide HTS, were between 0 and 14.9%. The average hit rate over all 10 libraries was 5.6%, compared with 4.8% for a library of randomly selected acids and 1.2% for a more diverse set of randomly chosen compounds. The authors noted that biological activity shown in a higher secondary screen may be a better way of assessing the success of a combinatorial library than the HTS hit rate. This is because an important lead structure can result from libraries with a low HTS hit rate, whereas those with a higher hit rate may

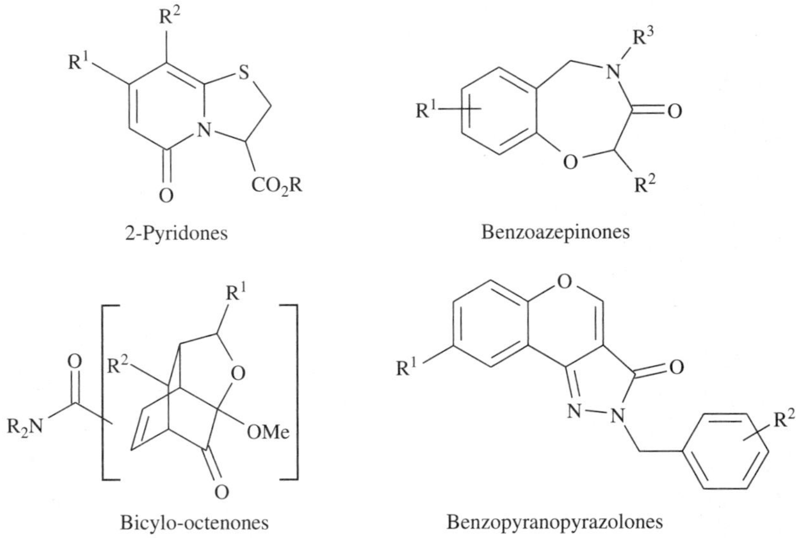

FIGURE 2.2.3 Examples of libraries inspired by privileged structures (top) and natural products (bottom).

not deliver any real leads at all. Indeed, in the current case, the two "best" libraries (11.6 and 14.9% hit rate) yielded no leads, whereas two "less good" libraries (3.3 and 3.8% hit rate) gave lead compounds good enough to justify follow-up chemistry. Other agrochemically relevant libraries utilizing privileged structures have been, for example, based upon fused 2-pyridones [29] and benzoazepinones (Fig. 2.2.3) [30]. Good examples based upon natural product-like structures have been reported by Smith et al. (bicyclo-octenones) [31] and Zhou and Yang (benzopyranopyazolones) (Fig. 2.2.3) [32].

2.2.5 FRAGMENT-BASED DESIGN

In Section 2.2.3 of this chapter we discussed the huge chemical space potentially available for library synthesis. However, if instead of 30 heavy atoms we restrict ourselves to just 11 C, N, or O atoms, then the chemical space shrinks enormously. Instead of 10^{60} possible molecules, the number drops to around 10^7 substances [33], meaning that it now becomes possible to sample a reasonable portion of the available chemical space with a screening library. This means that the chance of finding a ligand for a target protein in a library of low-molecular-weight substances is higher than for conventional HTS. However, a library of such small molecules is unlikely to yield high-potency protein ligands, since the number of protein–ligand interactions will be, by definition, much lower than in a typical drug or agrochemical product. The ensuing screening hits are best viewed as inhibitor fragments that need to be further grown, or combined with other fragments, to yield high-potency drug-like ligands. Such fragment-based approaches are arguably one of the most successful library design strat-

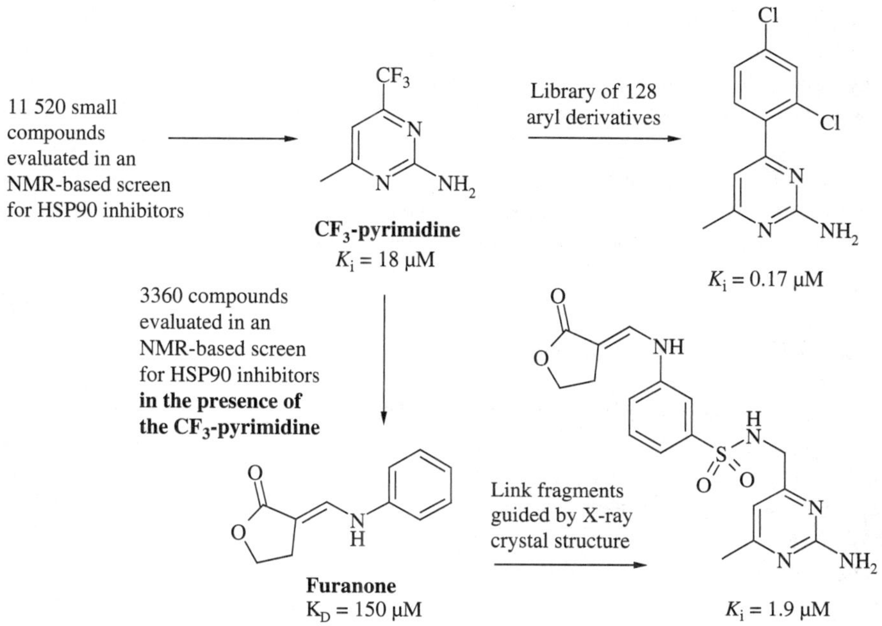

FIGURE 2.2.4 Fragment-based design and synthesis of HSP90 inhibitors.

egies currently available and have yielded several clinical drug candidates in recent years [34]. The approach starts with the identification of low-molecular-weight fragments that bind to the target protein, often with quite low mM levels of affinity. This is achieved by using functional or direct biochemical assays, mass spectrometric methods, NMR (nuclear magnetic resonance)-based screening, or X-ray crystallography-based methods [35]. Obviously, the assay method used must be sensitive enough to reliably identify weak-binding ligands, and the fragments themselves must be soluble at relatively high concentrations to be detected. The value of such inhibitor fragments is considerably enhanced if structural information about their binding mode is available. In this regard, the direct screening of fragments using NMR [36] or X-ray crystallography [37] can offer distinct advantages.

Inhibitors of heat shock protein 90 (HSP90) are of interest as anticancer drugs and potential agents to help plants cope with heat stress [38]. Huth and coworkers have reported a fragment-based NMR screening approach to identify novel HSP90 inhibitors [39]. They started by screening 11,520 low-molecular-weight compounds in an NMR-based HSP90 screen. This led to the identification of the CF_3-pyrimidine lead shown in Figure 2.2.4 ($K_i = 18\,\mu M$). Subsequently, X-ray crystallographic structural information was used to direct the synthesis of a library in which the CF_3 group was replaced by 128 different aryl groups. After biochemical screening, this gave rise to a 100-fold more potent inhibitor with $K_i = 0.17\,\mu M$ (Fig. 2.2.4). In an attempt to identify more novel substituents for the lipophilic aryl binding site, a second round of NMR screening was then conducted, this time in the presence of saturating amounts of the CF_3-pyrimidine. The furanone shown in Figure 2.2.4 was the most potent hit identified, with a K_D of 150 μM in the presence of the CF_3-pyrimidine but >5000 μM in its absence. In order to guide the linking of these two cooperatively binding fragments, the structure of the ternary complex was solved using X-ray crystallography. Based upon information from this structure, a methylsulfonamide linker was incorporated between the pyrimidine and furanone to give a new inhibitor that bound 10-fold better than the CF_3-pyrimidine ($K_i = 1.9$ μM) (Fig. 2.2.4). The aforementioned example nicely illustrates how structure-based design input can provide important guidance for optimizing and growing fragment structures.

2.2.6 LIGAND-BASED DESIGN

Ligand-based drug design is based upon knowledge about a set of molecules and their efficacies with respect to the biological action of interest. In this definition, the biological action is not restricted to the binding of a ligand to a biological target protein. Consequently, the term "ligand" refers to any molecule showing a particular biological activity. The ligands are characterized by a set of descriptors, for example, the chemical structure, three-dimensional atomic positions, and other experimental or calculated properties. Ligand-based design aims to build a model from these descriptors that explains the efficacies of known molecules in order to predict the efficacies of novel ones.

2.2.6.1 Quantitative Structure–Activity Relationship (QSAR)

Quantitative structure–activity relationship (QSAR) is a popular ligand-based design method that seeks to find a mathematical expression (usually a linear combination) which predicts the biological efficacies of molecules directly from a set of molecular properties (descriptors) [40]. Historically, the descriptors used in this approach were derived from (or are identical to) physical–chemical quantities that had been experimentally determined. As early as 1937, Hammett found a quantitative relationship between the rate of hydrolysis of substituted benzoic acid esters and a substituent-specific factor σ, which was derived from the acidity constants of the corresponding benzoic acids [41]. The substituent constant σ turned out to be a good measure for assessing the relative electronic effect of aromatic substituents in aromatic molecules in general. Hansch and coworkers introduced a substituent parameter π representing the hydrophobic contribution of a substituent, which was calculated from the octanol–water partition coefficient. They found that the biological activities (A_i) of a set of aromatic compounds could be estimated with moderate but surprising accuracy on a logarithmic scale using a linear combination of the σ and π parameters [42] (Eq. 2.2.2).

$$\log(A_i) = a + b \cdot \sigma_i + c \cdot \pi_i \tag{2.2.2}$$

The index i in Equation 2.2.2 refers to one out of a set of molecules. In this simplified equation, the molecules differ only at one substituent position. The coefficients a, b, and c are determined by a least-squares procedure, for example, a linear regression. Hansch [43] achieved an even better correlation when he added a quadratic term to the equation (Eq. 2.2.3). This change better accounts for the observation that the biological activity within a series of compounds often goes through an optimum with respect to hydrophobicity.

$$\log(A_i) = a + b \cdot \sigma_i + c \cdot \pi_i + d \cdot \pi_i^2 \qquad (2.2.3)$$

One of the problems associated with applying QSAR equations to novel molecules lies in the nonavailability of the descriptors. Therefore, various computational methods have been developed that are able to calculate properties of a compound from its chemical structure. Well-known examples include the models for predicting the octanol–water partition coefficient or its logarithm, log P [44]. Other easily accessible or computable properties are the PSA [10], surface area and volume of a molecule, molecular weight, and number of H-bond donor and acceptor atoms. These properties are also used as criteria to select compounds for large screening collections with the aim to optimize bioavailability. Many more descriptors can be calculated from the molecular structure and applied in a QSAR approach. For example, Adenot and coworkers [45] developed a discrimination model for the blood–brain barrier permeation of 1700 molecules from the World Drug Index. Apart from the aforementioned descriptors, they included molar refractivity, the number of rotational bonds, moments of inertia, quantum mechanical (QM) properties, as well as topological descriptors in their analysis. Topological properties of a molecule, such as branching, ring size, and atom counts, belong to a group of structural 1D and 2D descriptors that have been successfully deployed in establishing QSAR models. Based only upon molecular topology descriptors and a set of 33 thiobenzanalides, Garcia-Domenech and coworkers, for example, derived a QSAR for antifungal activity [46]. Subsequently, they used their QSAR model to screen a virtual library of thiobenzanalide analogs and were able to propose new chemical structures that were predicted to have improved antifungal activity.

Calculating QM properties can be a very time-consuming process, even with modern high-powered computers. Consequently, QM descriptors are not usually the first choice for a QSAR-based selection or library design procedure. However, Goeller et al. [47] showed that by performing QM calculations on a semiempirical level using appropriate (but not the highest level) hardware, QM descriptors could be derived with reasonable computational effort. They created a QSAR solubility model based upon QM and other descriptors by applying a neural network algorithm. The model was designed to handle large data sets of several tens of thousands of compounds in virtual screening-based library design.

2.2.6.2 Similarity-Based Design

As already mentioned in Section 3, similarity is an important principle in molecular design. It is based upon the hypothesis that chemical (structural) similarity and biological similarity are related. In order to measure chemical similarity, structural

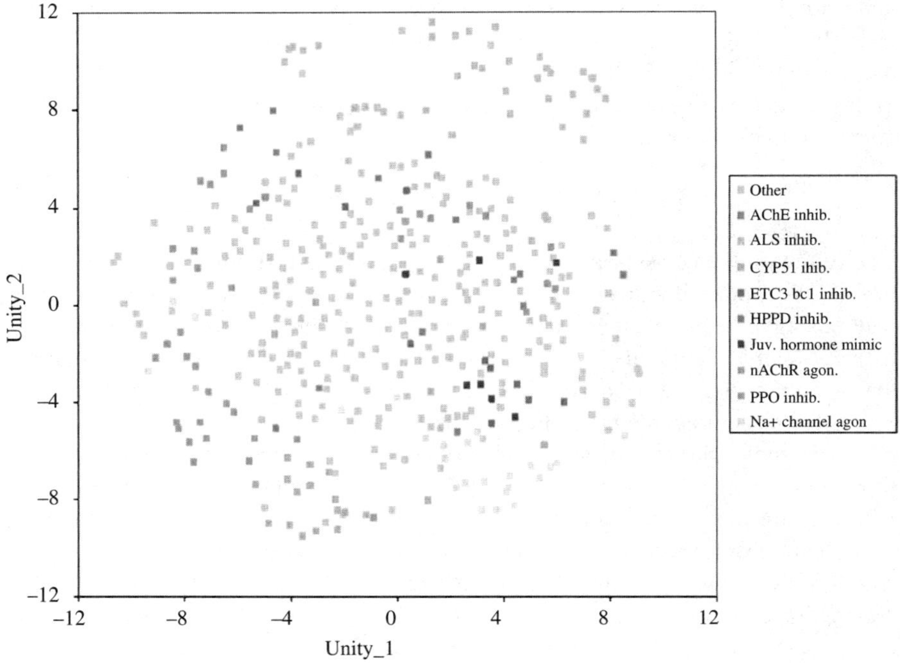

FIGURE 2.2.5 A set of 500 agrochemicals with known modes of action mapped into a 2D chemical space obtained using a PCA-NLM transformation of the Unity fingerprints. For color detail, please see color plate section.

features are often encoded in bit strings, also referred to as fingerprints or keys. Daylight fingerprints [48], Unity fingerprints [49], BCI keys [50], and MDL keys [51] are four of the best-known examples. The similarity score of a molecule with respect to another molecule based on bit strings is usually expressed as a Tanimoto coefficient or index [52], a number between 0 (no similarity at all) and 1 (identical fingerprints). Fingerprints are often said to span a multidimensional chemical space. For the purpose of visualization but at the cost of information loss, the dimensionality of this space can be reduced to 2D or 3D by transforming the bit strings using, for instance, principal component analysis (PCA) or multidimensional scaling (MDS) [53]. In Figure 2.2.5, we have mapped a set of 500 agrochemicals with known modes of action into a 2D chemical space obtained by a PCA-NLM [54] transformation of the Unity fingerprints. The modes of action with the highest population are color-coded showing a clear, though not unambiguous, relationship between chemical similarity and biological mode of action. Apart from fingerprints, various other similarity descriptors have been proposed and applied. For example, Pearlman introduced the so-called BCUT values [55] which are obtained by combining topological descriptors and molecular properties using a rather complex algorithm. Rarey and coworkers proposed the concept of feature trees [56] to describe molecular structures in a similarity space. In all of these approaches, the distance between two compounds in a descriptor space, expressed as a euclidean distance or Tanimoto index, represents a similarity score. Such similarity scores, with respect to one or more reference molecules with known biological efficacy, can be used to estimate a novel molecule's potency. For example, Willet and coworkers applied various fingerprint-based similarity analysis methods to a pesticide data set comprising 132,784 compounds containing 7127 biologically active molecules [57]. The data set was divided into a training set and a test set. They compared the results of classical similarity searches with those obtained from machine learning algorithms, such as the support vector machine (SVM) [58] and binary kernel discrimination (BKD) [59]. The classical similarity score was defined to be the fingerprint-based Tanimoto similarity of a compound to the most similar active within the training set. The SVM scores and BKD scores were obtained from models that had been trained using the training data set and then optimizing the intrinsic SVM and BKD parameters. The scores were then used to rank the compounds in the test set. All three scoring models were able to enrich actives in the high-ranking compounds; however, SVM and BKD achieved significantly better results than the classical similarity ranking.

Fischer and coworkers have presented a software tool called "LoFT," which comprises a set of algorithms using feature tree descriptors among others, to enable similarity-driven multi-objective focused library design [60]. The components of fingerprints and property vectors such as BCUTs are often used directly as descriptors for a QSAR model.

2.2.6.3 Pharmacophore Approach

In another ligand-based approach, a set of selected molecules with experimentally determined efficacies is used to identify a pharmacophore [61], which contains those structural features that are necessary in order to show the desired biological activity. For molecules that are known to act on a particular molecular target, these features may be exactly the ones that dominate the interaction with the active site of the target. In the simplest case, a pharmacophore is a list of functional groups and steric and electronic features, the presence of which is essential for activity. In addition, a well-defined pharmacophore model contains three-dimensional information such as the mutual distances between the functional groups, hydrophobic centers, etc. The model may also be complemented by features that are disadvantageous, for example, an excluded volume. In a variant of the pharmacophore approach, molecules are described by their molecular fields rather than by their atomistic representation. The classical ComFA approach, originally introduced by Cramer et al. [62], calculates electrostatic and van der Waals potentials on three-dimensional grid points surrounding the molecules. The grid points most relevant for biological efficacy are identified by statistical analyses similar to those used in the aforementioned QSAR methods. A more recent molecular field approach, called FieldTemplater, was introduced by Cresset BioMolecular Discovery [63]. Ligands are represented by a set of field points that directly reflect the molecules' ability to interact with a molecular target via electrostatic, van der Waals, and H-bonding forces. A good example has been reported by Harris et al. [64] who used FieldTemplater for the design of libraries targeted toward ligand-gated ion channels.

López-Ramoz and coworkers used a pharmacophore approach for the selection of potential 4-hydroxyphenylpyruvate dioxygenase (HPPD) inhibitors from a library [65]. This study gives an excellent overview of the design process of enzyme inhibitors in general and HPPD herbicides in particular. The authors applied various drug and library design methods, both in parallel to directly compare their performance and in a consecutive manner to use them as a series of filters. In the final step, after the number of molecules to be tested had already been reduced, they applied the most time-consuming method and performed a structure-based docking procedure. Interestingly, they came to the conclusion that the pharmacophore filter was much more efficient than the more complex docking experiment in terms of retrieving the most potent compounds. However, they also state that the success of one or the other method greatly depends upon the data set to be analyzed. In general, it can be said that ligand-based methods tend to show an intrinsic structural bias toward the compounds used to create a particular model and, therefore, rarely reveal really novel ligand structures.

2.2.7 STRUCTURE-BASED DESIGN

Structure-based design methods make use of the three-dimensional structure of the molecular target of a drug. The structure of the target, usually a protein but sometimes a nucleic acid, is determined experimentally by X-ray crystallography or NMR spectroscopy [66]. Alternatively, a homology-based model of a protein derived from an experimental structure of a related protein may be used, although this generally gives less reliable results. The publicly accessible Protein Data Bank (PDB) [67] currently contains the 3D coordinates of more than 70,000 proteins. In many cases, these structures were solved with a bound substrate or inhibitor, thus providing additional important information on features of the binding site. The potency of a drug candidate is estimated by predicting its affinity to the binding site of the molecular target. At first glance, this approach seems to be straightforward and, due to the amount and quality of initial information available, the most promising library design method discussed so far. However, there are still several hurdles that must be overcome to make structure-based design a truly powerful library design tool. Proteins are not rigid structures as presented in the PDB, and the binding of a ligand to a protein is accompanied by significant conformational changes within the protein as well as to the ligand itself. Protonation and tautomeric states play a significant role in the binding process and may even change upon binding. Moreover, affinity does not only depend on the interaction of a ligand with a protein interface but is greatly influenced by solvation effects before and after binding. Taking all these aspects appropriately into account is still a big challenge even with modern computers. Nonetheless, it is a highly attractive technique, especially for lead finding, since in contrast to ligand-based methods, it does not directly depend upon known bioactive molecules and is, therefore, potentially able to identify molecules with a completely novel chemical structure.

The first step in a virtual affinity test is placing the potential ligand into the binding site of the target, taking the conformational flexibility of the ligand into account. This step is usually referred to as "docking." The objective is to identify those ligand conformations and orientations, the so-called poses, which fit sterically best within the binding site and show the most energetically favorable interaction patterns. This may be accomplished using force-field methods [68] or other empirical approaches [69]. Due to the extensive computational effort, higher-level theoretical approaches such as quantum chemical methods, which in principle could give more accurate results, are usually not feasible in the context of virtual screening or library design. Reddy et al. [70] divided the various commercially available and academic docking software packages into three categories according to the algorithms used to account for conformational flexibility. Programs like GLIDE [71], QXP [72], GOLD [73], or AutoDock [74], for example, optimize ligand conformations and orientations simultaneously utilizing stochastic (e.g., Monte Carlo, genetic algorithms) or deterministic (e.g., steepest decent, trial-and-error algorithms) optimization strategies. A second category of programs, including the packages FRED [75] and SLIDE [76], generates an ensemble of low-energy conformations of each ligand prior to their placement within the binding site. This is much less time-consuming, provided that only a relatively small number of conformations are sampled. However, it is also less accurate because the selected ligand conformations are not obtained as a response to the interaction with the protein and may be misleading. Fragment-based approaches (e.g., FlexX [77] and DOCK 4.0 [78]) represent the third category of program and have turned out to offer a good compromise between speed and accuracy. Ligands are usually broken at rotational bonds into fragments. These are rigidly docked into the binding site while taking the geometrical constraints imposed by the bonds between the fragments into account. Usually a base fragment is chosen, either automatically or by the user's intervention, and several favorable docked poses are used as starting points to "grow" the molecule within the binding site. In all of the procedures just described, the protein is treated as a rigid entity.

In the second step of the virtual affinity test, the best poses are assessed using so-called scoring functions to estimate the free energies of binding. Scoring is used not only to select the best ligand structures from a set of candidate molecules but also to select the correct pose(s) of a particular ligand. It is noteworthy that in most approaches the objective function of the docking step and the scoring function are not identical. However, the two functions must be compatible; otherwise, one may miss potent candidate ligands or, potentially even worse, one is confronted with a plethora of allegedly good but false-binders. Scoring functions can be divided into force-field-based and empirical methods. Force-field functions comprise contributions from van der Waals and electrostatic interactions as well as H-bonding energies between the ligand and the binding site. In order to take entropic and solvation effects into account, while avoiding time-consuming simulation techniques including explicit solvent molecules, the solvent (i.e., water) is usually treated implicitly using continuum solvation models. The MM/PBSA [79] and GBSA [80] procedures are the most prominent examples. Empirical scoring functions [81] are comprised of parameterized terms that attempt to take the relevant contributions to the binding energy into account. The empirical terms include, for instance, H-bond energies, contact surface, so-called buriedness, and the number of rotatable bonds. The parameters are obtained by a fitting procedure using a reference data set so as to optimally reproduce experimental free energies of binding. Lange et al. have developed a scoring function called HYDE [82] which differs from other empirical approaches in that it contains no parameters that are fitted to experimental binding energies or experimental structures. It is based upon the concept that, in the first instance, solvated ligand molecules cause a perturbation of the pure water H-bonding network. The HYDE score is determined by calculating the water-accessible surface area and the number of made and unmade hydrogen bonds before and after ligand–protein complexation. The so-called hydrophobic effect is an intrinsic part of the HYDE score. HYDE parameters are obtained using a fitting procedure including experimental octanol–water partition coefficients rather than experimental protein–ligand binding energies. Thus, HYDE has no intrinsic bias toward certain proteins or ligand classes and is very generally applicable.

Wang et al. [83] used the docking software packages DOCK and AutoDock to perform a virtual screening campaign on the ACD 3D database [84], containing approximately 164,000 small organic molecules, against the enzyme AHAS [85]. The protein structure used in this study originates from the co-crystallization of AHAS from *Arabidopsis thaliana* with a herbicidal sulfonylurea inhibitor. Based upon the virtual screening results, Wang and coworkers selected 14 compounds that were purchased and tested for *in vitro* AHAS inhibition and *in vivo* herbicidal activity. Three of these compounds showed interesting *in vitro* activity including the dyestuff indigo carmine (Fig. 2.2.6). Although this compound was herbicidally inactive, its structure differed considerably from known AHAS inhibitors, which inspired Wang and coworkers to screen a follow-up library of 32 isatin derivatives against AHAS [86]. This led to the discovery of the novel isatin-like inhibitors shown in Figure 2.2.6, which pleasingly also exhibited herbicidal activity, albeit at weaker levels than commercial sulfonylurea herbicides.

In order to identify new fungicide candidates that could overcome the resistance of fungi to sterol 14α-demethylase (CYP51) inhibitors, Zhang et al. [87] subjected more than 100,000 commercially available compounds to a high-throughput virtual screening process. The screening protocol included a Lipinski filter step and a two-step docking procedure. Since they are integral membrane proteins, there is no 3D structure of a fungal CYP51 enzyme available to date. Therefore, the authors constructed a homology model of the fungal *Penicillium digitatum* CYP51 based upon the PDB crystal structure of bacterial *Mycobacterium tuberculosis* CYP51 [88]. To make the process more efficient, they constructed a pharmacophore model from known azole fungicides and incorporated this into the primary docking run using FlexX-pharm [89], which produced docking poses that fitted both the binding site and fulfilled the pharmacophore constraints. The resulting poses were scored with the PMF [90] and GOLD [73] scoring functions. The 13,000 best compounds were then docked into the binding site model without constraints in order to cross-validate the binding modes. Based upon the information from the two docking steps, 500 compounds were selected for further evaluation. A similarity analysis as well as a visual inspection further reduced the number of candidates down to 30. From these, seven compounds with considerable experimental inhibitory activity and that do not belong to the traditional azole-type inhibitors were identified as novel lead structures.

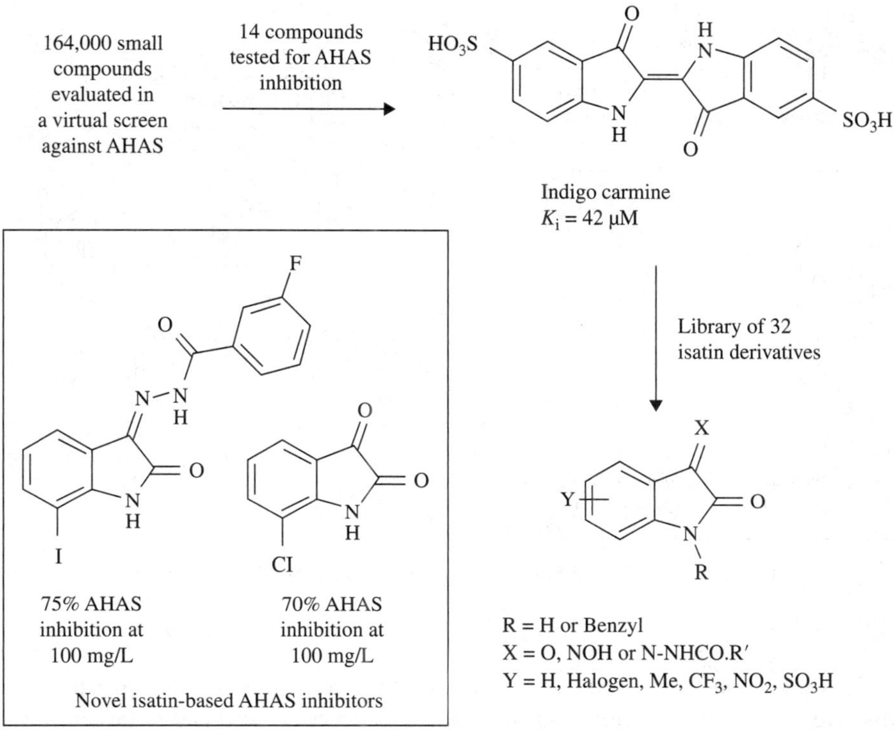

FIGURE 2.2.6 Virtual screening as an aid to the design of a library of AHAS inhibitors.

A structure-based design approach to the discovery of new inhibitors of the herbicidal target enzyme HPPD was successfully integrated into the library design process by Camper and Parker [91]. An important goal of their work was to identify inhibitors showing high selectivity toward plant HPPD in order to avoid toxicity concerns associated with inhibition of the mammalian enzyme. Molecular modeling studies, using the plant (*Arabidopsis*) and rat HPPD protein crystal structures, indicated that substituted tetramic acids had a good chance of binding selectively to the plant enzyme. In particular, examination of the binding pocket suggested that a lysine tail functionalized as an amide or urea would be able to make beneficial hydrogen bonding interactions with a glutamine residue in a previously unexploited binding channel. In order to estimate the desired selectivity, Camper et al. docked the members of virtual combinatorial tetramic acid libraries into the 3D structures of HPPD from both *A. thaliana* and rat using the software package FlexX. Subsequently, a small solid-phase library of lysine tetramic acids was synthesized as detailed in Figure 2.2.7. Biochemical testing revealed novel compounds showing a high degree of potency and selectivity toward the plant enzyme, but unfortunately the *in vivo* activity was not compelling.

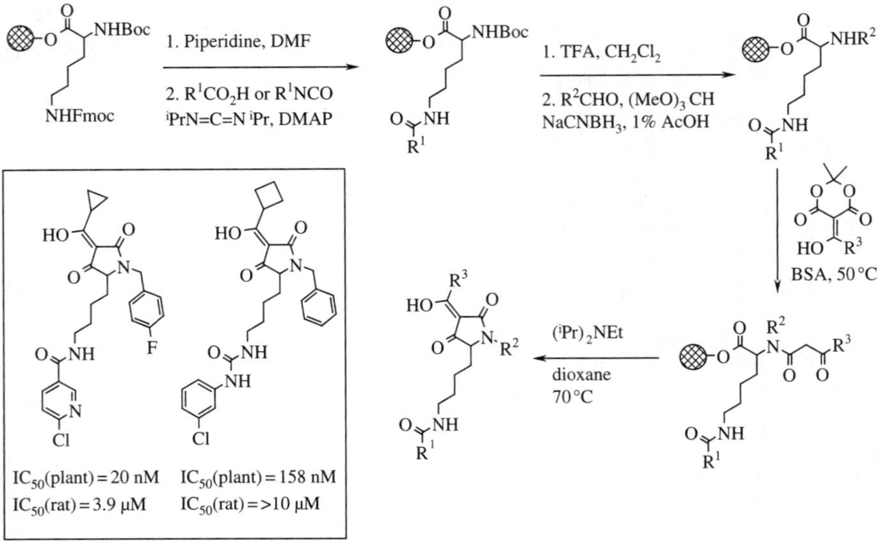

FIGURE 2.2.7 Synthesis of a library of potential HPPD inhibitors.

In some cases, especially during the early phase of drug and agrochemical discovery, it can be advantageous to synthesize libraries designed to target a protein family, rather than a specific protein. The concept of SoftFocus© library design was introduced by the BioFocus group to describe these more generic libraries [64]. The first step in this target-family approach is the identification of scaffolds capable of binding to multiple family members. Selectivity between family members is achieved in a second step in which recognition elements are attached to the scaffold. The approach has been nicely illustrated for the protein kinase family, which are important targets for anti-inflammatory and anticancer drugs as well as for fungicides. In order to keep computational effort to a manageable level, just seven kinases were selected from the very large number of publicly available kinase protein structures. These were chosen to be representative in terms of protein conformations and ligand binding modes. Various minimally substituted scaffold structures were docked into the seven protein structures and accepted or rejected according to their estimated affinity and ability to bind to more than one kinase. Appropriate substituents were selected based upon the environment and size of the binding site pocket available around an accepted scaffold in its particular binding mode. In this way, 36 SoftFocus kinase libraries were designed, synthesized, and screened against 17 different kinases at BioFocus. Nearly 90% of the libraries showed higher hit rates than a diverse compound collection (hit rate 1.7%), and half of the libraries showed hit rates that were three or more times higher (up to 31%). Figure 2.2.8 shows three library compounds, each designed to favor one of the three known kinase binding modes, which were identified as lead structures by other research groups.

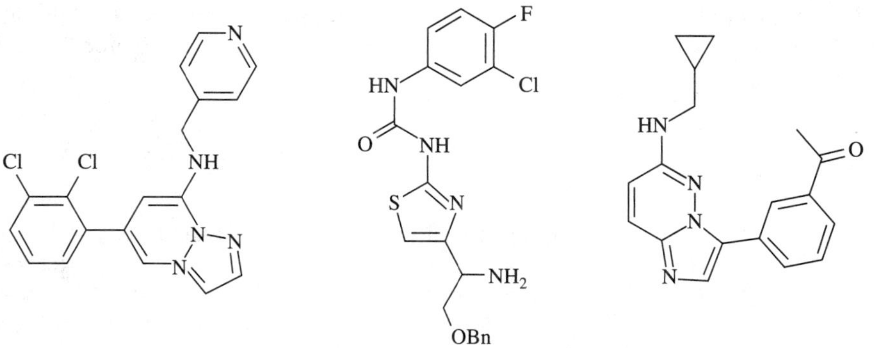

Compound A
Library SFK 03
CDK2 inhibitor
Hingle binding
PDP code 2R3F

Compound B
Library SFK 48
p38 α inhibitor
DFG-out binding
PDP code 3IW8

Compound C
Library SFK 33
PIM-1 inhibitor
Invariant lysine binding
PDP code 2C3I

FIGURE 2.2.8 Examples of kinase inhibitors from SoftFocus libraries designed to target different kinase binding modes.

2.2.8 CONCLUSION AND SUMMARY

The challenge faced by any library design method is how to best rationally select molecules that will address a specific biological problem from a practically limitless chemical diversity space. Early libraries attempted to address this challenge by synthesizing large numbers of compounds with a high peripheral diversity. Although these libraries often gave hits in HTS screens, the properties of the molecules did not lend themselves to optimization, and only very few real progressable leads were discovered. Modern library design methods concentrate on selecting out smaller highly focused libraries that probe a comparatively small, well-defined area of chemical space. Concepts such as bioavailability, diversity-oriented synthesis, privileged structures, fragment-based design, pharmacophore and other ligand-based methods, and structure-based design are all routinely used to help target chemical library synthesis. Computer-based methods are of increasing importance, since virtual compounds are cheap to generate and screen in virtual screens. The aim is to conserve valuable research resources by selecting out molecules with an increased probability of biological activity for synthesis and real screening. Many of these very same library design tools can also be effectively used for selecting out interesting compounds for screening from large collections of commercially available or proprietary compounds.

Large, full combinatorial chemistry matrices, in which all starting materials are reacted with all derivatizing reagents, are still valuable as virtual libraries, but are no longer routinely synthesized. One could even go so far as to say that true combinatorial chemistry, at least in terms of small-molecule discovery, now exists only in the virtual world. In the real world, it has been replaced by automated parallel synthesis methodologies that allow quick and efficient access to the multiple, small, targeted libraries required for success in modern biological and biochemical screening campaigns.

REFERENCES

1. (a) Evans, D. A. How can technology feed the world safely and sustainably? in "*Pesticide Chemistry and Bioscience: The Food-Environment Challenge*," Brooks, G. T.; Roberts, T. R., eds., Royal Society of Chemistry Special Publication 233. Royal Society of Chemistry, Cambridge, UK, **1999**, pp. 3–24. (b) Hay, J. V. Herbicide discovery in the 21st century — a look into the crystal ball, in "*Pesticide Chemistry and Bioscience: The Food-Environment Challenge*", Brooks, G. T.; Roberts, T. R., eds., Royal Society of Chemistry Special Publication 233. Royal Society of Chemistry, Cambridge, UK, **1999**, pp. 55–65.
2. Tietjen, K.; Drewes, M.; Stenzel, K. High throughput screening in agrochemical research. *J. Comb. Chem. High Throughput Screen.*, **2005**, *8*, 589–594
3. (a) Lahana, R. How many leads from HTS? *Drug Disc. Today*, **1999**, *4*, 447–448.
4. (a) Smith, S. Combinatorial chemistry in the development of new crop protection products, *Pesticide Outlook,* February **2003**, *14*, 21–26. (b) Pasteris, R. J. *New Discoveries in Agrochemicals*, ACS Symposium Series 892. American Chemical Society, Washington, DC, 2005, pp. 109–118.(c) Lindell, S. D.; Scherkenbeck, J. Prospects for combinatorial chemistry in the agrosciences, *Comb. Chem. High Throughput Screen.*, **2005**, *8*, 555–562.(d) Scherkenbeck, J.; Lindell, S. Applications of combinatorial chemistry in the agrosciences, *Comb. Chem. High Throughput Screen.*, **2005**, *8*, 563–576.(e) Smith, S. C.; et al. Targeting chemical inputs and optimising HTS for agrochemical discovery, *Comb. Chem. High Throughput Screen.*, **2005**, *8*(7), 577–587.(f) Lindell, S.; Pattenden, L. C.; Shannon, J. Combinatorial chemistry in the agrosciences, *Bioorg. Med. Chem.*, **2009**, *17*, 4035–4046.
5. (a) Tempest, P. A.; Armstrong, R. W. Cyclobutenedione derivatives on solid support: Toward multiple core structure libraries, *J. Am. Chem. Soc.*, **1997**, *119*, 7607–7608. (b) Sauer, W. H. B.; Schwarz, M. K. Size doesn't matter: Scaffold diversity, shape diversity and biological activity of combinatorial libraries, *Chim. Int. J. Chem.*, **2003**, *57*, 276–283.(c) Burke, M. D.; Schreiber, S. L. A planning strategy for diversity-oriented synthesis, *Angew. Chem. Int. Ed.*, **2004**, *43*, 46–58.
6. Lipinski, C. A.; et al. Experimental and computational approaches to estimate solubility and permeability in drug discovery and development settings, *Adv. Drug Deliv. Rev.*, **1997**, *23*, 3–25.
7. Briggs, G. G. "Predicting uptake and movement of agrochemicals from physical properties," presentation at SCI Meeting: "Uptake of Agrochemicals and Pharmaceuticals," Belgrave Square, London, England, December 9, **1997**.
8. Abraham, M. H.; et al. Hydrogen bonding. 32. An analysis of water–octanol and water–alkane partitioning and the delta log P parameter of seiler. *J. Pharm. Sci.*, **1994**, *83*, 1085–1100.
9. Briggs, G. G; Desbordes, P.; Genix, P. "Are there limits to the physical properties of fungicides?" poster presentation at 10th IUPAC International Congress on the Chemistry of Crop Protection, Basel, Switzerland, August 4–9, **2002**.
10. Ertl, P.; Rohde, B.; Selzer, P. Fast calculation of molecular polar surface area as a sum of fragment-based contributions and its application to the prediction of drug transport properties, *J. Med. Chem.*, **2000**, *43*, 3714–3717.

11. Briggs, G. G. Theoretical and experimental relationships between soil adsorption, octanol-water partition coefficients, water solubilities, bioconcentration factors, and the parachor, *J. Agric. Food Chem.*, **1981**, *29*, 1050–1059.
12. Hartley, G. S.; Graham-Bryce, I. J. "Physical Principles of Pesticide Behaviour: The Dynamics of Applied Pesticides in the Local Environment in Relation to Biological Response," Academic Press, London, New York, **1980**, *vol. 2*, pp. 883–886.
13. Clarke, E. D.; Delaney, J. S. Physical and molecular properties of agrochemicals: An analysis of screen inputs, hits, leads, and products, *Chimia International Journal for Chemistry*, **2003**, *57*, 731–734.
14. (a) Oprea, T. I.; et al. Is there a difference between leads and drugs? A historical perspective, *J. Chem. Inf. Comput. Sci.*, **2001**, *41*, 1308–1315. (b) Proudfoot, J. R. Drugs, leads, and drug-likeness: An analysis of some recently launched drugs, *Bioorg. Med. Chem. Lett.*, **2002**, *12*, 1647–1650.
15. (a) Clarke, E. "Bioavailability: An agrochemical research perspective," presentation at QSAR and Chemoinformatics Group Meeting, Jealotts Hill, Bracknell, England, April 25, **2001**. (b) Clarke, E. http://www.ukqsar.org/2011/07/12/spring-meeting-2001/. Accessed on July 2, 2013.
16. (a) Tice, C. M. Selecting the right compounds for screening: Does Lipinski's Rule of 5 for pharmaceuticals apply to agrochemicals? *Pest. Manag. Sci.*, **2001**, *57*, 3–16. (b) Tice, C. M. Selecting the right compounds for screening: Use of surface-area parameters, *Pest. Manag. Sci.*, **2002**, *58*, 219–233.
17. (a) Bohacek, R. S.; McMartin, C.; Guida, W. C. The art and practice of structure-based drug design: A molecular modeling perspective, *Med. Res. Rev.*, **1996**, *16*, 3–50. (b) Walters, W. P.; Stahl, M. T.; Mureko, M. A. Virtual screening—An overview, *Drug Discov. Today*, **1998**, *3*, 160–178.
18. Bleicher, K. H.; et al. Hit and lead generation: Beyond high-throughput screening, *Nat. Rev. Drug Discov.*, **2003**, *2*, 369–378.
19. (a) Schnur, D. M.; Tebben, A. J.; Cavallaro, C. L. Library Design: Ligand and Structure-Based Principles for Parallel and Combinatorial Libraries, in "*Comprehensive Medical Chemistry II*", Mason, J., ed., Elsevier Ltd., Oxford, UK, **2007**, vol. *4*, *Computer-Assisted Drug Design*, pp. 307–336. (b) Schnur, D. M.; et al. Methods for combinatorial and parallel library design, in "*Chemoinformatics and Computational Chemical Biology*," Bajorath, J., ed., Methods in Molecular Biology 672. Humana Press, New York, **2011**, pp. 387–434.
20. Hackler, R. E.; et al. Chemistry and miticidal activity of fused pyrimidine derivatives of fenazaquin, in "*Advances in the Chemistry of Insect Control III*," Briggs, G. G., ed., Royal Society of Chemistry, Cambridge, UK, **1994**, pp. 70–84.
21. Seo, A.; et al. Flubendiamide, a new insecticide characterized by its novel chemistry and biology, in "*Pesticide Chemistry: Crop Protection, Public Health, Environmental Safety*," Ohkawa, H.; Miyagawa, H.; Lee, P. W., eds., Wiley-VCH GmbH & Co., Weinheim, **2007**, pp. 127–135.
22. (a) Cordier, C.; et al. Natural products as an inspiration in the diversity-oriented synthesis of bioactive compound libraries, *Nat. Prod. Rep.*, 2008, *25*, 719–737. (b) Mentel, M.; Breinbauer, R. Combinatorial solid phase natural product chemistry, *Top. Curr. Chem.*, **2007**, *278*, 209–241.
23. Evans, B. E.; et al. Methods for drug discovery: Development of potent, selective, orally effective cholecystokinin antagonists, *J. Med. Chem.*, **1988**, *31*, 2235–2246.

24. For reviews on privileged structures see: (a) Duarte, C. D.; Barreiro, E. J.; Raga, C. A. Privileged structures: A useful concept for the rational design of new lead drug candidates, *Mini-Rev. Med. Chem.*, **2007**, *7*, 1108–1119. (b) Constantine, L.; Barlocco, D. Privileged structures as leads in medicinal chemistry, *Curr. Med. Chem.*, **2006**, *13*, 65–85.(c) DeSimone, R. W.; et al. Privileged structures: Applications in drug discovery, *Comb. Chem. High Throughput Screen.*, **2004**, *7*, 473–493–456.
25. Arve, L.; Voigt, T.; Waldmann, H. Charting biological and chemical space: PSSC and SCONP as guiding principles for the development of compound collections based on natural product scaffolds, *QSAR Comb. Sci.*, **2006**, *25*, 449–456.
26. Noren-Muller, A.; et al. Discovery of protein phosphatase inhibitor classes by biology-oriented synthesis, *Proc. Natl. Acad. Sci. U. S. A*, **2006**, *103*, 10606–10611.
27. Turner, J. A.; et al. *New Discoveries in Agrochemicals, ACS Symposium Series 892*. American Chemical Society, Washington, DC, **2005**, pp. 119–129.
28. Tomlin, C. D. S. (ed.) "The Pesticide Manual" (14th ed.), British Crop Protection Council, Alton, UK, 2006
29. Ruiz J. M.; Lorsbach, B. A. *New Discoveries in Agrochemicals*, ACS Symposium Series 892. American Chemical Society, Washington, DC, **2005**, pp. 99–108.
30. Carreras, I.; Scherkenbeck, J.; Paulitz, C. Communication: Polymer-assisted solution-phase synthesis of 4,5-dihydro-1,4-benzoxazepin-3(2H)-ones, *Comb. Chem. High Throughput Screen.*, **2005**, *8*, 643–647.
31. Smith, S. C.; et al. Synthesis and agrochemical screening of a library of natural product-like bicyclo[2,2,2]octenones, *Comb. Chem. High Throughput Screen.*, **2005**, *8*, 607–615.
32. Zhou, Z.-Z.; Yang, G.-F. Insecticidal lead identification by screening benzopyrano[4,3-c] pyrazol-3(2H)-ones library constructed from multiple-parallel synthesis under microwave irradiation, *Bioorg. Med. Chem.*, **2006**, *14*, 8666–8674.
33. Fink, T.; Bruggesser, H.; Reymond, J.-L. Virtual exploration of the small-molecule chemical universe below 160 Daltons, *Angew. Chem. Int. Ed.*, **2005**, *44*, 1504–1508.
34. Congreve. M.; et al. Recent developments in fragment-based drug discovery, *J. Med. Chem.*, **2008**, *51*, 3661–3680.
35. Erlanson, D. A.; McDowell, R. S.; O'Brien, T. Fragment-based drug discovery, *J. Med. Chem.*, **2004**, *47*, 3463–3482.
36. (a) Shuker, S. B.; et al. Discovering high-affinity ligands for proteins: SAR by NMR, *Science*, 1996, *274*, 1531–1534. (b) Hajduk, P. J.; et al. High-throughput nuclear magnetic resonance-based screening, *J. Med. Chem.*, **1999**, *42*, 2315–2317.
37. Hartshorn, M. J.; et al. Fragment-based lead discovery using X-ray crystallography, *J. Med. Chem.*, **2005**, *48*, 403–413.
38. McLellan, C. A.; et al. A rhizosphere fungus enhances Arabidopsis thermotolerance through production of an HSP90 inhibitor, *Plant Physiol.*, **2007**, *145*, 174–182.
39. Huth, J. R.; et al. Discovery and design of novel HSP90 inhibitors using multiple fragment-based design strategies, *Chem. Biol. Drug Des.*, **2007**, *70*, 1–12.
40. Kubinyi, H. QSAR and 3D QSAR in drug design. Part 2: Applications and problems, *Drug Discov. Today*, **1997**, *2*, 538–546.
41. Hammett, L. P. The effect of structure upon the reactions of organic compounds. Benzene derivatives, *J. Am. Chem. Soc.*, **1937**, *59*, 96–103.
42. Hansch, C.; et al. Correlation of biological activity of phenoxyacetic acids with Hammett substituent constants and partition coefficients, *Nature*, **1962**, *194*, 178–180.

43. Hansch C. Quantitative approach to biochemical structure–activity relationships, *Acc. Chem. Res.*, **1969**, *2*, 232–239.

44. (a) Chou, J. T.; Jurs, P. C. Computer-assisted computation of partition coefficients from molecular structures using fragment constants, *J. Chem. Inf. Comput. Sci.*, **1979**, *19*, 172–178. (b) Ghose A. K.; Crippen G. M. Atomic physicochemical parameters for three-dimensional structure-directed quantitative structure–activity relationships I. Partition coefficients as a measure of hydrophobicity, *J. Comput. Chem.*, **1986**, *7*, 565–577.

45. Adenot, M.; Lahana, R. Blood-brain barrier permeation models: Discriminating between potential CNS and non-CNS drugs including P-glycoprotein substrates, *J. Chem. Inf. Comp. Sci.* **2004**, *44*, 239–248.

46. García-Domenech, R.; et al. QSAR by molecular topology of 2,4-dihydroxythiobenzanilides. A virtual screening approach to optimize the antifungal activity, *Ind. J. Chem.*, **2002**, *41B*, 2376–2384.

47. Goeller, A. H.; et al. In silico prediction of buffer solubility based on quantum-mechanical and HQSAR- and topology-based descriptors, *J. Chem. Inf. Model.*, **2006**, *46*, 648–658.

48. Daylight Chemical Information Systems Inc. http://www.daylight.com. Accessed on July 2, 2013.

49. Tripos Inc. http://www.tripos.com. Accessed on July 2, 2013.

50. Barnard Chemical Information Ltd. http://www.digitalchemistry.co.uk/. Accessed on July 30, 2013.

51. Durant, J. L.; et al. Reoptimization of MDL keys for use in drug discovery, *J. Chem. Inf. Comput. Sci.*, **2002**, *42*, 1273–1280.

52. Rogers, D. J.; Tanimoto, T. T. A computer program for classifying plants, *Science*, **1960**, *132*, 1115–1118.

53. (a) Shaw P. J. A. "Multivariate Statistics for the Environmental Sciences," Hodder-Arnold, London, **2003**. (b) Borg, I.; Groenen, P. "Modern Multidimensional Scaling: Theory and Applications" (2nd ed.), Springer, New York, **2005**, pp. 207–212.

54. Lee, J. A.; Verleysen M. "Nonlinear Dimensionality Reduction," Springer, New York, **2007**.

55. Pearlman, R. S.; Smith, K. M. Metric validation and the receptor-relevant subspace concept, *J. Chem. Inf. Comp. Sci.*, **1999**, *39*, 28–35.

56. Rarey, M.; Dixon, J. S. Feature trees: A new molecular similarity measure based on tree matching, *J. Comput.-Aided. Mol. Des.*, **1998**, *12*, 471–490.

57. Wilton, D. J.; Harrison, R. F.; Willet, P. Virtual screening using binary kernel discrimination: analysis of pesticide data, *J. Chem. Inf. Model.*, **2006**, *46*, 471–477.

58. Burges, C. J. A tutorial on support vector machines for pattern recognition, *Data Min. Know. Discov.*, **1998**, *2*, 121–167.

59. Harper, G.; et al. Prediction of biological activity for high-throughput screening using binary kernel discrimination, *J. Chem. Inf. Comput. Sci.*, **2001**, *41*, 1295–1300.

60. Fischer, J. R.; Lessel, U.; Rarey, M. LoFT: Similarity-driven multiobjective focused library design, *J. Chem. Inf. Model.*, **2010**, *50*, 1, 1–21.

61. Yang, S.-Y. Pharmacophore modeling and applications in drug discovery: Challenges and recent advances, *Drug Discov. Today*, **2010**, *15*, 444–450.

62. Cramer, R. D., III; Patterson, D. E.; Bunce, J. D. Comparative molecular field analysis (CoMFA). 1. Effect of shape on binding of steroids to carrier proteins, *J. Am. Chem. Soc.*, **1988**, *110*, 5959–5967.

63. (a) http://www.cresset-group.com/products/fieldtemplater/. Accessed on July 2, 2013. (b) Cheeseright, T.; et al. Molecular field technology applied to virtual screening and finding the bioactive conformation, *Expert Opin. Drug Discov.*, **2007**, *2*, 131–144.
64. Harris. C. J.; et al. The design and application of target-focused compound libraries, *Comb. Chem. High Throughput Screen.*, **2011**, *14*, 521–531.
65. López-Ramoz, M.; Perruccio, F. HPPD: Ligand- and target-based virtual screening on a herbicide target, *J. Chem. Inf. Model.*, **2010**, *50*, 801–814.
66. Hus, J. C.; Marion, D.; Blackledge, M. De novo determination of protein structure by NMR using orientational and long-range order restraints, *J. Mol. Biol.*, **2000**, *298*, 927–936.
67. Protein Data Bank. http://www.rscb.org/pdb. Accessed on July 2, 2013.
68. Banks, J. L.; et al. Integrated Modeling Program, Applied Chemical Theory (IMPACT), *J. Comput. Chem.* **2005**, *26*, 1752–1780.
69. Shoichet, B. K.; Kuntz, I. D.; Bodian, D. L. Molecular docking using shape descriptors, *J. Comput. Chem.*, **2004**, *13*, 380–397.
70. Reddy, A. S.; et al. Virtual screening in drug discovery – A computational perspective, *Curr. Protein Pept Sci.*, **2007**, *8*, 329–351.
71. (a) GLIDE, Schrödinger Inc. http://www.schrodinger.com. Accessed on July 2, 2013. (b) Friesner, R. A.; et al. Glide: a new approach for rapid, accurate docking and scoring. 1. Method and assessment of docking accuracy, *J. Med. Chem.*, **2004**, *47*, 1739–1749.
72. McMartin, C.; Bohacek, R. S. QXP: Powerful, rapid computer algorithms for structure-based drug design, *J. Comput.-Aided Mol. Des.*, **1997**, *11*, 333–344.
73. Jones, G.; et al. Development and validation of a genetic algorithm for flexible docking, *J. Mol. Biol.*, **1997**, *267*, 727–748.
74. Morris, G. M.; et al. Distributed automated docking of flexible ligands to proteins: parallel applications of AutoDock 2.4, *J. Comput.-Aided Mol. Des.*, **1996**, *10*, 293–304.
75. (a) FRED, OpenEye Scientific Software. http://www.eyesopen.com. Accessed on July 2, 2013. (b) McGann, M. FRED pose prediction and virtual screening accuracy, *J. Chem. Inf. Model.*, **2011**, *51*, 578–596.
76. Schnecke, V.; Kuhn, L. A. Virtual screening with solvation and ligand-induced complementarity, *Drug Des. Discov.*, **2000**, *20*, 171–190.
77. Lengauer, T.; Rarey, M. Computational methods for biomolecular docking, *Curr. Opin. Struct. Biol.*, **1996**, *6*, 402–406.
78. Ewing, T. J. A.; Kuntz, I. D. Critical evaluation of search algorithms for automated molecular docking and database screening, *J. Comp. Chem.*, **1997**, *18*, 1175–1189.
79. (a) Honig, B.; Nicholls, A. Classical electrostatics in biology and chemistry, *Science*, **1995**, *268*, 1144–1149. (b) Sitkoff, D.; Sharp, K. A.; Honig, B. Accurate calculation of hydration free energies using macroscopic solvent models, *J. Phys. Chem.*, **1994**, *98*, 1978–1988.
80. Jayaram, B.; Sprous, D.; Beveridge, D. L. Solvation free energy of biomacromolecules: Parameters for a modified generalized born model consistent with the AMBER force field, *J. Phys. Chem. B*, **1998**, *102*, 9571–9576.
81. (a) Böhm, H. J. The computer program LUDI: A new method for the de novo design of enzyme inhibitors, *J. Comp.-Aided Mol. Design*, **1992**, *6*, 61–78. (b) Sotriffer C. A.; Gohlke, H.; Klebe, G. Docking into knowledge-based potential fields: A comparative evaluation of DrugScore, *J. Med. Chem.*, **2002**, *45*, 1967–70.

82. Reulecke, I.; et al. Towards an integrated description of hydrogen bonding and dehydration: Decreasing false positives in virtual screening with the HYDE scoring function, *ChemMedChem*, **2008**, *3*, 885–897.

83. Wang, J.-G.; et al. Identification of some novel AHAS inhibitors via molecular docking and virtual screening approach, *Bioorg. Med. Chem.*, **2007**, *15*, 374–380.

84. ACD **2005**.1, Accelrys Inc. http://www.accelrys.com. Accessed on July 2, 2013.

85. Duggleby, R. G.; Pang, S. S. Acetohydroxyacid synthase, *J. Biochem. Mol. Biol.*, **2000**, *33*, 1–36.

86. Wang, J.; et al. Chemical synthesis, in vitro acetohydroxyacid synthase (AHAS) inhibition, herbicidal activity, and computational studies of isatin derivatives, *J. Agric. Food Chem.*, **2011**, *59*, 9892–9900.

87. Zhang, Q.; et al. Structure-based rational screening of novel hit compounds with structural diversity for cytochrome P450 sterol 14alpha-demethylase from *Penicillium digitatum*, *J. Chem. Inf. Model.*, **2010**, *50*, 317–325.

88. Podust, L. M.; Poulos, T. L.; Waterman, M. R. Crystal structure of cytochrome P450 14α-sterol demethylase (CYP51) from *Mycobacterium tuberculosis* in complex with azole inhibitors, *Proc. Natl. Acad. Sci. U. S. A.*, **2001**, *98*, 3068–3073.

89. Hindle, S. A.; et al. Flexible docking under pharmacophore type constraints, *J. Comput.-Aided Mol. Des.*, **2002**, *16*, 129–149.

90. Muegge, I.; Martin, Y. C. A general and fast scoring function for protein-ligand interactions: a simplified potential approach, *J. Med. Chem.*, **1999**, *42*, 791–804.

91. Camper, D. L.; Parker, M. H. A structure-based design approach to plant selective 4-hydroxyphenylpyruvate dioxygenase inhibitors, in "*ACS Symposium Series, Synthesis and Chemistry of Agrochemicals VII*", Lyga, J. W.; Theodoridis, G., eds., **2007**, *vol. 948*, pp. 105–117.

2.3

NATURAL PRODUCT-BASED LIBRARIES

Alan L. Harvey

2.3.1 INTRODUCTION

High-throughput screening (HTS) is the dominant approach used by pharmaceutical companies to find leads for new therapeutic agents for human diseases [1]. High-throughput screening can also be used to find pharmacological tools for dissecting physiological pathways (see References [2–4] for recent reviews). The same strategy and the same chemical collections can be applied to problems in plant chemical biology. This chapter outlines some of the recent work related to collections of natural products that can be used for screening, for either drug discovery or chemical biology purposes.

The success of HTS campaigns depends on the quality of the assay (e.g., the validity of the target in the context of the disease, the robustness of the assay to minimize artifacts) and the quality of the chemical collection used in the assay (overall chemical diversity, absence of "frequent-hitter" compounds and "promiscuous inhibitors") [5]. In terms of chemical diversity, there are many publications on how to estimate the molecular diversity of collections of compounds and how to compare different collections (e.g., see References [6–8] and references therein). It has been calculated that the theoretical number of compounds with up to 11 heavy atoms from C, N, O, and F is 26 million [9]. Given that a typical compound screening deck in a large pharmaceutical company may have one or two million compounds, it is perhaps surprising that any hits are found in HTS assays [6]. One possible answer is that the collections of compounds in existence are consciously or unconsciously biased toward metabolites and other natural products that

Plant Chemical Biology, First Edition. Edited by Dominique Audenaert and Paul Overvoorde.

tend to be biologically active [6]. An analysis of screening campaigns undertaken by Novartis revealed that the plates in the screening collection containing natural products gave a much higher hit rate than plates containing compounds from combinatorial chemistry or from traditional medicinal chemistry methods [8]. A recent review about compound collections written from an industrial perspective [10] emphasized the need to include natural product-like diversity in such collections, and there are compelling reasons to return to natural products for modern drug discovery [11].

The continuing contributions of natural products to the development of new medicines have been comprehensively reviewed (see, e.g., References [12–14]). Examples include antibacterial (e.g., retapamulin, a semisynthetic derivative of pleuromutilin; telavancin, a modified vancomycin), antifungal (e.g., the echinocandin derivative, anidulafungin), and anticancer agents (e.g., romidepsin, a microbial natural product; eribulin, a synthetic analog modeled on halichondrin). In 2010, 50% of the small-molecule new chemical entities that were approved were natural products or derivatives [14].

Natural products are also proving to be successful as probes for disturbing protein–protein interactions, and this ability will prove to be critical for further studies in chemical biology. Bauer et al. [15] give examples of unusual compounds with interesting activities relating to protein–protein interactions. It is notable that the chemical scaffolds of these compounds are underrepresented in traditional screening collections.

2.3.2 PLANT-BASED COLLECTIONS

Much of the research on the bioactivity of compounds from plants is based on focused investigations of traditionally used medicinal plants. There are relatively few readily available collections of plant extracts or compounds isolated from plants. In part, this may be due to the expectations regarding access and benefit-sharing arrangements under the terms of the United Nations Convention on Biological Diversity [16].

The French company Greenpharma offers a plant extract library in a multi-well plate format (www.greenpharma.com); in the United States, TimTec can provide 2600 extracts from plants from Eastern Eurasia (www.timtec.net), while AMRI offers a broad range of extracts from different sources, including plants (www.amriglobal.com); in Singapore, MerLion Pharmaceuticals also can provide a wide range of extracts, including some from almost 30,000 plant species (www.merlionpharma.com).

From noncommercial sources, the National Cancer Institute in the United States maintains a natural products repository that can be accessed for screening purposes, and this contains plant extracts as well as samples of marine origin ([17]; http://dtp.nci.nih.gov/branches/npb/repository.html). The University of Strathclyde in the United Kingdom has built up a screening collection through its network of collaborating institutions throughout the world (http://www.sulsa.ac.uk/facilities/SNPL). It contains extracts of 5000–6000 plant species that cover 90% of recognized plant families. The extracts have been prepared by exhaustive methanolic extraction followed by detannification and freeze-drying. Samples are reconstituted in DMSO and stored in 96-well plates. The collection is used for in-house drug discovery projects, and it can be supplied under license to external collaborators.

Plant extracts have been suggested to be more compatible with cell-based assays than with molecular assays [11, 18]. Certainly, extracts have been found to yield potent hits in cell-based assays. For example, Nesterov et al. [19] screened 6732 plant extracts on an assay in which a gene reporter was linked to pathways activated by cholecystokinin CCK1 receptor stimulation; 387 hits were found, although only two were selective when tested on a counter screen. Fractionation of one of these hit extracts gave a compound with an IC_{50} of 13 pM. Mechanism of action studies revealed that the compound did not interfere with the receptor itself, but on the G-protein coupled to the receptor.

Plant extracts typically contain hundreds of different compounds. Hence, using extracts in screening assays has the obvious disadvantage of not knowing the identity of the compound or compounds responsible for any bioactivity detected. There is also the possibility that activity could result from the combined effects of two or more constituents of the extract. Since it is generally not feasible to isolate all of the compounds in an extract in advance of screening and to test them singly, there have been various attempts to simplify the hit identification process by preparing less complex fractions from the starting extract. These have involved solid-phase extraction systems [20] or HPLC methods [21]. More recently, a method for rapidly producing fractions of plant extracts in screening-ready formats has been published [22]. Detannified extracts were fractionated into 24 separate fractions by HPLC and then dried. Importantly, the weight of each fraction was recorded so that the fractions could be reconstituted in multi-well plates at a standard concentration.

Plant extracts have also been processed to enrich the proportion of drug-like molecules on the basis of their physicochemical properties. By using solid-phase chromatography with appropriate properties, fractions could be obtained with log $P<5$, which is typical of drugs [23, 24]. These fractions could be subsequently sub-fractionated to give samples that were convenient for use in bioassays.

Another approach to making plant extracts more compatible with screening assays is to treat the extracts with chemicals that will combine with some of the more reactive chemical groupings on the natural products and make them more drug-like [25]. This could be done in a combinatorial fashion ("chemical conditioning") to enhance the drug-like diversity within extracts or in a more targeted manner ("chemically engineered extracts") using single reagents at a time [26]. Typically, this involves adding amine functions through carboxyl groups, introducing sulfur atoms via hydroxyl and amino groups, and bromination of aromatic rings. Such approaches have been shown to result in novel compounds with bioactivities not found in the starting materials [26].

2.3.3 MICROBIAL-DERIVED SAMPLES

Natural products from microbial sources have long been recognized for their potential as drugs. Undoubtedly, similar sources could be a rich source of compounds for studies in chemical biology. As is the case with plant extracts, there is also a lack of readily accessible microbial extracts for screening. Although a mainstay of drug discovery efforts in the pharmaceutical industry for many years, collections of microbial samples tend to remain proprietary [27]. Companies such as MerLion and AMRI will provide microbial samples, and there are smaller specialist companies such as Hypha Discovery that can provide libraries of fractions and extracts from microbial cultures (www.hyphadiscovery.co.uk). In the United States, Magellan BioScience has a collection of over 55,000 fungal samples (http://www.magellanbioscience.com/); in the United Kingdom, BioFocus offers access to a large collection of subfractions prepared from cultures of fungi and actinomycetes (http://www.biofocus.com/). The nongovernmental organization CABI has a very extensive collection of bacteria and fungi, and it can provide services including the production of samples for screening (http://www.cabi.org/). The Japanese research organization RIKEN has created a "Natural Products Depository" that contains microbial metabolites and that is available to collaborators (http://www.riken.go.jp/engn/r-world/research/lab/asi/cb-cp/index.html). This is being expanded to include partially purified fractions and pure compounds [28].

As it is well recognized that a very small proportion of bacterial species can be readily cultured in the laboratory, there is considerable effort directed at expanding the range of small molecules that can be obtained from "unculturable" species (see References [29, 30] for recent reviews). Techniques range from the traditional (culturing in a range of media, using chemicals to elicit the production of secondary metabolites [31]) to the genetic (metagenomics in order to sample environmental DNA extracted from soil samples through expression in readily culturable host organisms; see, e.g., References [32, 33]). However, these techniques do not yet appear to have been used to create large-scale collections of samples for use in screening campaigns.

2.3.4 SAMPLES OF MARINE ORIGIN

Marine organisms are very productive sources of unusual natural products that have been found to have a wide variety of biological activities [34]. There are organizations that have collected macro- and microorganisms from different environments, and a few collections are available for third-party screening. The Spanish company PharmaMar offers access to its very extensive collection of marine samples (http://www.pharmamar.com/partnering.aspx), and the U.S. company Magellan BioScience makes available a collection of samples based on marine microorganisms. As previously mentioned, the National Cancer Institute in the United States has an extensive collection of samples prepared from marine animals and plants [17]. In the United Kingdom, workers at the University of Aberdeen are building a collection of marine natural products that will be available for screening (http://www.abdn.ac.uk/ncs/chemistry/research/mbc/).

To make more productive use of scarce marine natural products, methods have been developed to produce libraries of pre-fractionated extracts [35, 36]. Other approaches to improving access to natural products of marine origin are based on the recognition (or assumption) that much of the small-molecule chemical diversity is produced not by the marine macroorganisms but by some of the symbiotic bacteria that colonize them. Here again, there is the difficulty of getting the marine bacteria to grow in laboratory conditions and also to produce the same range of small molecules as synthesized *in situ*. While there has been progress in modifying culture conditions for marine bacteria, a promising approach is the metagenomic one, whereby clusters of biosynthetic genes are cloned from marine bacteria and engineered into more convenient host organisms (see Reference [29] for a review). Advances in the understanding of the production of polyketides by bacteria associated with sponges have been reviewed [37].

2.3.5 FUTURE PERSPECTIVES

Large collections of pure natural products that are readily available for screening are rare: BioFocus offers 500 compounds of mainly microbial origin (http://www.biofocus.com/); Enzo Life Sciences can provide 500 pure compounds that have mainly been derived from plants (http://www.enzolifesciences.com/); TimTec offers 720 pure natural products (http://www.timtec.net/); and AnalytiCon Discovery can provide 3400 compounds from plants and 1300 from microbial sources (http://www.ac-discovery.com/). An alternative approach has been to use natural product chemical scaffolds as the basis for creating libraries of synthetic or semisynthetic analogs (see, e.g., References [38–41]). AnalytiCon Discovery offers 20,000 synthetic compounds based on 60 naturally occurring chemical scaffolds (http://www.ac-discovery.com/).

While these methods provide samples of known structures, they are still labor intensive. There may be more benefit from processing natural products to enrich the original samples with compounds of drug-like physicochemical properties followed by rapid separation into single compounds of fractions containing a few different compounds. The focus on compounds with particular log P values has been mentioned earlier [23, 24], and there are developments of chromatographic resins that would separate compounds according to their key functional groups, such as amines, carboxyls, and hydroxyls [42].

Alternatively, there could be further developments of techniques that can directly detect interactions of natural products with their targets. This has been done through electrospray ionization-Fourier transform ion cyclotron resonance mass spectrometry for binding to RNA [43] and a model protein target [44].

2.3.6 CONCLUDING REMARKS

Natural products have been exceptionally useful in drug discovery [12, 14], and they have also been useful in providing tools for dissecting a wide variety of physiological pathways. Many academic groups and a few specialist companies have interests in finding novel biologically active compounds from different sources. There could be scope for chemical biology programs to screen broad collections of natural products in order to find new probes for specific targets. However, this will depend on collaborations developing. Additionally, there will continue to be the serendipitous discovery of useful probes for chemical biology based on observations of unusual phenotypic effects of natural products and subsequent investigations of the underlying mechanism. This can be exemplified by the recognition that the microbial compound FR901464, originally noted for its effects on cancer cells *in vitro*, is a specific inhibitor of SF3b, a protein critical for splicing mRNA [45].

REFERENCES

1. Maccaron R, et al. (2011). Impact of high-throughput screening in biomedical research. *Nature Reviews Drug Discovery 10*, 188–195.
2. Carlson EE (2010). Natural products as chemical probes. *ACS Chemical Biology 5*, 639–653.
3. La Clair JJ (2010). Natural product mode of action (MOA) studies: a link between natural and synthetic worlds. *Natural Product Reports 27*, 969–995.
4. Schmitt EK, et al. (2011). Natural products as catalysts for innovation: a pharmaceutical industry perspective. *Current Opinion in Chemical Biology 15*, 497–504.
5. Renner S, et al. (2011). Recent trends and observations in the design of high-quality screening collections. *Future Medicinal Chemistry 3*, 751–766.
6. Hert J, et al. (2010). Quantifying biogenic bias in screening libraries. *Nature Chemical Biology 5*, 479–483.
7. Singh N, et al. (2009). Chemoinformatic analysis of combinatorial libraries, drugs, natural products, and molecular libraries small molecule repository. *Journal of Chemical Information and Modeling 49*, 1010–1024.
8. Sukuru SC, et al. (2009). Plate-based diversity selection based on empirical HTS data to enhance the number of hits and their chemical diversity. *Journal of Biomolecular Screening 14*, 690–699.
9. Fink T, Reymond JL (2007). Virtual exploration of the chemical universe up to 11 atoms of C, N, O, F: assembly of 26.4 million structures (110.9 million stereoisomers) and analysis for new ring systems, stereochemistry, physicochemical properties, compound classes, and drug discovery. *Journal of Chemical Information and Modeling 47*, 342–353.
10. Drewry DH, Macarron R (2010). Enhancements of screening collections to address areas of unmet medical need: an industry perspective. *Current Opinion in Chemical Biology 14*, 289–298.
11. Carter GT. (2011). Natural products and pharma 2011: strategic changes spur new opportunities. *Natural Product Reports 28*, 1783–1789.
12. Harvey AL (2008). Natural products in drug discovery. *Drug Discovery Today 13*, 894–901.
13. Mishra BB, Tiwari VK (2011). Natural products: an evolving role in future drug discovery. *European Journal of Medicinal Chemistry 46*, 4769–4807.
14. Newman DJ, Cragg GM (2012). Natural products as sources of new drugs over the 30 years from 1981 to 2010. *Journal of Natural Products 75* (3), 311–335.
15. Bauer RA, Wurst JM, Tan DS (2010). Expanding the range of "druggable" targets with natural product-based libraries: an academic perspective. *Current Opinion in Chemical Biology 14*, 308–314.

16. Harvey AL, Gericke N (2011). Bioprospecting: creating a value for biodiversity. In *Research in Biodiversity – Models and Applications*, ed. IY Pavlinov. Rijeka, Croatia: InTech Open Access Publishing, pp. 323–338.
17. McCloud TG (2010). High throughput extraction of plant, marine and fungal specimens for preservation of biologically active molecules. *Molecules 15*, 4526–4563.
18. Rishton GM (2008). Natural products as a robust source of new drugs and drug leads: past successes and present day issues. *American Journal of Cardiology 101*, 43D–49D.
19. Nesterov A, et al. (2010). Screening a plant extract library for inhibitors of cholecystokinin receptor CCK1 pathways. *Journal of Biomolecular Screening 15*, 518–527.
20. Thiericke R (2000). Drug discovery from nature: automated high-quality sample preparation. *Journal of Automated Methods and Management in Chemistry 22*, 149–157.
21. Eldridge GR, et al. (2002). High-throughput method for the production and analysis of large natural product libraries for drug discovery. *Analytical Chemistry 74*, 3963–3971.
22. Tu Y, et al. (2010). Automated high-throughput system to fractionate plant natural products for drug discovery. *Journal of Natural Products 73*, 751–754.
23. Quinn RJ, et al. (2008). Developing a drug-like natural product library. *Journal of Natural Products 71*, 464–468.
24. Camp D, et al. (2012). Drug-like properties: guiding principles for the design of natural product libraries. *Journal of Natural Products 75*, 72–81.
25. Rishton GM (2008). Molecular diversity in the context of leadlikeness: compound properties that enable effective biochemical screening. *Current Opinion in Chemical Biology 12*, 340–351.
26. Ramallo IA, et al. (2011). Chemically engineered extracts: source of bioactive compounds. *Accounts of Chemical Research 44*, 241–250.
27. Wagenaar MM (2008). Pre-fractionated microbial samples – the second generation natural products library at Wyeth. *Molecules 13*, 1406–1426.
28. Kato N, et al. (2012). Construction of a microbial natural product library for chemical biology studies. *Current Opinion in Chemical Biology 16*(1–2), 101–108.
29. Liu X, et al. (2010). Bioprospecting microbial natural product libraries from the marine environment for drug discovery. *Journal of Antibiotics 63*, 415–422.
30. Davies J (2011). How to discover new antibiotics: harvesting the parvome. *Current Opinion in Chemical Biology 15*, 5–10.
31. Pettit RK (2010). Small-molecule elicitation of microbial secondary metabolites. *Microbial Biotechnology 4*, 471–478.
32. Banik JJ, et al. (2010). Tailoring enzyme-rich environmental DNA clones: a source of enzymes for generating libraries of unnatural natural products. *Journal of the American Chemical Society 132*, 15661–15670.
33. Craig JW, et al. (2010). Expanding small-molecule functional metagenomics through parallel screening of broad-host-range cosmid environmental DNA libraries in diverse proteobacteria. *Applied Environmental Microbiology 76*, 1633–1641.
34. Mayer AM, et al. (2010). The odyssey of marine pharmaceuticals: a current pipeline perspective. *Trends in Pharmacological Sciences 31*, 255–265.
35. Bugni TS, et al. (2008). Marine natural product libraries for high-throughput screening and rapid drug discovery. *Journal of Natural Products 71*, 1095–1098.
36. Bugni TS, et al. (2008). Fractionated marine invertebrate extract libraries for drug discovery. *Molecules 13*, 1372–1383.

37. Hochmuth T, Piel J (2009). Polyketide synthases of bacterial symbionts in sponges – evolution-based applications in natural products research. *Phytochemistry 70*, 1841–1849.
38. Cordier C, et al. (2008). Natural products as an inspiration in the diversity-oriented synthesis of bioactive compound libraries. *Natural Product Reports 25*, 719–737.
39. Grabowski K, Baringhaus KH, Schneider G (2008). Scaffold diversity of natural products: inspiration for combinatorial library design. *Natural Product Report 25*, 892–904.
40. Wilk W, et al. (2010). Principles, implementation, and application of biology-oriented synthesis (BIOS). *Biological Chemistry 391*, 491–497.
41. Thomas GL, Johannes CW (2011). Natural product-like synthetic libraries. *Current Opinion in Chemical Biology 15*, 1–7.
42. Odendaal AY, Trader DJ, Carlson EE. (2011). Chemoselective enrichment for natural products discovery. *Chemical Science 2*, 760–764.
43. Cummins LL, et al. (2003). Multitarget affinity/specificity screening of natural products: finding and characterizing high-affinity ligands from complex mixtures by using high-performance mass spectrometry. *Journal of Natural Products 66*, 1186–1190.
44. Vu H, Pham NB, Quinn RJ. (2008). Direct screening of natural product extracts using mass spectrometry. *Journal of Biomolecular Screening 13*, 265–275.
45. Kaida D, et al. (2007). Spliceostatin A targets SF3b and inhibits both splicing and nuclear retention of pre-mRNA. *Nature Chemical Biology 3*, 576–583.

PART THREE

IDENTIFICATION OF NEW CHEMICAL TOOLS BY HIGH-THROUGHPUT SCREENING

3.1

ASSAY DESIGN FOR HIGH-THROUGHPUT SCREENING

Frank W. An and Jose R. Perez

3.1.1 INTRODUCTION

High-throughput screening (HTS) emerged in the early 1990s [1, 2] as an approach to investigate a broad sampling of diverse compounds for the identification of modulators of a specific biology. High-throughput screening now typically refers to a screening campaign where 10,000–100,000 compounds are tested in a day. Compounds identified in these investigations could be used to further elaborate the biology being investigated or serve as a starting point for the development of novel therapeutic agents. High-throughput screening has been and is still a core function for new lead identification in pharmaceutical industry, with approximately 50–80% of new leads being derived from HTS [3]. Further, a number of recent FDA-approved drugs originated from an HTS, such as Maraviroc (Pfizer), Dasatinib (Bristol-Myers Squibb), and Eltrombopag (GlaxoSmithKline) [3, 4]. Growing interest in HTS has spread to the academic sector in recent years [5, 6], among the most notable being funding by the U.S. National Institute of Health (NIH) of the Molecular Libraries Screening Center Network as a pilot phase in 2004 and the Molecular Libraries Production Center Network (MLPCN) as a second phase in 2008. The goal of this effort was to support HTS and structure–activity relationship (SAR) studies in the public sector for the development of small-molecule probes (http://mli.nih.gov/mli/mlpcn/).

To achieve the high throughput, HTS relies on assorted instrumentations including liquid handlers (dispensers, washers, micro-volume compound addition tools), multimode plate readers, incubators, and plate sealers/peelers. Some of

Plant Chemical Biology, First Edition. Edited by Dominique Audenaert and Paul Overvoorde.

these instruments, such as liquid dispensers and plate readers, can have plate stackers to work in a semiautomated mode. However, in a fully integrated system, a central automated robotic platform connects the diverse instrumentation and schedules all activities in a coordinated and concurring manner. In addition, a large and diverse compound collection of several hundred thousand to several million compounds, often filtered to have "drug-like" chemicophysical properties conforming to the Lipinski Rule of Five [7], is typical among large screening centers. These large compound collections sometimes are supplemented by smaller-sized (in thousands) "focused libraries" that target a specific target class, such as GPCR, kinase, or ion channels.

Although sophisticated instrumentation and compound collection are necessary "hardware" for HTS, a successful execution of an HTS campaign foremost requires an appropriate and robust primary assay. This chapter focuses on the discussion of critical considerations and practices during assay design and development, from conception to execution of the HTS process.

3.1.2 APPROACHING ASSAY DEVELOPMENT

Typically, developing a primary assay suitable for HTS is the first step after the initial biological characterization and validation of the target or pathway, followed by adaptation to automation, the actual HTS campaign, and subsequent confirmatory and secondary assays that verify and establish the activity, potency, and selectivity of hits. In this section, we focus on a number of critical considerations that need to be thought through before investigators embark on actual bench work and develop an assay.

3.1.2.1 Biological Relevance

Probably the most far-reaching question to ask yourself in planning an assay is its biological relevance, namely, the assay that is developed should be the best possible model for the biology one wants to target. This sounds overly simplified but it is not always that easy. If the assay does not accurately mirror the biology, then subsequent efforts, no matter how successful, may not identify molecules that will modulate the biology of interests. A few assay considerations to ensure that the assay mirrors the target biology are as follows. First, the assay's context needs to be as physiological as possible. One example of this would be the study of cytokine IL-1 production in primary monocytes [8, 9].The expression and processing of IL-1 could not be accurately modeled in an HEK 293 or HeLa (ATCC) cell system. Similarly, the cholesterol biosynthesis pathway is best studied in a hepatic cell lineage such as HepG2, Huh-7 cells. In developing *in vitro* enzyme assays, the Investigator may need to choose between using a truncated enzyme versus full length (e.g., kinase domain or full-length kinase). The use of full-length protein would be preferred, but often the full-length protein is difficult to express or is not active while the truncated protein may not model the protein's native conformation. Second, an assay format/technology should facilitate targeting the biological endpoint. For example, a cell-surface receptor binding assay using a labeled ligand only enables the identification of compounds that disrupt binding to that target site, while a cell-based functional assay (e.g., monitoring Ca^{2+} mobilization or resulting secondary messengers) will allow the identification of modulators with diverse modes of actions. In addition, a target may have two related biological activities, such as conformational changes of nuclear receptor upon ligand binding and subsequent translocation into nucleus [10]. Other times, a single readout might not be sufficient to accurately describe the biology, such as in the case of predicting compound hepatotoxicity in a cell-based system [11]. In these scenarios, a multiple-readout assay, such as the high-content, image-based technology, would be not only preferable over a single-readout plate reader format but also necessary [10–12]. Finally, assay conditions need to be biased for the identification of compounds with the desired mode of activity. For example, if the allosteric site of a kinase is the target, ATP concentrations several times the K_m concentrations would help bias the assay against competitive inhibition at the ATP-binding pocket. In case where the experimental conditions cannot be easily manipulated, secondary assays that verify the desired mode of action and/or rule out the undesired mode of action need to be designed in advance. It is recognized that often investigators choose an approach based on simplicity, familiarity, reagent availability, or some other non-biology-driven rationales, such as timelines. However, in our opinion, the best and most successful practice is to always push toward disease-relevant biology.

3.1.2.2 Prior Arts

It has been our experience that often assays or even full HTS campaigns for the target of interest have been reported in literature. Investigators will want to investigate these previous efforts for two perspectives. First, are any of the assays reported in the literature technically and economically superior to the one being proposed for the current project? A small improvement in protocol or procedure may make a large positive difference in an HTS campaign. A good example is the evolution of calcium mobilization detection kits used on Fluorometric Imaging Plate Reader (FLIPR). The first generation of no-wash kit eliminated the dye wash step and greatly reduced assay failures, enhanced signal-to-background ratio, and increased throughput [13, 14]. Further follow-on improvement led to additional no-wash commercial kits that employed different technologies that proved to be better in performance [15]. Second, how did the published researchers run the assay? What were the researchers looking for? Did the assay serve their purpose? How does your own biology differ? For instance, despite earlier reports of multiple biochemical assays with purified phosphodiesterases (PDEs), investigators developed cell-based PDE4 assays in order to reproduce the physiological environment of PDE4 function and to eliminate compounds of poor cell permeability [16]. A comprehensive search of prior work will inform the investigator if an HTS will be run with the assurance that the best and most fit assay technology will be utilized.

3.1.2.3 Throughput

When approaching assay development, a researcher will need to consider what throughput he or she needs versus what is achievable based on the biology being investigated. Primary screens that test hundred thousands or millions of compounds require high throughput. The trade-off for speed and throughput is often information content. For a focused library screen or early hits validation of a few hundred to a few thousand compounds, a medium-throughput assay will satisfy the requirement. For a focused SAR survey, where only a few dozen compounds need to be tested each time, low-throughput assay is not only acceptable but sometimes better in providing a more authenticated method, richer information, or more physiological context. As an example of this, GSK3b activity can be measured by kinase activity in a high-throughput ADP-Glo biochemical assay for an HTS, followed by a medium-throughput microfluidics or β-catenin nuclear translocation assay (microscope based) for hit validation. Downstream of these assays, compounds can be tested in disease-relevant cell-based Tau phosphorylation assay [17] or the GSK3b specific substrate CRMP2 phosphorylation assay [18] (low-throughput ELISA or Western) to prioritized scaffold series. This transition from higher-throughput to higher-resolution or more biologically relevant assays is often evident in probe development project design.

3.1.2.4 Manual versus Automation

High-throughput screening does not necessarily mean fully automated screening on larger robotic platforms. Fully automated screen at our facility refers to software-controlled, robot-assisted, multiple instrument-integrated screens. Certain semi-automated or manual operations can achieve as good or better throughput by being more flexible and nimble. Semiautomated or manual operation here refers to researcher transporting plates between stand-alone instruments at predetermined time points. Sometimes, semiautomated or manual operation is also referred as workstation or walkup mode. As an example, one of the authors of this chapter has completed an HTS of DNA repair protein against 360,000 compounds in walkup mode in 7 h with the help of a colleague. The same screen on a fully automated system would have taken approximately 3 days. Having thought about the mode of screening helps the researcher take advantage of the unique features associated with each mode. The major advantages for automation-based screening are a significant reduced need for human intervention, extended work hours, and consistency in procedure timing. The major disadvantages for automated screening are availability of automation systems when multiple campaigns need to be run, more stringent requirement on reagent stability, robotic failures, reliance on the existing capability of scheduler software and hardware specifications to improve throughput. In contrast, the one distinctive advantage for walkup workstation screening is flexibility in choice of instruments, scheduling, equipment maintenance, and availability. One major disadvantage for manual screening is the variability introduced in the timing of each step as each instrument processes plates at slightly different rates. However, this can be minimized by handling plates in smaller batches within tolerable time variation specific to given assays. We often practice a combined mode of execution in our screening facility to take advantage of both manual and automated screening, such as seeding and culturing cells manually offline and using the automated system for liquid dispense, compound addition, and reading.

3.1.2.5 Data Format and Informatics

High-throughput screening generates an enormous amount of data. Screening centers need to build a large informatics infrastructure and computational capacity for data capture, storage, association, retrieval, and analysis (including but not limited to data normalization and correction, hit calling, cherry-picking dose responses, and structural analysis). The procedures and software packages involved therein, once developed, become standard for the institution and are often difficult to accommodate deviations from the standard input data format or data processing work flow, at least not at the screen practitioner level without particular help from informatics expert. Screen practitioners are therefore reminded to think about these existing data processing structure and its limitations beforehand to ensure that the HTS data can be readily captured and analyzed. This is usually less of an issue for plate reader data unless they utilize a less common plate format, such as 96-well plates for a secondary assay in a center that runs almost exclusively 1536- or 384-well plate formats. It can be a significant issue for high-content screening where multiple readouts and data output formats need to be transformed to fit existing data processing pipeline. In any nonstandard data generation situation or whenever in doubt, screen practitioners can save tremendous time and frustration at the post-HTS data analysis stage if they work with their informatics support team before deciding the assay format and jointly map out a path for data capture and analysis.

3.1.2.6 Other Practical Factors

Timeline is a common consideration in HTS practices. Often a project only has a specific window of time to get ready for screening due to system availability or an inflexible deadline. Assay developers will need to ask themselves the question whether they have time to work on a more information-rich, but more complex assay, and balance out information content and throughput against time spent. For example, if a cell-based assay works well in a 384-well format and one has to deliver the screen by an approaching deadline, one may opt not to miniaturize it further to a 1536-well format because the time saved in conducting a 1536-well format assay may not be enough to compensate the extra time spent on developing a robust assay in this higher-density format. In addition to timeline, reagent and consumable cost, reagent availability, and familiarity with specific assays are also factors that have considerable impact on assay design in practice. In conceptualization and design of assays for HTS, investigators need to balance these practical considerations and the need to deliver an assay that is biologically relevant, high-throughput-compatible, and operationally efficient and economical.

3.1.3 ASSAY DEVELOPMENT

Assay development refers to the process where a specific biological event, from a biochemical reaction to a whole cell phenotypic change, is modeled in a well and transformed into a quantifiable, HTS-executable format with acceptable robustness, throughput, and cost.

3.1.3.1 Characteristics of a Good Assay

A frequently used measurement for HTS assay robustness is Z′ factor, which is a mathematical descriptor developed by Zhang et al. [19] as follows:

$$Z' = 1 \quad \frac{3 \times (\text{std1} + \text{std2})}{|\text{mean1} - \text{mean2}|}$$

std1, 2: standard deviation for neutral and positive controls, respectively. Mean 1, 2: the average of neutral and positive controls, respectively.

Z′ factor measures the degree of separation of the neutral and positive control populations. $Z' \geq 0.5$ indicates a robust assay. $Z<0.5$ but greater than 0 suggests an assay of borderline quality. Screening in duplicate may be needed to call hits with confidence. $Z' < 0$ denotes an assay of poor quality likely resulting in a high false-positive rate. To achieve a high Z′ values, one wants to find a minimum of a five fold signal-over-background ratio with percentage coefficient of variation (% CV, the ratio of standard deviation and mean) less than 15% of neutral and positive controls. Improvements in the Z′ value can be achieved through increases in the signal-to-noise ratio or decreases in variation or both. Because standard deviations contribute significantly to Z′ values, Z′ is also a uniformity indicator of the assay. Typically, it is easier to achieve higher Z′ for homogeneous biochemical assays than for cell-based assays due to higher variation in cell growth and physiology.

3.1.3.2 Simplicity of Assays

Regardless of the type of HTS campaign, automated or walkup, simplicity is one of the keys to achieve high data quality and process efficiency in this instrument-based process. Many assays captured in the literature are not readily suitable for HTS. They may use larger reaction volumes or contain a number of reagent additions, transfers, washes, or centrifugations. While these steps are quite readily manageable for small numbers of samples, they are not readily adaptable to HTS and would make the process extremely cumbersome and error prone. A good HTS protocol has the minimum number of steps and is as close as possible to the ideal "mix-and-read" paradigm, where all biological reagents are dispensed in one or two steps, compounds are predispensed in wells or added in one step, the reaction is incubated for a short period of time, and the plate is read. Investigators are encouraged to take a prototype assay and eliminate as many steps as possible without affecting the assay outcome, such as removing spinning or shaking and eliminating or minimizing number of washes. In some extreme cases, the existing assay format or technology needs to be completely redeveloped for HTS. An example would be a multi-wash step ELISA protocol to investigate protein–protein interaction; this protocol in our center was redeveloped to a no-wash alpha screen (a bead-based proximity technology developed by Perkin Elmer).

3.1.3.3 Miniaturization

Execution of an HTS requires that the assay be adapted to the microtiter plate format. Microtiter plates are available in multiple well densities ranging from 6 to 3456 wells per plate. Currently, most HTS assays use either the 384- or 1536-well formats, with reaction volumes typically between 20 and 80 μL for standard 384-well plates, 10 and 20 μL for low-volume 384-well plates, and 2 and 10 μL for 1536-well plates. Most biochemical assays can be readily miniaturized directly from assay tubes to a 384-well format. Cell-based assay formats have become more adaptable to the 384-well format. Usually, miniaturization to 384 requires two steps, first being the verification of the conservation of the signal-to-noise ratio and % CV in a smaller volume in a 384-well plate format (using single or multichannel pipette) and second being measuring assay performance with liquid handling instruments (dispense, wash, and transfer). It is common to observe a loss in signal-to-noise ratio as the assay is miniaturized to higher-density formats. In general, when miniaturizing from 96- to 384-well formats, a 50% loss in signal is observed; however, a 50–75% reduction in reagent use is realized. The loss in signal-to-noise ratio is less of a problem with a robust assay but can become a limiting factor in miniaturization to higher-density formats. Worth noting, miniaturization of cell-based assays can alter the biological response and may require re-optimization of the assay conditions. Positional effects may also manifest themselves at this stage. A positional effect refers to a phenomenon where the readout is consistently stronger or weaker at certain locations of a microtiter plate, usually at the corners, at the edges, or as a gradient across the plate. Most of the time, the positional effect is caused by environmental factors, which can be difficult to identify, and the solution is often empirically determined. Use of ambient temperature for biochemical incubations and the use of efficient circulating incubator for cellular assays tend to mitigate much of the positional effect(s). For these reasons, it is recommended that assay development and miniaturization experiment uses the same type of instruments that will be used for HTS to minimize this problem. Some HTS data analysis software such as Genedata (Genedata Inc.) can correct, to a degree, the positional effect computationally.

Miniaturization to a 1536-well format represents an additional distinct challenge than that to a 384-well format. Liquid handling and compound transfer becomes more difficult as an assay is transferred into the 1536-well format, and requires equipment specially designed with sub-microliter volume and positional accuracy for a 1536-well format. Most assay run in the 1536-well format are addition-only homogeneous assays as filtration and liquid transfer out of the 1536 microtiter plate present a significant technical challenge. Since the volume is considerably smaller in the 1536 format than that in the 384 format, and throughput nearly four times higher, it is recommended that biochemical assays be miniaturized, or at least attempted, to save reagents and time. To achieve compound transfer into the 1536 plates, compounds in DMSO are often added to a dry plate using an acoustic transfer instrument prior to running of the assay. These plates are referred to as "Assay-Ready Plates" (ARP). These ARPs can be prepared up to six months in advance without significant loss in activity of the compound. Cell-based assays in the 1536-well format have been reported [16, 20, 21], but such assays, particularly those involving adherent cells, generally are considered difficult, given very small surface areas, small number of cells, small volume, and their impact on cell growth and stochastic response.

3.1.3.4 Importance of Controls

It is important to include positive controls, as well as neutral (vehicle) controls, while developing and executing an assay. Positive controls can be small-molecule modulators or biological agents or conditions (e.g., no enzyme for inhibitor screens). The main difference between compound and biological controls is that small-molecule controls when used judicially validate the sensitivity of the assays by showing the detectability of active compounds under assay conditions. In contrast, biological controls would not provide such information. Therefore, a small-molecule compound control is always preferred to a biological control. Biological controls can be convenient in cases when small-molecule controls are not available. It is a good practice to have both positive controls and neutral controls on the same assay plate so that data generated on each plate can be scaled to their respective controls, minimizing the impact of inter-plate variations and permitting across-plate analysis of all plates in a campaign.

3.1.3.5 Reproducibility

A good assay is reproducible. Reproducibility means the signal-to-background, raw or scaled values are constant across the plate, plate to plate, and day to day. Reproducibility depends on not only consistent performance of reagents and instrumentation but also the assay being less susceptible to variation due to factors that are hard to predict or control, such as variation in ambient temperature, frequent temperature changes with opening and closing of incubator doors, and microenvironment of individual wells. Often if extraordinary care is needed to make an assay work, it could be a warning sign that the assay tends to suffer variability during HTS where human involvement is intentionally reduced to a minimum. A well-developed, robust assay will not be susceptible to these factors and will readily adapt to an automated HTS.

3.1.3.6 Reagent Stability

Reagent stability needs to be carefully established during assay development, even more so for runs on automated systems. Instability of reagents can result in loss of signal over time or lead to inconsistent signal levels beyond normalization across the run. Such reagent instability can be managed by multiple reagent additions to the reservoirs during the HTS run, but has the drawback in that it is taxing for screen team staff and limits throughput in an automated screen. Stability for enzymes sometimes can be enhanced by the addition of mild dilute detergent such as 0.01–0.001% Brij 35 and 0.1% BSA or in the presence of one substrate. Lowering temperature for reagent reservoir is another option, and the reaction temperature can be restored by running the tubing through a water bath set at the desired temperature.

3.1.3.7 Reagents and Consumables Availability

At the assay development stage, investigators also need to consider the supply of necessary reagents and consumables. Availability of sufficient volume or mass of custom reagents, such as an enzyme, can affect the start timing of campaign. In evaluating and comparing commercial consumable items, vendor facilitation is a consideration. Well-planned HTS campaigns always have the necessary stocks in-house or have a guaranteed delivery of consumables on a predetermined schedule, in particular the items that require long lead time to procure, such as bar-coded plates and large quantity of enzymes. Not having readily available consumables can lead to not only an interruption in the specific HTS campaign but also the missing of scheduled instrument time and disruption of the larger facility screening portfolio.

3.1.3.8 Ruling Out Artifacts by Counter Screens

With all detection techniques, there is a risk of false-positive artifacts. This applies to, both, biochemical screens where coupled enzymes are included and cell-based reporter assays that utilize enzymatic reactions. A common example would be luciferase inhibitors or activators from luciferase coupled or luciferase reporter assays [22, 23]. Counter screens, where only the coupled enzyme or the reporter enzyme is present, are needed to rule out such false positives. Another type of false-positives is compound-mediated as seen in fluorescence intensity (FI) assays by autofluorescence (for gain of signal assays) or quench (for loss of signal assays) of test compounds. A technique used to eliminate the positive FI contribution of a compound is to take multiple reads and use the difference as the change in FI. Natively fluorescent compounds have also been found to interfere with fluorescent polarization (FP). These can be identified by looking at the raw data in either the S or P polarized plain and will appear as a large increase in overall FI. When building a screening funnel, reagent artifact counter screens and orthogonal assays should be included to verify the activity observed in the primary assay.

3.1.4 ASSAY TYPES AND CONSIDERATIONS

3.1.4.1 Overview of Biochemical and Cell-Based Assays

Assays tend to fall into two categories: biochemical (i.e., target based) or cell based. The biochemical assays utilize purified recombinant or native proteins to recapitulate biological process in a well. Such assays assess enzymatic activities [24, 25], receptor–ligand binding [26], or protein–protein interactions [27] in cell-free environment. This approach has been a standard mainstream approach in the pharmaceutical industry for decades. In the last decade, cell-based assays have become a significant portion of HTS portfolio [28–30], and for the first time more first-in-class small-molecule drugs were originated from cell-based phenotypic assays than target-based biochemical assays [30].

The advantage of a defined target is that it enables investigators to identify direct modulators of the specific target. Since the target is defined, available function and structural information can be leveraged for this particular target or this class of targets. Compounds identified using this approach can readily be optimized for potency and subsequently desired selectivity through intelligent SAR studies. A major drawback for biochemical assays is uncertainty of translation of efficacy into cell- or animal-based model systems due to the simplicity of the test systems. Further, cellular permeability and to a degree solubility are not monitored in this type of assay. Structure–activity relationship studies based solely on biochemical assays run the risks of generating a lead molecule that fails the subsequent cellular or animal model due to permeability or pharmacokinetic considerations. Therefore, SAR study including parallel cell-based assays is recommended.

In contrast to biochemical assays, cell-based assays model disease-relevant biology in the context of complex biology of the cells, including redundancy and cross talk between various biological signaling pathways. Cell-based assays include functional assays (e.g., secondary messenger generation and intracellular posttranslational protein modification), reporter gene assays (luciferase, beta-galactosidase, and beta-lactamase), and phenotypic assays that monitor cellular processes such as cell migration, cytokinesis, and viability [31–34]. Unlike biochemical assays, cell-based assays often do not require a priori knowledge or hypothesis of a target. In fact, all relevant cellular pathways can be interrogated in cell-based assays, thus increasing the repertoire of targets and possible chemical scaffolds for small-molecule leads [29, 35]. Cell-based assays also select for cell-permeable compounds and against those that result in acute toxicity. One major disadvantage with cell-based assays is the black box aspect where the target or targets that produced the readout are unknown. Target identification becomes a significant challenge that requires extensive effort, despite recent advances in this area [36–39]. In addition, some cellular readouts may be indirect and too downstream from the biological event of interest to complicate interpretation of small-molecule actions in the context of many intersecting signaling pathways. Decisions on which type of assays to use mostly depend on the specific biology and availability of purified reagents.

3.1.4.2 Biochemical Assay Conditions/Considerations

Biochemical or target-based assays usually measure enzyme activity, small-molecule binding, or protein–protein interaction. Functional assays, such as a receptor activation assay, usually are the first choice in that they are most direct, distinguish activity between agonist and antagonist, and permit detection of allosteric modulators. Receptor binding assays cannot satisfy the latter two objectives. However, binding assays are relatively simpler and easier to develop and are often good choice when the functional assays are difficult to set up or no functional readout is readily available (e.g., transcriptional factors). In addition to using site-specific isotope or fluorescent molecule-labeled tracers, thermal shift assays using nonselective binding dyes have gained popularity recently despite certain limitations [40]. The following is an illustration on how to approach target-based assay development for HTS using enzyme with known substrate(s) as an example.

Detection Technology. The first task is to decide on a detection platform. Many technologies exist to measure enzymatic activity. For example, an enzyme that uses NADPH as a cofactor to reduce or NADP to oxidize its substrate can be monitored by absorbance or FI of NADPH. In another example, kinase reactions can be monitored by phosphorylation of a peptide, production of ADP, or depletion of ATP, using radiometric (e.g., scintillation proximity assay (SPA), fluorescence polarization (FP), fluorescence resonance energy transfer (FRET)) or bioluminescence technologies. Sensitivity, compatibility to existing plate reader instrument, and cost are some of the factors to consider. If a coupled enzyme detection assay is the choice, a counter screen for coupled enzyme inhibition will be needed to rule out false positives.

K_m *and* V_{max} *Determination.* After an appropriate detection platform has been selected and all of the reagents have been secured, the investigator can start testing conditions to optimize the assay. One of the first experiments that should be performed is K_m and V_{max} determination using the Michaelis–Menten kinetics [41]. Data generated from a substrate titration experiment performed under constant enzyme conditions are graphed on an initial velocity [v] versus substrate concentration [S] plot yielding a hyperbolic curve. The K_m is equal to the substrate concentration at which one half maximal enzyme reaction velocity is achieved. It should be noted that the accurate measurement of K_m according to the Michaelis–Menten equation requires measuring the initial velocity (i.e., under substrate concentration-dependent first-order kinetics), which is recorded typically under conditions of less than 10% of substrate conversion; under 30% conversion is usually acceptable before observing the negative effects of substrate depletion. For multi-substrate reactions, the K_m value for one substrate is usually measured with the second substrate under saturating conditions. V_{max} is observed under saturating substrate conditions (i.e., zero-order kinetics) and in practice is equivalent to the maximum signal observed under given reaction conditions. Room temperature is preferred for biochemical HTS assays because it reduces temperature changes as the plates are being moved in and out of incubators, particularly for short incubations of less than 30 min.

Substrate Concentration. With K_m value(s) known, the investigator can set the concentrations of the substrates around K_m for an assay with maximum sensitivity for detecting inhibitors of all modalities including competitive, noncompetitive, and uncompetitive. One exception is if the biology intent changes, the aforementioned rule needs to be adjusted. For example, if the investigator does not wish to find ATP-competitive inhibitors, then a much higher ATP concentration would be desirable to bias the reaction to finding noncompetitive or uncompetitive inhibitors.

Enzyme Concentration. Enzyme titration and time course experiments need to be performed at the desired substrate concentration(s) to determine the optimal amount of enzyme and time that will be required to produce sufficient increase of signal over background. Typically, linear plots in enzyme versus product and time versus product experiments suggest a sensitive and robust reaction. Deviations of linearity, also referred to as shouldering, could suggest substrate depletion or enzyme stability issues. With the optimal substrate and enzyme concentrations and reaction time determined, the Z′ factor, which was described earlier, can then be determined using appropriate controls.

Reagent Stability. Finally, stability of the enzyme and other reaction components needs to be checked. Stability of these reaction components ideally needs to be on the order of 8–12 h at ambient temperature to enable an automated HTS to go overnight and maximize throughput. If a reagent is found to be unstable at room temperature over time, the investigator can consider multiple cycles of reagent replenishing per run or executing the HTS in walkup mode using smaller numbers of plates per batch.

3.1.4.3 Cell-Based Assay Considerations

Although the general set of considerations as discussed in Section 3.1.4.2 of this chapter applies to assay development for both biochemical and cell-based assays, cell-based assay development introduces additional factors that were not addressed in the biochemical assay development.

Cell Type and Cell Lines. It is logical that the biology being studied dictates the species or tissue types of cells to use. However, when it comes to the source of cells for the specific species or tissue, one needs to balance the relevance of the biology, availability, ease of operation, and cost in the context of HTS. While human primary cells are the most physiologically relevant for human disease-related studies, they are difficult to maintain and expensive to procure, and in some cases the desired response may diminish in the short term (e.g., [42]). Established cell lines, either transformed or immortalized, are the most commonly used in the HTS setting. Clonal difference between cell lines is common, and selection of a particular line to use is based on biology, annotation, conventional/common use in a certain field, and empirical preferences. Recently, induced pluripotent stem cell technology provides the promise of delivering large quantity of human cells with physiology close to that of human primary cells [43, 44].

Cell State. Cells in culture are living, replicating elements. Cell density and growth phase can have dramatic effects on responsiveness in terms of magnitude and directionality. The biology being studied will have a direct bearing on what growth conditions the cells should be in. An example would be studying signal transduction pathways. To study these pathways, the background is often reduced by growing the cells to higher density and then replacing the media with a low serum formulation. This process minimizes the background by minimizing growth and serum-induced signals. Under these quiescent conditions, the observed reporter activity is more directly dependent on pathway stimulation such as the addition of a receptor agonist. This is very different from a cell proliferation assay where to maximize the sensitivity to the effects of test agents on cell proliferation, the cells should be actively growing.

Cell Culture. The most basic underlying factor for robust cell-based assays is cell culture. Using cells under certain passage numbers to minimize phenotypic drift, consistent split frequency, uniform dispersion of cell aggregates, and quality control of media components are all important to keep the cells consistently in the desired physiological states. Attention to details and excellent record keeping during cell culturing are strongly recommended because most of the assay variations can often be traced back to issues with cell culturing.

Cell Density. Cell density will directly affect robustness of response in selected assays. Cells are often titrated to achieve the optimal signal-to-background ratio and lowest variation possible. All being equal, smaller number of cells is preferred to reduce reagent and human cost and space requirement.

Adherent and Suspension Cells. Both classes of cells have been used for HTS. Attention is needed when assays require medium exchanges, which is common to many cell-based assays. Instrument settings need to be carefully adjusted during aspiration to avoid cell loss, and during dispense to not disturb the monolayer of adherent cells.

Stable Cell Lines and Transient Transfections. Both stable lines and transiently transfected cell have been used in HTS. Stable cell lines take longer to establish, but are preferred in our opinion in the HTS setting as they produce much less variability as compared to transient transfection cell systems. The needed expression levels may drive the selection of one method over the other as a transient transfection model will provide a more robust response but will diminish over 72–96 hours. Precautions against very high expression level of exogenous protein are twofold. First, it risks introducing nonphysiological artifact of the target if the endogenous target protein is usually expressed at low to medium level, and second, it risks affecting other normal cellular processes.

Incubation Time. Some cellular responses take minutes to be measurable, such as phosphorylation that is often maximal in 15 minutes, while others may take days, such as luciferase reporter gene assays. It is a balance to minimize the time cells are exposed to compounds to reduce nonspecific toxicity while achieving a maximal signal.

DMSO Titration. Cell-based assays tend to be more sensitive to DMSO concentrations than biochemical assays. A titration of up to 1% DMSO is recommended. From the experience in our laboratory, 1% is usually the maximally tolerated concentration of DMSO in assays extending beyond a few hours.

Positional/Edge Effect. Cell-based assays are more susceptible to positional effect, which often happens along an edge of the assay plate. This is often associated with uneven warming of the plate in the incubator causing changes in cell growth conditions and uneven evaporation. Maintaining a maximal level of humidity and forced air circulation within the incubators tends to help reduce the severity of positional effect.

3.1.5 AUTOMATION ADAPTATION AND VALIDATION

Adaptation to automation is the next step in launching an HTS. After the assay has been developed and optimized manually, the assay protocol needs to be translated into a linear process adaptable to the HTS automation. The goal at this stage is to ensure that the quality of the assay is maintained on the actual instrumentation, that the HTS protocol is robust, and that an acceptable throughput is achieved. Homogeneous biochemical assays can typically be adapted quickly (1–2 weeks), while cell-based assays can take much longer depending on the assay biology. Certain mechanical steps during the automation process, such as frequent opening and closing of incubator doors or the timing of scheduled robotic plate movements, can negatively impact the assay performance, so the automation protocol will need to be validated experimentally.

The robustness and throughput of the automation protocol are first evaluated through computer simulations. Accurate timing of robotic movements is captured during an abbreviated dry run executing the new automation assay protocol with the required peripheral instrumentation in position. Such simulations will reveal any potential conflicts due to simultaneously occurring robotic moves. Adjustment of start time (such as adding delays to the timing when an assay plate enters the run) and assignment of proper prioritization of particular steps based on knowledge of assay biology (such as low robotic priority to enforce absolutely equal incubation period during a long incubation step) are often needed to provide flexibility to the scheduler software.

Several operational factors can also impact an automation protocol's efficiency. First, as mentioned before, a mixed operation of manual and automated screening may increase throughput. Procedures that involve long incubations, such as seeding cells or extended (>48 h) compound incubations, may be performed offline and brought back onto the automated platform at the end of specific incubation periods. Second, assuming flexibility of the automation platform, peripheral instruments need to be arranged to minimize robotic movement distances during consecutive steps. Third, screen runs need to be appropriately staffed for reagent replenishment and recovery from instrument errors, which are not uncommon during automated runs. For semiautomated screening runs, duration and the scheduling of instrument time on each separate workstation are key determining factors for batch size and the number of batches that can be processed in a day.

Validation of the automation protocol should be initially performed with a small number of compound plates. The authors routinely perform initial validation runs of seven 384-well plates or equivalent (approximately, 2000 compounds) in duplicate and repeated independently to assess the robustness of the signal and reproducibility across plates and runs. Next a scaled-up validation run of 40–100 plates is typically performed before commitment to a full-scale HTS run. This run size will enable the identification of potential issues not seen with a small number of screened plates such as the effect of temperature fluctuations due to frequent opening and closing of incubator doors or reagents settling or precipitating in supply lines over a long period of time. If validation results are not satisfactory, the assay conditions and/or the automation protocol will need to be adjusted to correct assay performance. Although data analysis software can sometimes help correct positional effects due to instrumentation-introduced variability or errors, experimental correction is always preferred over computational approaches.

3.1.6 CLOSING

HTS has become an integral part of drug and small-molecular probe discovery. Combining biology, a robust and correctly developed assay, and instrumentation to screen compounds en mass can provide an investigator early-stage chemical tools with which to further explore the targeted biological system or possibly seed drug discovery projects. A well-designed HTS assay, biochemical or cell based, targets the desired biology, delivers a statistically robust throughput of over tens of thousands of compounds screened per day, and is cost-effective to run. Adapting these assays to automation frees the investigator to pursue more fundamental biological questions or complex disease-relevant models and greatly enhances throughput. These technologies and instrumentations that were once mostly found in the industrial sectors have now reached the academic sector and can be accessed by all investigators to advance their science.

ACKNOWLEDGMENT

The authors are supported by NIH grant 1 U54HG005032-1.

REFERENCES

1. Mayr, L. M., and Fuerst, P. (2008) The future of high-throughput screening. *Journal of Biomolecular Screening* **13**, 443–448.
2. Paslay, J. W., Morin, J. E., and Harrison, R. K. (2010) High Throughput Screening in the Twenty-First Century Lead-Seeking Approaches. *Lead-Seeking Approaches* (Hayward, M. M. ed.), Springer, Berlin, Heidelberg. pp. 175–212.
3. Macarron, R., et al. (2011) Impact of high-throughput screening in biomedical research. *Nature Reviews Drug Discovery* **10**, 188–195.
4. Pereira, D. A., and Williams, J. A. (2007) Origin and evolution of high throughput screening. *British Journal of Pharmacology* **152**, 53–61.
5. Frearson, J. A., and Collie, I. T. (2009) HTS and hit finding in academia – from chemical genomics to drug discovery. *Drug Discovery Today* **14**, 1150–1158.
6. Roy, A., et al. (2010) Open access high throughput drug discovery in the public domain: a Mount Everest in the making. *Current Pharmaceutical Biotechnology* **11**, 764–778.
7. Lipinski, C. A., et al. (1997) Experimental and computational approaches to estimate solubility and permeability in drug discovery and development settings. *Advanced Drug Delivery Reviews* **23**, 3–25.
8. Perregaux, D. G., et al. (2000) ATP acts as an agonist to promote stimulus-induced secretion of IL-1β and IL-18 in human blood. *Journal of Immunology* **165**, 4615–4623.

9. Perregaux, D. G., et al. (2001) Identification and characterization of a novel class of interleukin-1 post-translational processing inhibitors. *The Journal of Pharmacology and Experimental Therapeutics* **299**, 187–197.
10. Jones, J. O., and Diamond, M. I. (2008) A cellular conformation-based screen for androgen receptor inhibitors. *ACS Chemical Biology* **3**, 412–418.
11. O'Brien, P. J., et al. (2006) High concordance of drug-induced human hepatotoxicity with in vitro cytotoxicity measured in a novel cell-based model using high content screening. *Archives of Toxicology* **80**, 580–604.
12. Jones, J. O., An, W. F., and Diamond, M. I. (2009) AR inhibitors identified by high-throughput microscopy detection of conformational change and subcellular localization. *ACS Chemical Biology* **4**, 199–208.
13. Schlag, B., and Dunlop, J. (2010) *Molecular Devices, FL IPR application note #53*. http://www.moleculardevices.com/Documents/general-documents/mkt-appnotes/flipr-appnotes/FLIPR%20App%20Note%2053%20%28Calcium%203%29%20rev%20B.pdf. Accessed on July 29, 2013.
14. Zhang, Y., et al. (2003) Evaluation of FLIPR calcium 3 assay kit—a new no-wash fluorescence calcium indicator reagent. *Journal of Biomolecular Screening* **8**, 571–577.
15. Xin, H., et al. (2007) Evaluation of no-wash calcium assay kits: enabling tools for calcium mobilization. *Journal of Biomolecular Screening* **12**, 705–714.
16. Titus, S. A., et al. (2008) A cell-based PDE4 assay in 1536-well plate format for high-throughput screening. *Journal of Biomolecular Screening* **13**, 609–618.
17. Martin, L., et al. (2011) The new indirubin derivative inhibitors of glycogen synthase kinase-3, 6-BIDECO and 6-BIMYEO, prevent tau phosphorylation and apoptosis induced by the inhibition of protein phosphatase-2A by okadaic acid in cultured neurons. *Journal of Neuroscience Research* **89**, 1802–1811.
18. Soutar, M. P. M., et al. (2010) Evidence that glycogen synthase kinase-3 isoforms have distinct substrate preference in the brain. *Journal of Neurochemistry* **115**, 974–983.
19. Zhang, J.-H., Chung, T. D. Y., and Oldenburg, K. R. (1999) A simple statistical parameter for use in evaluation and validation of high throughput screening assays. *Journal of Biomolecular Screening* **4**, 67–73.
20. Maffia, A. M., Kariv, I., and Oldenburg, K. R. (1999) Miniaturization of a mammalian cell-based assay: luciferase reporter gene readout in a 3 microliter 1536-well plate. *Journal of Biomolecular Screening* **4**, 137–142.
21. Seitz, P. M., et al. (2010) Development of a high-throughput cell-based assay for superoxide production in HL-60 cells. *Journal of Biomolecular Screening* **15**, 388–397.
22. Auld, D. S., et al. (2008) Characterization of chemical libraries for luciferase inhibitory activity. *Journal of Medicinal Chemistry* **51**, 2372–2386.
23. Auld, D. S., et al. (2008) Specific mechanism for nonspecific activation in reporter-gene assays. *ACS Chemical Biology* **3**, 463–470.
24. Burns, S., et al. (2006) Identification of small-molecule inhibitors of protein kinase B (PKB/AKT) in an AlphaScreen™ high-throughput screen. *Journal of Biomolecular Screening* **11**, 822–827.
25. Sudo, K., et al. (2005) High-throughput screening of low molecular weight NS3-NS4A protease inhibitors using a fluorescence resonance energy transfer substrate. *Antiviral Chemistry & Chemotherapy* **16**, 385–392.

26. Allen, M., Reeves, J., and Mellor, G. (2000) High throughput fluorescence polarization: a homogeneous alternative to radioligand binding for cell surface receptors. *Journal of Biomolecular Screening* **5**, 63–69.
27. Kenny, C. H., et al. (2003) Development of a fluorescence polarization assay to screen for inhibitors of the FtsZ/ZipA interaction. *Analytical Biochemistry* **323**, 224–233.
28. An, W., and Tolliday, N. (2010) Cell-based assays for high-throughput screening. *Molecular Biotechnology* **45**, 180–186.
29. Johnston, P. A., and Johnston, P. A. (2002) Cellular platforms for HTS: three case studies. *Drug Discovery Today* **7**, 353–363.
30. Swinney, D. C., and Anthony, J. (2011) How were new medicines discovered? *Nature Reviews Drug Discovery* **10**, 507–519.
31. Chambers, C., et al. (2003) *Combinatorial Chemistry & High Throughput Screening* **6**, 355–362.
32. Eggert, U. S., et al. (2004) Parallel chemical genetic and genome-wide RNAi screens identify cytokinesis inhibitors and targets. *PLoS Biology* **2**, e379.
33. Li, X., et al. (2007) Functional characterization of cell lines for high-throughput screening of human neuromedin U receptor subtype 2 specific agonists using a luciferase reporter gene assay. *European Journal of Pharmaceutics and Biopharmaceutics* **67**, 284–292.
34. Yarrow, J. C., et al. (2005) Screening for cell migration inhibitors via automated microscopy reveals a Rho-kinase inhibitor. *Chemistry & Biology* **12**, 385–395.
35. Krejci, P., Pejchalova, K., and Wilcox, W. (2007) Simple, mammalian cell-based assay for identification of inhibitors of the Erk MAP kinase pathway. *Investigational New Drugs* **25**, 391–395.
36. Chan, J. N., Nislow, C., and Emili, A. (2010) Recent advances and method development for drug target identification. *Trends in Pharmacological Sciences* **31**, 82–88.
37. Cong, F., Cheung, A. K., and Huang, S.-M. A. (2012) Chemical genetics-based target identification in drug discovery. *Annual Review of Pharmacology and Toxicology* **52**, 57–78.
38. Kasper, A. C., et al. (2009) The end game of chemical genetics: target identification. *Future Medicinal Chemistry* **1**, 727–736.
39. Smith, C. (2004) Drug target identification: a question of biology. *Nature* **428**, 225–231.
40. Senisterra, G., Chau, I., and Vedadi, M. (2012) Thermal denaturation assays in chemical biology. *Assay and Drug Development Technologies* **10**, 128–136.
41. Menten, L., and Michaelis, M. I. (1913) Die kinetik der invertinwirkung. *Biochemistry Zeitung* **49**, 333–369.
42. Hengstler, J. G., et al. (2000) Cryopreserved primary hepatocytes as a constantly available in vitro model for the evaluation of human and animal drug metabolism and enzyme induction. *Drug Metabolism Reviews* **32**, 81–118.
43. Das, A. K., and Pal, R. (2010) Induced pluripotent stem cells (iPSCs): the emergence of a new champion in stem cell technology-driven biomedical applications. *Journal of Tissue Engineering and Regenerative Medicine* **4**, 413–421.
44. Zhou, H., and Ding, S. (2010) Evolution of induced pluripotent stem cell technology. *Current Opinion in Hematology* **17**, 276–280.

PART FOUR

USE OF CHEMICAL BIOLOGY TO STUDY PLANT PHYSIOLOGY

4.1

USE OF CHEMICAL BIOLOGY TO UNDERSTAND AUXIN METABOLISM, SIGNALING, AND POLAR TRANSPORT

KEN-ICHIRO HAYASHI AND PAUL OVERVOORDE

4.1.1 INTRODUCTION

The class of plant hormones categorized as auxins play crucial roles in multiple aspects of plant development. These molecules exert their effects on cell division, cell elongation, and cell differentiation. In concert with other plant hormones, auxin mediates such diverse processes as embryogenesis, hypocotyl elongation, phyllotaxy, apical dominance, vascularization, and lateral root formation [1]. Auxin also participates in a plants response to its environment. Consequently, auxin-mediated events control the dynamic architecture of plants, and efforts during the past century to elucidate its mode of action have a rich and storied history [1–3].

Our current understanding of auxin biology has been enhanced by the approaches that meld synthetic chemistry with biological assays. In this chapter, we describe natural and synthetic molecules that function as auxins and provide an overview of how these compounds are transported in a polar fashion *in planta*. We also review ways that novel compounds have been used to define aspects of auxin metabolism, auxin transport, and auxin signaling.

Plant Chemical Biology, First Edition. Edited by Dominique Audenaert and Paul Overvoorde.

4.1.2 NATURALLY OCCURRING AUXINS

The term "auxin," which is derived from the Greek work "auxein" (= to increase, to help), was originally associated with molecules that promoted cell elongation in *Avena* (oat) coleoptiles. Citing the work of Berger and Avery along with that of Haagen Smit, Frits Went commented in a 1945 review [4], "Chemical isolation and identification of indole acetic acid from vascular plants has been accomplished. This makes it necessary to use the term 'auxin' as a generic name for all substances, produced by a plant as growth hormones or as a correlation carriers, which give a response in the *Avena* test." This physiological definition has complicated a chemical classification since structurally diverse molecules promote cell elongation and are therefore considered auxins. Furthermore, some of these molecules are naturally occurring and can be identified from plant extracts, and others, as described later, are synthetic and are not produced by plants.

Three of the four categories that make up the suite of naturally occurring auxins contain indole-3-acetic acid (IAA) as the core structure (Fig. 4.1.1**a** and **b**, Table 4.1.1). Although first identified as a growth-promoting substance found in human urine [5], IAA was subsequently identified as a metabolite in maize [6]. Since this time IAA has been isolated from a range of plants, including moss, fern, gymnosperms, and angiosperms, and has been isolated from a number of fungi. IAA is widely recognized as the predominant naturally occurring auxin in land plants. The halogenated IAA molecule, 4-chloroindole-3-acetic acid (4-Cl-IAA, Fig. 4.1.1**a** and **b**), is also active in bioassays and has been identified in several legumes and *Pinus sylvestris*, but not in *Arabidopsis* [7]. 4-Cl-IAA displays higher activity than IAA in several bioassays. The increased potency of 4-Cl-IAA could result from greater chemical stability or be due to inefficient removal through conjugation or oxidation. The detailed physiological roles of 4-Cl-IAA remain unclear [8]. A third category of IAA-based auxin is indole-3-butyric acid (IBA). Initially classified as a synthetic auxin, this molecule has been found as a metabolite in several plant species [9]. The working model is that IBA serves as a storage form of IAA. Strong genetic evidence indicates that through a pathway that resembles peroxisomal beta-oxidation of fatty acids, IBA is converted to IAA [10]. Local metabolism of IBA in the root cap contributes to the control of lateral root formation, likely through its conversion to IAA [11]. The biochemical characterization of enzymes involved in IBA metabolism should reveal regulatory aspects of IBA-IAA homeostasis [12]. The fourth example of a naturally occurring auxin is phenylacetic acid (PAA, Fig. 4.1.1**a** and **b**), which has been found in *Tropaeolum* [13]. PAA elicits less of a response than IAA, and little is known about its synthesis, perception, or signaling in physiological responses [8].

4.1.3 AUXIN BIOSYNTHESIS

Despite its central role, defining the biochemical pathways plants use to synthesize the predominant auxin, IAA continues to be an active area of research [14]. Two sets of biosynthetic pathways have been proposed—one that is tryptophan dependent and the other

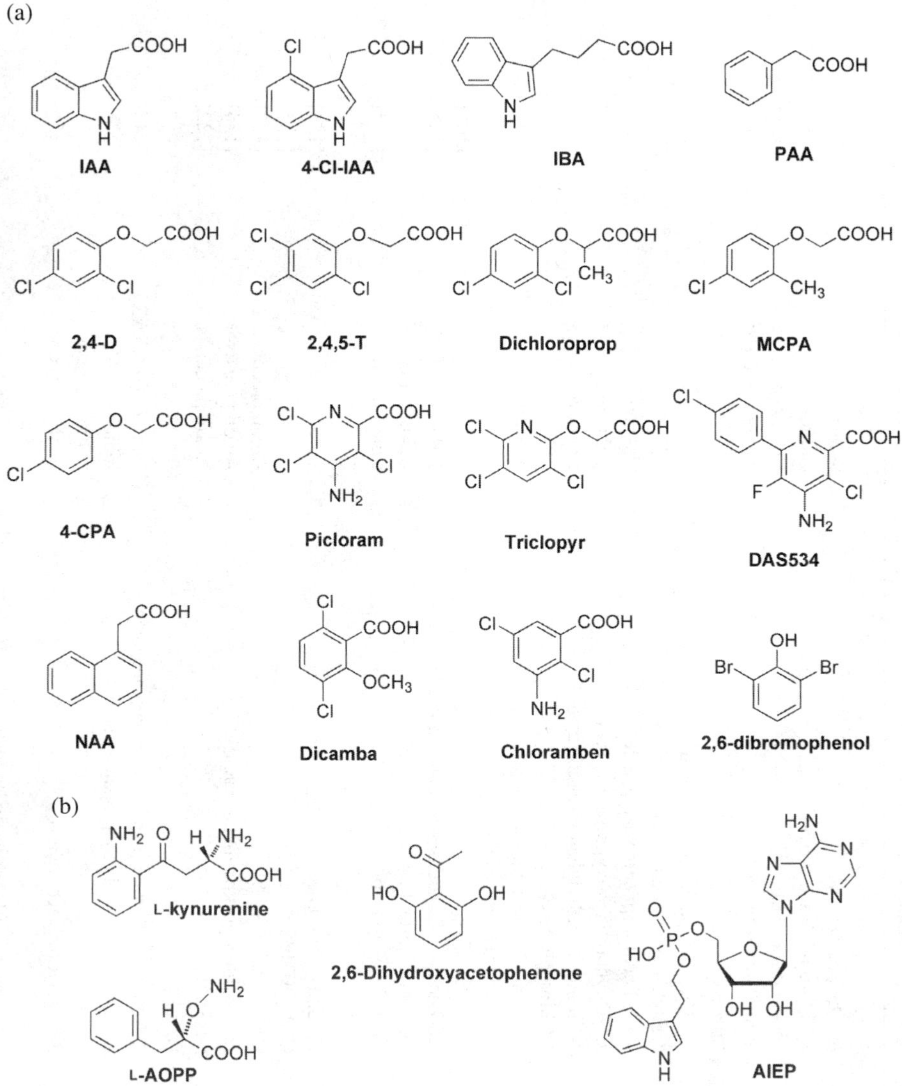

FIGURE 4.1.1 Overview of compounds used as auxins or that alter auxin metabolism. (**a**) Chemical structures of natural and synthetic auxins. See Table 4.1.1 for an explanation of acronyms. (**b**) Chemical structures of molecules that alter auxin metabolism. See Table 4.1.1 for an explanation of the acronyms and targets, where known.

that is tryptophan independent [15]. Recently, the convergence of work from several groups demonstrated that the tryptophan-dependent pathway consists of a simple two-step metabolic pathway [16–18]. First, tryptophan is converted to indole-3-pyruvic acid (IPyA) by amino transferases encoded by the three-member *TRYPTOPHAN AMINOTRANSFERASE OF ARABIDOPSIS1/TRYPTOPHAN AMINOTRANSFERASE*

TABLE 4.1.1 Chemizcal tools for auxin biology

Type name	Chemical name or CAS registry	Activity/target	References
Endogenous auxins			
IAA	Indole 3-acetic acid	$SCF^{TIR1/AFB}$	[77]
4-Cl-IAA	4-Chloroindoleacetic acid	Unknown	[7]
PAA	Phenylacetic acid	Unknown	[13]
IBA	Indole 3-butyric acid	Auxin precursor	[1]
Synthetic auxins and herbicides			
2,4-D	2,4-Dichloro-phenoxyacetic acid	$SCF^{TIR1/AFB}$	[77]
NAA	Naphthalene-1-acetic acid	$SCF^{TIR1/AFB}$	[77]
2,4,5-T	2,4,5-Trichlorophenoxyacetic acid	Unknown	
Dichloroprop	2,4-Dichlorophenoxypropanoic acid	Unknown	
MCPA	4-Chloro-2-methylphenoxyacetic acid	Unknown	
4-CPA	4-Chlorophenoxyacetic acid	Unknown	
Picloram	4-Amino-3,5,6-trichloropicolinic acid	AFB4/5	[80]
Triclopyr	2-((3,5,6-trichloropyridin-2-yl)oxy)acetic acid	unknown	
DAS534	CAS: 496851-93-5	AFB5	[104]
Dicambia	3,6-Dichloro-2-methoxybenzoic acid	TIR1 and AFB5	[105]
Chloramben	3-Amino-2,5-dichlorobenzoic acid	Unknown	
2,6-Dibromophenol	2,6-Dibromophenol	Unknown	[153]
Synthetic "proauxins"			
Compound 602	CAS: 313700-05-9		[123]
Compound 533	CAS: 303138-07-0		[123]
Sirtinol	CAS: 410536-97-9	Auxin precursor	[127]
2,4-DB	2,4-Dichlorophenoxybutyric acid	Auxin precursor	[154]
W7	CAS: 91192-94-8	Auxin precursor	[122]
Alters auxin metabolism			
L-kynurenine	2-Amino-4-(2-aminophenyl)-4-oxobutanoic acid	Inhibits TAA/TAR	[24]
L-AOPP	2-(Aminooxy)-3-phenylpropanoic acid	Inhibits IAA biosynthesis	[27]
2,6-DHA	2,6-Dihydroxyacetophenone	Inhibits IAA conjugation	[33]
AIEP	Adenosine-5′-(2-indoleethyl)phosphate	Inhibits GH3	[34]

Inhibitors of auxin transport			
NPA	2-(Naphthalen-1-ylcarbamoyl)benzoic acid	PIN and PGP	[140]
TIBA	2,3,5-Triiodobenzoic acid	Vesicle trafficking	[149]
PBA	2-(1-Pyrenoyl) benzoic acid	Vesicle trafficking	[149]
1-NOA	1-Naphtoxy acetic acid	Auxin influx transport	[146]
Gravacin	188438-05-3	ABCB19/PGP19	[134]
BUM	5809-23-4	ABCB transporters	[144]
Brefeldin A (BFA)	20350-15-6	PIN recycling	[148]
5-Alkoxy-IAA	Benzyloxy-IAA: 4382-53-0 Hexyloxy-IAA: 1270002-54-4	PIN, ABCB and AUX1	[147]
7-Alkoxy-NAA	Benzyloxy-NAA: 1270002-60-2 Hexyloxy-NAA: 1270002-59-9	PIN and ABCB	[147]
Inhibitors of auxin regulated gene expression			
PCIB	*p*-Chloro-phenoxy-isobutyric acid	$SCF^{TIR1/AFB}$ pathway	[155]
MG132	Z-Leu-Leu-Leu-al	$SCF^{TIR1/AFB}$ pathway	[66]
Yokonolide B	171423-45-3	$SCF^{TIR1/AFB}$ pathway	[115]
Terfestatin A	CAS: 804556-26-1	$SCF^{TIR1/AFB}$ pathway	[156]
Toyocamycin	CAS: 606-58-6	$SCF^{TIR1/AFB}$ pathway	[117]
BH-IAA	CAS: 1020201-82-4	$SCF^{TIR1/AFB}$ pathway	[129]
PEO-IAA	2-Indole-4-oxo-4-phenylbutanoic acid	$SCF^{TIR1/AFB}$ pathway	[128]
Auxinole	4-(2,4-Dimethylphenyl)-2-(indole)-4-oxobutanoic acid	$SCF^{TIR1/AFB}$ pathway	[129]
Compound A	CAS: 350845-04-4	Unknown	[120]
Compound B	CAS: 301682-74-6	Unknown	[120]
Compound C	CAS: 54754-44-8	Unknown	[120]
Compound D	CAS: 420811-23-0	Unknown	[120]
2-FAA	2-Furylacrylic acid	Unknown	[121]
IAA-Trp	CAS: 57105-53-0	Unknown	[40]
TFIBA	4,4,4-Trifluoro-3-(2-indolyl)butyric acid	Unknown	[157]
	5,7-Dichloroindole-3-isobutyric acid	Unknown	[158]
	Indole 3-isobutyric acid	Unknown	[47]
	Maleic hydrazide	Unknown	[159]
	trans-Cinnamic acid	Unknown	[160]
	2,4-Dichloranisole	Unknown	[161]

RELATED (TAA1/TAR) gene family. Second, IPyA is converted to IAA by flavin monooxygenases encoded by the 11-member YUCCA gene family. With the knowledge of these components, future studies can build on earlier work that suggests transcriptional regulation of *TAA1/TARs* and *YUC* genes contribute to the spatial and temporal aspects of IAA biosynthesis required for auxin-mediated process such as shade-avoidance, root meristem maintenance, response to high temperature, and others [19–23].

Molecules that specifically target enzymes involved in the two-step, tryptophan-dependent pathway would allow components of the tryptophan-independent pathway to be discovered. Using a chemical genetics approach, He et al. [24] identified the molecule kynurenine (Kyn) as an effective and specific inhibitor of the TAA1/TAR aminotransferases. Interestingly, in animals, Kyn is formed during the degradation of tryptophan and is active as a signaling molecule [25, 26]. How Kyn is formed and what its *in planta* function might be remains to be established. Regardless of whether Kyn functions as a regulatory molecule in plants, this molecule represents a novel tool for manipulating the tryptophan-dependent synthesis of IAA.

A number of other potential inhibitors of auxin biosynthesis have been reported, but the enzymatic targets and modes of action have not been substantiated. For instance, Soeno et al. [27] showed that aminoethoxyvinylglycine (AVG) and L-amino-oxyphenylpropionic acid (AOPP) reduced free IAA levels in tomato seedlings. However, since these molecules likely affect multiple pyridoxyl-5-phosphate-dependent enzymes, including the ACC synthases involved in ethylene biosynthesis [28], defining the primary mode of action for these molecules represents a challenge. Similarly D-cycloserine and 6-fluoroindole reduce tryptophan-dependent IAA biosynthesis, but the enzymatic targets of these compounds have not been identified [29, 30].

4.1.4 AUXIN CONJUGATION AND RELEASE BY HYDROLYSIS

In addition to synthesis, IAA-conjugation, to either amino acids or sugars, controls the level of free-IAA in a plant cell [31]. The identification and characterization of enzymes involved in IAA conjugation and hydrolysis represent another area of IAA metabolism where a chemical biology approach has been beneficial. The GH3 protein family functions as IAA conjugating enzymes [32]. The amide conjugation of IAA to the amino acids aspartate (Asp) and glutamate (Glu) is not hydrolyzed to release free IAA, and as such is thought to represent intermediates in IAA catabolism. Lee and Starratt [33] reported that the stems of *Teucrium canadense* L. (Canadian germander) efficiently conjugate [2-^{14}C]-IAA to aspartate and glutamate. They also observed that pretreating stems with 2,6-dihydroxyacetonphenone inhibited the formation of these conjugates, suggesting that this molecule inhibits enzymes involved in IAA breakdown. In 2012, Bottcher et al. [34] based the design of an inhibitor of *GH3* IAA-amido synthetases on the two-step mechanism that was revealed by kinetic analysis of the *Oryza sativa* (rice) OsGH3-8 enzyme performed by Chen et al. [35]. Bottcher et al. [34] generated adenosine-5′-[2-(1H-indol-3-yl)ethylphosphate] (AIEP) as a mimic of the adenylated intermediate formed by the GH3 proteins (Fig. 4.1.2). Using two GH3 isoforms from *Vitis vinifera* L. (grapevine), they

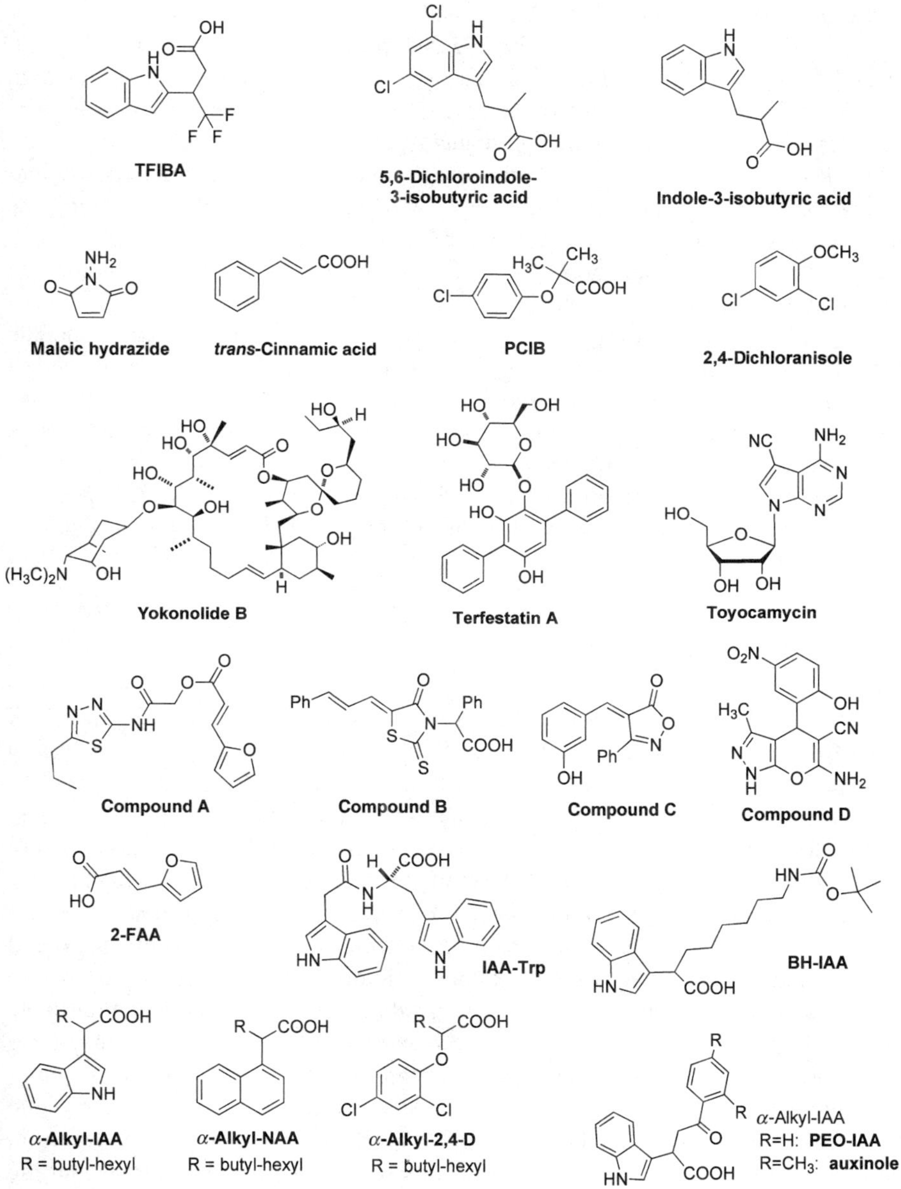

FIGURE 4.1.2 Overview of structures of potential antiauxins, auxin signaling inhibitors identified in screening, and rationally designed auxin signaling inhibitors. Compounds A–D are from Armstrong et al. (2004). See Table 4.1.1 for an explanation of the acronyms and targets, where known.

established AIEP as a potent competitive inhibitor of GH3 enzymes. Since AIEP competes with binding of MgATP and IAA to GH3, they speculate that this molecule will inhibit each of the six GH3 proteins found in grapevine. As such, AIEP represents a novel tool for studying auxin metabolism.

In contrast to stability of IAA-Asp and IAA-Glu, IAA conjugated to either alanine (Ala) or leucine (Leu) can be released through the activity of amidohydrolases [1, 31]. Supplying either IAA-Ala or IAA-Leu to plants yields high-auxin responses, including shortening of the primary root [36]. This observation was used in several screens to identify mutant plants that were resistant to exogenous IAA-Ala or IAA-Leu (e.g., [36–38]). Cloning of the mutant loci has identified IAA-amino acid amidohydrolases as well as other novel and characterized proteins that may be involved in the hydrolysis of IAA-Ala and IAA-Leu [1]. The release of phytotoxic molecules from IAA conjugates has also been proposed as a strategy to define the role of amidohydrolases in plants [39].

Beyond mediating auxin homeostasis, at least one IAA-amino acid conjugate appears to function as a signaling molecule. Staswick [40] documented a physiological role for IAA conjugated to Tryptophan (IAA-Trp). IAA-Trp causes agravitropic root growth in seedlings, and treatments reduce both auxin-responsive gene expression and lateral root formation [41] (Fig. 4.1.2). However, IAA-Trp does not prevent IAA binding to the TIR1 receptor. Therefore, IAA-Trp might serve as an endogenous inhibitor of an as yet unidentified pathway that regulates auxin action.

4.1.5 SYNTHETIC AUXINS

The discovery of IAA prompted the screen for synthetic compounds that elicit auxin-like responses [42]. During the 1940s through 1970s, many structurally diverse molecules were identified and their activities were documented. Based on their chemical cores, these synthetic molecules can be binned into five major groups: (i) indole-containing molecules, for example, halogenated IAA; (ii) naphthalene-containing molecules, for example, naphthaleneacetic acid (NAA); (iii) phenyl- and acetic acid-containing molecules, for example, 2,4-dichlorophenoxyacetic acid (2,4-D); (iv) benzoic acid-containing molecules, for example, dicamba; and (v) picolinate-type molecules, for example, picloram (Fig. 4.1.1**a** and **b**). These synthetic auxins disrupt normal auxin-regulated plant growth and thus have been widely used as herbicides. Despite the diversity of molecular structure, these synthetic auxins share a common pharmacophore features. Each of these molecules consists of a planner aromatic ring system separated from a carboxylic acid moiety by approximately 0.5 nm [43–46]. During the analysis of structural variants of some synthetic auxins, molecules such as *p*-chlorophenoxyisobutylic acid (PCIB) [47], indole-3-isobutyric acid [48] and α-(5,7-dichloroindole-3-) isobutyric acid [49] were found to antagonize auxin-mediated responses and were termed "antiauxins." These antiauxins were hypothesized to compete with auxin for binding to a receptor, thus blocking auxin-mediated responses (Table 4.1.1, Fig. 4.1.2). Such hypotheses foreshadowed the mechanistic description of the auxin-signaling mechanism and subsequent development of tools to examine and manipulate this process.

4.1.6 POLAR AUXIN TRANSPORT

In the late nineteenth century, Charles Darwin and his son, Francis, deduced from a series of experiments that a substance produced in the coleoptile tip of oats (*Avena*) can elicit a plant's bending toward a light source (known as phototropic response) [50]. Subsequently, Peter Boysen-Jensen [51] showed that the mobile substance could move into a gelatin block, and this observation formed the basis to show that the growth-promoting molecule that moved into the block could induce the bending of decapitated shoots in the absence of a light [42]. These observations led Nicolai Cholodny and Frits Went to independently posit that the asymmetrical distribution of auxin in response to a light leads to differential growth on the two sides of the elongating shoot leading to the shoot bending. The asymmetric distribution of auxin has also been used to explain a plants response to gravity (e.g., gravitropism) and touch (e.g., thigmotropism). The identification of multiple transport proteins coupled with advances in visualizing protein movement has enabled the formation of contemporary models that account for these physiological outcomes.

IAA produced in shoot can be directionally transported toward the root tips. Auxin-carrier proteins facilitate the directional auxin movement between cells, which is referred to as polar auxin transport [1, 52, 53]. The polar stream of auxin mediates proper plant development such as vascular formation, lateral root foundation, embryo and root patterning, and apical dominance. The polar auxin transport system controls the local auxin distribution leading to auxin gradient between neighboring cells. The local auxin distribution is achieved by the asymmetric cellular localization of auxin transporters at the plasma membrane in response to stimuli such as light, gravity, and touch. The ensuing asymmetric auxin gradient elicits tropic growth that allows plants to adapt to environmental condition. For gravitropism, the gravity signal perceived in collumella cells regulates the asymmetric localization of auxin transporters on plasma membrane and the asymmetric distribution of auxin results in differential root growth toward gravity. The precise localization of auxin transporters in a cell plays a crucial role in the auxin-regulated differential growth and development [52, 54].

Two types of transporters, auxin influx carriers and auxin efflux carriers, mediate polar auxin transport. Proteins belonging to the AUXIN RESISTANT (AUX)/Like-AUX1 (LAX) family proteins serve as auxin influx transport proteins. The *AUX1/LAX* gene family encodes transmembrane proteins that serve as proton gradient-driven symporters, similar to amino acid permeases [55]. In Arabidopsis, mutants that lack AUX1 show agravitropic and auxin-resistant root phenotypes [56]. The membrane-permeable synthetic auxin, NAA, rescued the agravitropic phenotype of *aux1* mutant [55]. Subsequent experiments show that AUX1 transports IAA and 2,4-D into cells, but not NAA [57].

For auxin export, the PIN efflux carriers and ABCB/PGP transporters coordinately control the direction and the rate of intercellular auxin movement [53, 58]. PIN and ABCB proteins show polar localization at the plasma membrane and, therefore, determine the direction and rate of auxin export. Among eight *Arabidopsis* PIN proteins, five (PIN1, PIN2, PIN3, PIN4, and PIN7) are localized at the plasma membrane, are expressed in overlapping or distinct cell types, and function as auxin efflux carriers [53]. This cell-type-specific expression and subcellular localization of PIN proteins ultimately determine the auxin gradient profile within a tissue. Among 21 members of ABCB subgroup in *Arabidopsis*, three ABCB proteins—ABCB1, ABCB4, and ABCB19—have been identified and characterized as the auxin transporters so far [59]. Single and multiple loss-of-function mutations in these PIN and ABCB proteins exhibit severe defects in embryogenesis, shoot and root tropisms, and lateral organ development [56].

4.1.7 CURRENT MODELS OF AUXIN SIGNALING

4.1.7.1 Auxin-Regulated Gene Expression

In the 1980s, with advances in molecular biology came evidence that auxin rapidly (e.g., within 5 min) and specifically upregulates the transcription of genes (reviewed by Abel and Theologis [3], Guilfoyle and Hagen [60]). More recent whole genome expression studies reveal that auxin regulates the expression of hundreds of genes [61].

Genes encoding three multimembered protein families—AUXIN/INDOLE-3-ACETIC ACID (*Aux/IAA*), GRETCHEN HAGEN (*GH3*), and SMALL AUXIN UP RNA (*SAUR*)—are among those most rapidly and strongly induced by auxin. The biochemical roles have been established for the GH3 and Aux/IAA proteins, and the activity of the SAUR proteins is beginning to be elucidated. Originally the RNAs for GH3s were identified in soybean, but they have been found in all plant species examined [62]. The GH3 protein family functions as IAA conjugating enzymes [32], and their role in IAA metabolism has been discussed earlier. The SAUR genes are found only in plants. Genome analysis shows that the *SAURs* are present as large gene families (e.g., 18 in *Physcomitrella*, 58 in rice, and 78 in *Arabidopsis*) and that these genes encode small proteins (e.g., 86–189 amino acids in *Arabidopsis*). Although a precise mechanistic role for these proteins remains to be determined, results from work on two separate subclades of the SAURs indicate that these proteins play a role in cell expansion [63, 64]. Finally, the *Aux/IAA* genes have also been found in all plants that have been examined. The 29-member gene family in *Arabidopsis* is the best characterized. Multiple reports demonstrate the central role of Aux/IAA proteins in auxin signaling [1].

Aux/IAAs encode short-lived, nuclear proteins that are repressors of auxin-mediated gene expression. The majority of Aux/IAA proteins contain four highly conserved domains (I, II, III, and IV). Domain I facilitates interaction with the transcriptional corepressor TOPLESS [65]. Domain II mediates interaction with IAA and the TIR1, which is a component of an ubiquitin E3 ligase [66]. Domain II is required for auxin-mediated degradation of the Aux/IAA proteins and many gain-of-function mutants with changes in domain II have been described [60]. Domains III and IV facilitate homo- and heterodimerization among the Aux/IAA proteins and with the auxin response factor (ARF) family of transcription factors [67, 68].

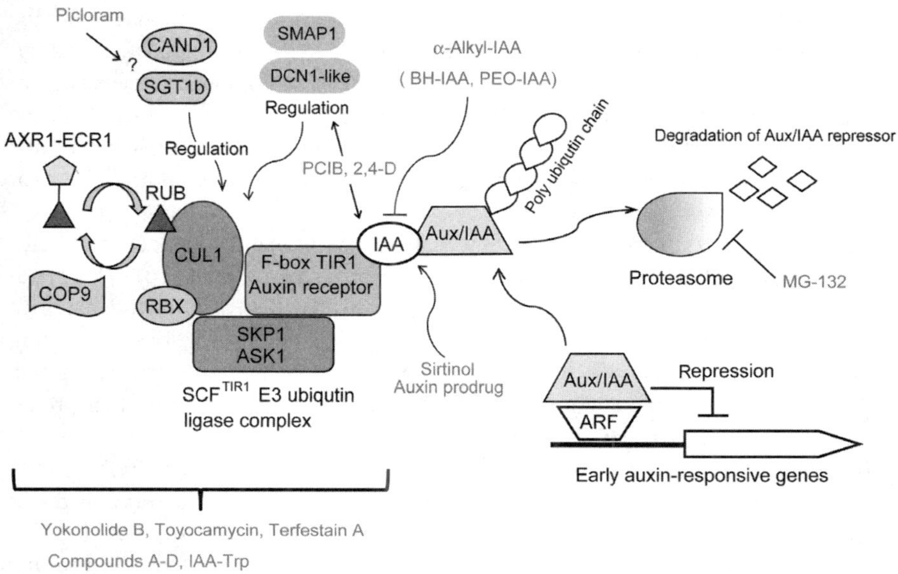

FIGURE 4.1.3 Auxin signaling and hypothesized sites of action for small molecule probes. The small molecule probe targeting SCFITIR1 pathway is indicated to show its target site or possible (indicated by ?) cellular target.
Auxin binds to TIR1, an F-box protein auxin receptor that forms E3 ubiquitin ligase complexes with the Skp1 (ASK1) and Cullin (CUL1) proteins. In the absence of auxin, Aux/IAA heterodimerize with ARFs that activate transcription to repress the transcriptional activation of auxin-responsive genes. Auxin binding promotes the ubiquitination and degradation of Aux/IAA repressor via proteasome pathway, which enables ARF function in promoting gene expression. The small molecule probe targeting SCF^{TIR1} pathway is indicated to show its target site.

The ARF proteins, which are encoded by 23 genes in *Arabidoposis*, can bind auxin-responsive elements (AuxRE) located in auxin-responsive promoters to either activate or repress transcription [69].

Figure 4.1.3 shows the current model for ARF proteins that activate transcription. When the ARF interacts with an Aux/IAA protein, transcriptional activation of auxin-responsive genes is blocked. In the presence of auxin, the Aux/IAA proteins become ubiquitinated and are degraded by the 26S proteasome. This enables the activating ARFs to bind to AuxREs and mediate transcription of genes, which includes the *Aux/IAAs*. Production of Aux/IAA protein resets the system for another pulse of auxin. The role of Aux/IAA-ARF dimers that involve ARF proteins that repress transcription has not been established.

The ubiquitin-conjugating activity of the SCF^{TIR1} complex plays a central role in mediating auxin-regulated gene expression. This complex emerges from the interactions of Skp1 (ASK1), Cullin (CUL1), and the F-box protein (TIR1) and functions as an E3 ubiquitin ligase [70, 71]. The SCF^{TIR1} complex mediates the ubiquitination of Aux/IAA proteins, and auxin enhances this activity by promoting the interaction between TIR1 and the Aux/IAA proteins [70, 71]. Modification of the cullin, CUL1, by the ubiquitin-related protein RUB1/NEDD8 regulates the SCF^{TIR1} function [72]. The components of the RUB-conjugation pathway, including AXR1, ECR1, and RCE1, are implicated in auxin-regulated SCF^{TIR1} E3 ligase activity, and mutations of these components cause auxin-resistant phenotypes and auxin-related developmental defects [73]. Other defective mutants in auxin-response, *cand1* (cullin-associated and neddylation-dissociated) and *sgt1b* [74, 75] were shown to have altered SCF^{TIR1}activity. The molecular mechanism for how SCF^{TIR1} function is regulated is not understood.

Biochemical and structural studies show that auxin promotes the interaction of TIR1 with the domain II motif (GWPPV) found in most Aux/IAA proteins [76, 77]. Crystal structures show auxin trapped in a small cavity formed by the two proteins that form this coreceptor. The auxin molecule extends the hydrophobic interactions, suggesting that it acts as a "molecular glue" to hold the two proteins together. These structural data explain why point mutations in domain II of Aux/IAA proteins prevent their proper auxin-dependent degradation, which leads to pleiotropic phenotypes [60, 72].

In *Arabidopsis*, there are four additional proteins, called "AUXIN SIGNALING F-BOX" (AFB), that share sequence similarity with TIR1 and that serve as auxin receptors. AFB1, AFB2, and AFB3 share 60–70% amino acid identity with TIR1 [78, 79] and provide overlapping auxin receptor activity. In contrast, AFB4 and AFB5 form a subclade that is more diverged from TIR1. AFB4 and AFB5 share only 50% amino acid identity with TIR1. AFB4 and AFB5 work as auxin receptors but serve to negatively regulate of auxin action [80]. Interestingly, AFB5 displays a high affinity for picloram, which AFB4 lacks.

Recently, S-phase kinase-associated protein 2A (SKP2A) was shown to be another F-box auxin receptor. SKP2A regulates the proteolysis of cell cycle transcription factors, such as E2FC and DPB. Auxin binding to SKP2A was confirmed by a ligand-binding assay using radioactive IAA. Auxin enhanced the degradation of SKP2A protein, and the binding of IAA to SKP2A promoted the degradation of E2FC and DPB to trigger cell division [81, 82].

4.1.7.2 Auxin-Binding Protein1 (ABP1) Regulates Cell Division and Cell Size

Auxin-binding protein 1 (ABP1) is believed to perceive auxin on plasma membrane that is distinct from the nucleus-localized TIR1 receptor. ABP1 was identified as a soluble 22 kDa glycoprotein that specifically binds to auxin (reviewed by Braun et al [83], Tromas et al. [84]). ABP1 is localized predominantly within the endoplasmic reticulum, but in small quantities at the plasma membrane. Initially, ABP1 functions were thought to be implicated in the rapid regulation of membrane potential and ion fluxes at the plasma membrane since the auxin receptor activity of ABP1 was largely determined by electrophysiological experiments such as auxin-induced plasma membrane hyperpolarization in protoplasts [85, 86]. ABP1 is also involved in the auxin-induced cell swelling of protoplasts in a Pea and *Arabidopsis* hypocotyls [87, 88]. The ectopic expression of ABP1 induces auxin-dependent epinastic curvature of young tobacco leaf [89] and the suppression of ABP1 function reduces cell division rates in tobacco cultured cell [90, 91].

In *Arabidopsis*, a single *ABP1* gene was found in genome and a homozygous null mutant showed embryo lethality [92]. Recently, the conditional knockdown of ABP1 function in *Arabidopsis* through the use of an anti-ABP1-monoclonal antibody demonstrated that ABP1 plays a role in regulating cell elongation and cell division, as well as directly or indirectly impacting auxin-responsive gene expression [93, 94]. Plants heterozygous (*abp1/ABP1*) for the knockout showed defects in auxin-related responses, such as longer hypocotyls, agravitropic roots and hypocotyls, and reduced apical dominance [95]. The heterozygous line also displayed altered auxin-induced gene expression and reduced basipetal auxin transport in young roots. These results imply that ABP1 regulates polar auxin transport, which controls local auxin gradients [95].

PIN proteins function as auxin efflux carriers and participate in polar auxin transport. The localization of PIN protein in the plasma membrane determines the direction of auxin flow to control local auxin distribution. ABP1 could regulate the intracellular recycling of PIN transporters between the plasma membrane and endosomes [96]. Auxin signaling via ABP1 inhibits clathrin-mediated internalization of PIN proteins [97]. Clathrin is a protein that participates in the formation of coated vesicles during the process of endocytosis. ABP1 plays a positive role in recruiting clathrin to the plasma membrane, which in turn facilitates PIN endocytosis. The binding of auxin to ABP1 inhibits clathrin-mediated endocytosis of PIN proteins, resulting in a reduced internalization of PIN proteins, thereby promoting the efflux of auxin.

The Rho GTPases, ROP2 and ROP6, coordinately regulate the interdigitated growth of epidermal pavement cells. Auxin rapidly activates the expression of *ROP2* and *ROP6* within 30 s. Auxin signal mediated by ABP1 is required for the activation of these ROP proteins [98]. Activated ROP signal might modulate the PIN activity regulating local auxin concentration that determines the interdigitated growth pattern. ROP2 also positively regulates the initiation and tip growth of root hairs [99]. Multiple components downstream of ABP1 have been described, and their regulation remains under investigation.

4.1.8 APPLICATION OF AUXIN-RELATED MOLECULES IN CHEMICAL GENETIC APPROACH

Using the reference plant, *Arabidopsis thaliana*, as a genetic chassis, our understanding of auxin biology has been advanced by mutant screens that use synthetic auxin molecules. For example, the initial applications of a chemical genetic screen involved looking for AUXIN-RESISTANT (*axr1*) mutants that grew in the presence of the synthetic phenoxyacetic acid-type auxin, 2,4-D [100]. The *axr1* mutant exhibited defects in a multiple auxin responses. The *AXR1* gene was later shown to encode a subunit of the heterodimeric ubiquitin E1-like enzyme complex in the RUB-conjugation pathway that regulates SCF^{TIR1} activity [73, 101]. Similar types of screens led to the identification of dominant mutants such as *axr3/iaa17* and *axr2/iaa7*, which result in changes in domain II that disrupt Aux/IAA-TIR1 interactions [72, 102].

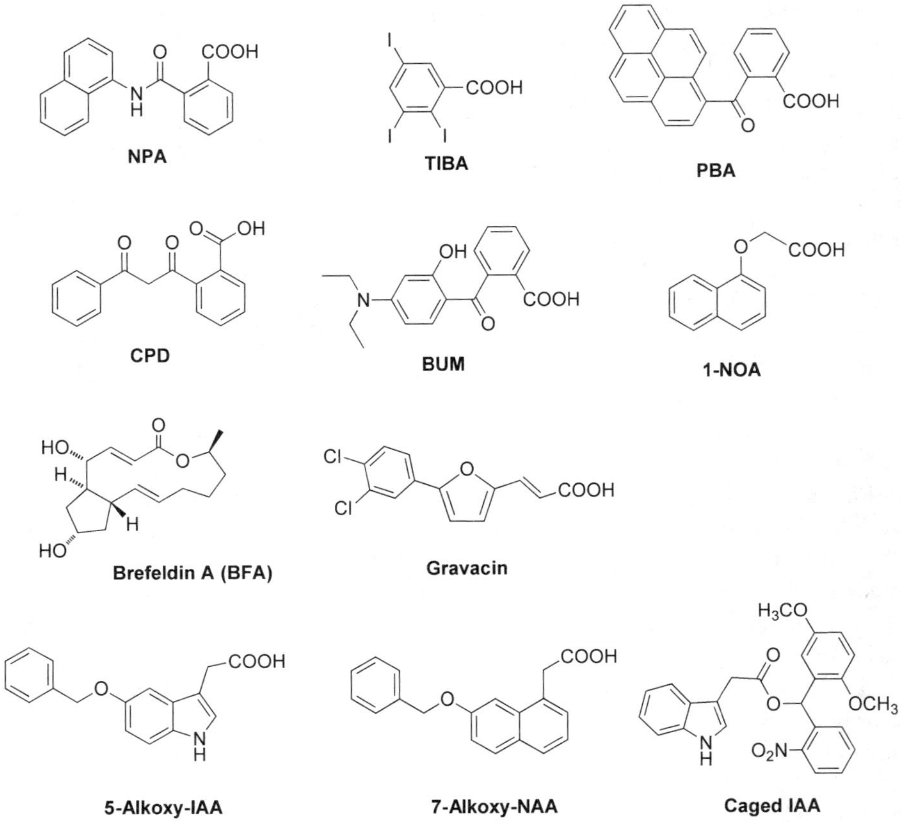

FIGURE 4.1.4 Chemical structures of auxin transport inhibitors and caged auxin. See Table 4.1.1 for an explanation of the acronyms and targets, where known.

Combining genetics with small molecules that alter distinct auxin-related process, such as polar auxin transport, has extended the impact of the genetic approach. Mutants resistant to auxin transport inhibitors (Fig. 4.1.4), such as 1-naphthylphthalamic acid (NPA) and 1-(2′-carboxyphenyl)-3-phenylpropane-1,3-dione (CPD), have been identified and described as *TRANSPORT INHIBITOR RESPONSE* (*TIR*) mutants [103]. As described earlier, *TIR1* encodes a F-box protein that functions as a co-receptor for auxin-induced changes in gene expression [70, 71]. Recent reports indicate that *TIR2* encodes a tryptophan aminotransferase, TAA1, that catalyzes the conversion of tryptophan to indole-3-pyruvic acid in auxin biosynthetic pathways [14, 23]. *TIR2/TAA1* is also involved in the local auxin biosynthesis that controls a positive regulatory loop required for root gravitropism.

The specificity of TIR1 or the related AFB proteins for distinct auxins has been revealed through similar types of genetic screens. For example, mutants resistant to DAS534, a synthetic analog of the auxinic herbicide picloram, have been identified. One of these mutants results from the disruption of AFB5, which falls in the AFB4 subclade of TIR1-related genes [104]. Interestingly, loss of *AFB5* confers resistance to picolinate auxins, but not to 2,4-D or IAA. The selective interaction of this subclade of AFB proteins with picloram was further demonstrated by the highly selective resistant to picloram of the *afb4 afb5* double mutant [80]. Another example is the way that Dicamba, a benzoic acid-type synthetic auxin, has less growth inhibiting effects on the roots of *tir1* and *afb5* mutant seedlings compared to other *afb* mutants. Furthermore, this Dicamba resistance was enhanced in the *tir1 afb5* double mutant [105]. The distinct sensitivity of *tir1*, *afb4*, and *afb5* mutants to different synthetic auxins indicates that the molecular recognition of auxins is different among the auxin receptor family. As mentioned, *Arabidopsis* has 29 Aux/IAA family proteins that interact with one of 23 ARF transcriptional factors. Thus, pleiotropic auxin responses at specific cells might be regulated by the distinct affinity among the receptors, together with the multiple combinations of the 6 TIR1/AFB coreceptors, the 29 AUX/IAA repressors, and the 23 ARF proteins [79, 72, 106]. Further complicating the interpretation of these data is the observation that AFB4 functions as negative regulator of auxin signaling, indicating additional regulatory mechanisms may be functioning in this system [80].

Various compounds that antagonize auxin action have been described as "antiauxins" [47] (Fig. 4.1.2). Auxin-related molecules that compete for binding to the receptor or that interfere with subsequent signaling steps without activating downstream events would be expected to antagonize auxin activity. *p*-Chlorophenoxyisobutyric acid (PCIB) is an analog of the synthetic auxin, 4-chlorophenoxyacetic acid, that exhibits antiauxin activity in classical auxin bioassays [107]. Oono et al. [108] demonstrated that PCIB stabilizes Aux/IAA proteins by acting at or upstream of the auxin receptor. Further, they showed that PCIB inhibited the early auxin-responsive gene expression that is mediated by the SCF^{TIR1} pathway. Consistent with these molecular data, PCIB reduced lateral root number and disrupted root gravitropism in *Arabidopsis*. PCIB did not, however, attenuate the inhibiting effect of auxin on primary root growth. The effect of other proposed antiauxins (Fig. 4.1.2), such as 2,4-dichloranisole (DCA), maleic hydrazide, *trans*-cinnamic acid, and 4,4,4-trifluoro-3-indole-3-butyric acid (TFIBA), had a minimal impact on auxin-induced reporter-gene expression [108]. The impact on gene expression of other auxin analogs, such as indole-3-isobutyric acid [48] and 5,7-dichloro-indole-3-isobutyric acid [49], that inhibit the auxin-induced elongation of *Avena* coleoptiles has not been described.

Rahman et al. [109] used PCIB-induced phenotypes to isolate PCIB-resistant mutants. The PCIB-resistant plants were expected to have a mutation in auxin perception or other downstream signaling component(s). The genetic screen identified alleles of two SCF^{TIR1} components, *tir1* and *cullin1*, as well as uncovering two novel loci, which they named *ANTIAUXIN-RESISTANT* (*AAR*). Both *aar1 and aar3* are resistant to PCIB and 2,4-D, yet they respond like wild type to other forms of auxin. *AAR1* encodes a SMALL ACIDIC PROTEIN 1 (SMAP1) involved in auxin-signaling pathway [109]. The other locus, *AAR3*, encodes a DEFECTIVE in CULLIN NEDDYLATION (DCN1)-like protein that regulates the auxin response in *Arabidopsis* roots [110]. Collectively, these examples show the power of combining genetic screens with different natural and synthetic auxin molecules.

4.1.9 CHEMICAL PROBES ON AUXIN SIGNALING FROM CHEMICAL LIBRARY AND NATURAL SOURCES

The identification of compounds that either activate or attenuate auxin signaling shares the common challenge faced by scientists who want to use chemical biology to target a particular aspect of plant biology [111]. Perhaps the most promising method for creating such specific probes would involve the rational design of compounds based on target protein structure; however, in most cases, the structure of the desired target has not been determined. An alternative approach, therefore, is necessary. Rapid, sensitive, and specific bioassay systems that allow large collections of molecules, generated either from diversity-oriented synthesis or from pools of natural products, to be interrogated represent one such advance [111]. For auxin signaling, two transgenic *Arabidopsis* auxin-responsive reporter systems, *BA3::GUS* and *DR5::GUS*, are well characterized and widely used [112, 113]. The *BA3::GUS* line contains the β-glucuronidase (*GUS*) reporter gene under the control of a promoter element derived from the pea *PS-IAA4/5* gene [114] while the *DR5::GUS* line expresses GUS from a synthetic auxin-responsive element [113]. As such, these reporter lines enable the impact of thousands of compounds on auxin signaling to be monitored in parallel and in within a few hours.

The use of reporter lines can be particularly powerful when assaying crude extracts of natural products where additional metabolites can lead to false-positives in classical phenotype-based assays, such as cell elongation and cell division. Using promoter-GUS reporters, two macrolide-type metabolites, yokonolide A and B (Fig. 4.1.2), from the soil bacteria *Streptomyces diastatochromogenes* were identified as auxin signaling inhibitors [115, 116]. Yokolonide B inhibits auxin-induced degradation of the Aux/IAA repressors without affecting proteasome activity, indicating that yokonolide B likely acts on SCF^{TIR1}-ubiqutination machinery or components that regulate its activity [115]. Similar screens for auxin signaling inhibitors from microbial metabolites have identified two additional auxin signaling inhibitors, terfestatin A and toyocamycin [117, 118] (Fig. 4.1.2). Both terfstatin A and toyocamycin inhibit auxin-induced Aux/IAA degradation and prevent auxin-induced gene expression, which results in auxin-related phenotypes *in planta*. The target(s) of these compounds, however, has not yet been determined.

Auxin signaling inhibitors from natural products have structural diversity and show potent inhibitory activity on auxin signaling. However, these compounds generally have optically active stereochemistry and complicated structures. In addition, technically challenging purification processes or synthetic schemes that require many steps pose problems in obtaining pure samples of these molecules. These features of natural products limit the large-scale use of this pool of molecules. Another source of molecules for the screening are commercially available compound libraries (see Chapter 2.1 for more details). These libraries generally consist of pure compounds organized in a manner that facilitates high-throughput screening. If a lead compound is simple and optically inactive, subsequent structure–activity analysis and structural optimization become more accessible. Diverse drug-like compound collections that meet "Lipinski's rule of five" [119] can be obtained from vendors such as ChemBridge and Maybridge.

Because the libraries consist of individual molecules, both reporter gene and phenotype-based approaches can be applied to identify and characterize novel auxin signaling modulators. The earliest approach to find auxin signaling inhibitors from a chemical library using auxin-responsive *BA3::GUS* reporter line revealed four nonauxin-like structures that inhibited auxin signaling [120] (Fig. 4.1.2). These compounds blocked auxin-regulated gene expression, Aux/IAA protein degradation, and auxin-related phenotypic responses. Among these molecules, the active pharm-core structure of compound A, a furyl acrylate ester of a thiadiazole heterocycle, was shown to be 2-furylacrylic acid (2-FAA) [121]. The active principle, 2-FAA, inhibited auxin-responsive gene expression and auxin-enhanced Aux/IAA protein degradation but failed to suppress the auxin-induced interaction between Aux/IAA repressor and TIR1 receptor proteins. Interestingly, *in planta*, 2-FAA antagonized the growth-inhibiting effects of auxin to restore primary root length. In contrast, PCIB, a classical antiauxin, does not counteract the inhibition of primary root growth by exogenous auxin nor does it inhibit the root growth [108]. Identification of the cellular target of 2-FAA has not yet been achieved.

Two phenotype-based screens led to the identification of a new series of molecules that function as "proauxins" (Fig. 4.1.5). First, a phenotype-based screen of a 10,000 compound chemical library revealed 13 compounds that inhibited the primary root growth of *Arabidopsis* seedlings [122]. One of the most potent active compounds (WH7) identified in this screen turned out to be an amide conjugate of the well-characterized synthetic auxin 2,4-D. Application of WH-7, or its derivatives, activated auxin-responsive gene expression and stimulated maize coleoptile elongation. The elongation activity of WH7 requires slight time-lag in comparison with IAA activity, leading the authors to suggest that WH7 works as prodrug of 2,4-D [122]. Second, Savaldi-Goldstein et al. [123] used a similar strategy to identify molecules in a 10,000 compound library that would stimulate *Arabidopsis* hypocotyl elongation. This screen identified 100 bioactive compounds. A subset of these auxin-related compounds functioned as auxin precursors (e.g., compound 533 and 602, Fig. 4.1.5), with amide derivatives of molecules such as phenoxyacetic acid being hydrolyzed by amidases to release active auxin in specific tissue (e.g., hypocotyl). Interestingly, this 2,4-D prodrug enhanced hypocotyl elongation, but exogenously supplied 2,4-D alone did not. This suggests that local temporal and spatial control of 2,4-D distribution plays a role in hypocotyl elongation.

Another avenue for identifying molecules that impact auxin-mediated physiological responses comes from exploring the impact on plants of compounds that are bioactive in other systems. The characterization of the molecule sirtinol represents an example of this approach. A screen in yeast for a chemical inhibitor of the sirtuin class of NAD-dependent histone deacetylases led to the identification of sirtinol [124] (Fig. 4.1.5). Interestingly, sirtinol did not cause changes in histone deacetylation when applied to mammalian cell lines, but it did cause auxin-related phenotypes in roots and vascular tissue of *Arabidopsis* seedlings [124]. Plants exposed to sirtinol resembled high-auxin phenotypes even though the structure of this compound was distinct from previously described auxins [124]. The identification of sirtinol-resistant mutants yielded to categories of plants. The first included plants that, in addition to being sirtinol resistant, were

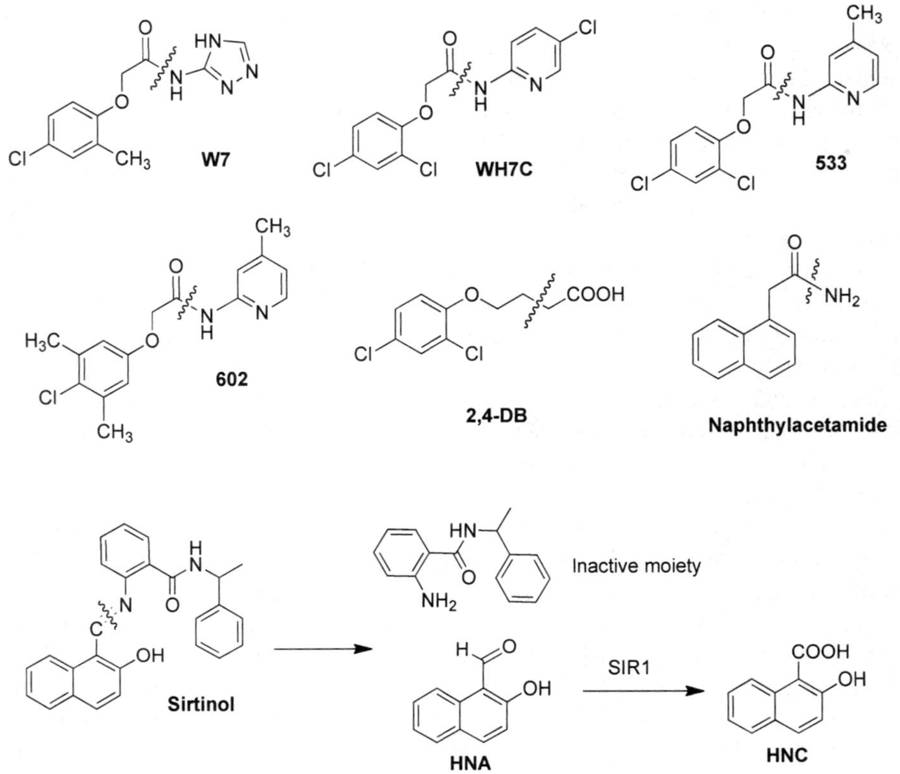

FIGURE 4.1.5 Chemical structures of auxin proauxins.
The absorption and distribution profile of auxin precursors is different from the parent molecules. The amide bond is likely cleaved by amidases to release the active auxin molecules. 2,4-DB was converted to the active 2,4-D though β-oxidation. The imine bond of sirtinol was hydrolyzed to give a HNA. This aldehyde was converted by the aldehyde oxidase, SIR1 to HNC, which is an NAA analog. Mutations in *SIR1* blocked the oxidation of HNA, which resulted in the sirtinol-resistant phenotype. For each molecule, the site of cleavage is indicated by the wavy line.

auxin resistant. This set of mutants included several known auxin-response mutants (e.g., *axr1*, *axr2*, *axr3*, and *axr6*) as well as a new locus, *AtCAND1*, which encodes a protein required for proper regulation of the SCF complex [74, 125]. The second category of mutants included novel SIRTINOL-RESISTANT (*sir*) plants that remained sensitive to auxin [126, 127]. Cloning of *sir1* showed that the *SIR1* gene encodes a protein that contains an ubiquitin-activating enzyme E1-like domain and a rhodanase-like domain. At the time, SIR1 was hypothesized to function as a novel component of auxin signaling, likely through regulation of Aux/IAA repressor degradation [126]. Subsequent studies revealed that sirtinol becomes metabolized to form a synthetic auxin [127]. The active-core moiety of sirtinol is 2-hydroxy-1-naphthaldehyde (HNA), which is enzymatically oxidized to 2-hydroxy-1-naphtoic acid (HCN), a structural analog of the synthetic auxin NAA [127] (Fig. 4.1.5). Each of the sirtinol-resistant mutants, *sir3*, *4*, and *5*, encodes enzymes that contain a molybdopterin cofactor and are involved in the oxidization of aldehydes to carboxylates. Thus, Dai et al. [127] showed that the imine bond of sirtinol is hydrolyzed to release the HNA and that the SIR proteins catalyze the ensuing oxidation of HNA to HNC. These data support the hypothesis that sirtinol is a pro-drug of the auxin molecule, HNC.

4.1.10 RATIONAL DESIGN OF AUXIN ANTAGONIST ON THE BASIS OF TIR1 STRUCTURE

The crystal structures of the TIR1-auxin complex illustrate the molecular recognition mechanism of auxin by the TIR1-Aux/IAA co-receptor [76, 77] (Fig 4.1.6). The molecular shape of TIR1 forms a mushroom-shaped structure, and auxin nestles on the floor of surface pocket formed by the LRR domain of TIR1. IAA is held in place by hydrophobic interactions of the aromatic ring and hydrogen bonding that form with NH groups in the indole structure. The binding is further stabilized by interactions between the carboxylate of IAA with the side chains of arginine and serine residues. Other synthetic auxins, 2,4-D and NAA, interact in a similar fashion with this TIR1-binding pocket. However, 2,4-D and NAA lack hydrogen donor capacity in their aromatic rings, which likely accounts for their lower affinity compared to IAA.

The Aux/IAA interacts with TIR1 only when auxin is present. Analysis of TIR1-auxin-Aux/IAA structure revealed a new molecular recognition mechanism for F-box proteins. In the crystal structure, the Aux/IAA7/AXR2 domain II peptide, which includes the GWPPV "degron" that is required for Aux/IAA protein degradation, interacts with auxin in the auxin-binding region of TIR1. The binding of the domain II peptide does not alter the interaction of TIR1 with IAA and does not shift the conformation of TIR1. These data indicate that TIR1 is not allosterically regulated. The crucial GWPPV motif in the domain II degron covers the IAA in the auxin-binding cavity of TIR1. The tryptophan and second proline residues found in this motif are positioned close to the aromatic ring of IAA and are stabilized by hydrophobic interactions. This observation explains why point mutations in the GWPPV motif of Aux/IAA proteins prevent their auxin-dependent ubiquitination and degradation by the SCF^{TIR1}–proteasome pathway, which leads to auxin-insensitive phenotype.

Furthermore, according to these data, Aux/IAA works as a co-receptor that forms the hydrophobic cavity in the auxin-binding site of TIR1, which stabilizes the TIR1-auxin-Aux/IAA complex. As such, auxin has been described as a "molecular glue" that holds the TIR1 and Aux/IAA proteins together [76, 77].

The molecular structure and mechanism of auxin perception opened the door for structure-based molecular design of auxins and antiauxins. The rational design of molecular probes can be powerful, especially, in the case of TIR1 receptor where the interaction between auxin, Aux/IAA, and TIR1 does not induce conformational or positional changes, even in the flexible side chains around the auxin-binding pocket. These observations imply that the molecular design of molecules that disrupt

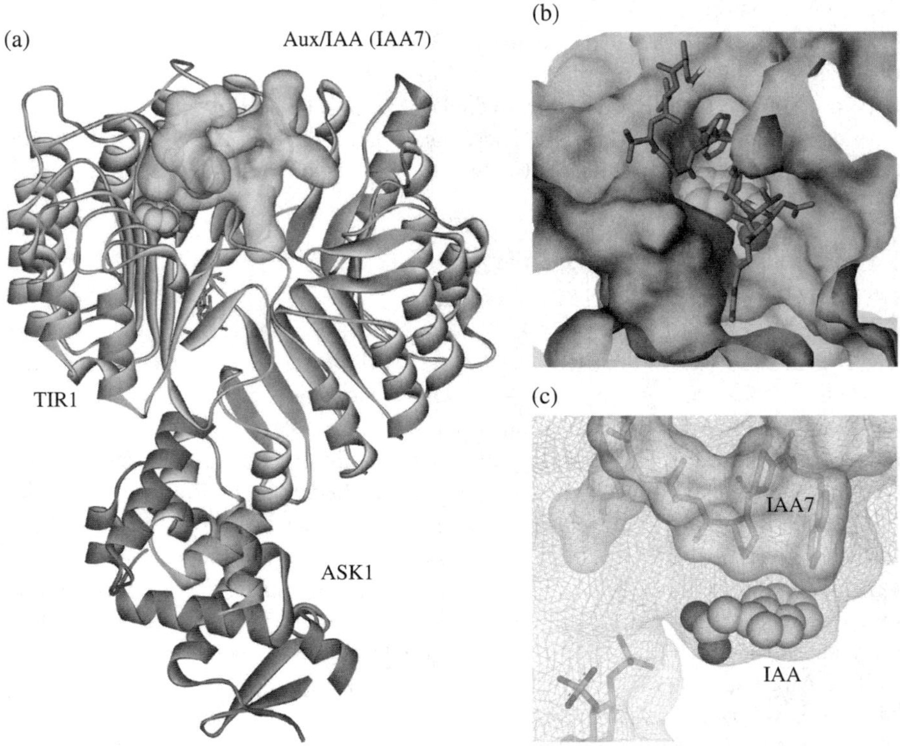

FIGURE 4.1.6 Structural highlights of the TIR1-IAA-Aux/IAA complex.
(**a**) Crystal structure of ASK1-TIR1-IAA-IAA7 complex. ASK1 is shown in olive green, TIR1 in blue, IAA in green and red, and the domain II portion of IAA7 is shown in pink. (**b**) View of IAA (green and red) serving as molecular glue in the TIR1 cavity, which is covered by the GWPPV motif of IAA7 peptide. (**c**) A slab view of the TIR1–IAA–IAA7 complex. The WPPV motif of IAA7 peptide (shown in pink) covers the auxin-binding site. Auxin fills a cavity between TIR1 and Aux/IAA that stabilizes the interactions among these three components. The molecular surface of TIR1 is shown as a light blue mesh. For color detail, please see color plate section.

Aux/IAA-TIR1 interactions involves a rigid auxin-binding site and can proceed without consideration of side chain rotations that might impact docking.

The rational design and subsequent testing of molecules for their ability to disrupt TIR1-Aux/IAA interaction show that substituting varying length alkyl groups at the α-position of IAA yields compounds that serve either as auxin agonists or antagonists [128]. For example, α-methyl-IAA or α-propyl-IAA simulates TIR1-Aux/IAA interaction, leading to enhanced Aux/IAA degradation and auxin-related phenotypes. In contrast, IAA molecules with butyl or longer chains still interact with TIR1, but prevent Aux/IAA docking, which means these compounds function as antiauxins (Fig. 4.1.2). Among the first generation of molecules examined, α-*tert*-butoxycarbonylaminohexyl-IAA (BH-IAA) was effective at antagonizing auxin action mediated by $SCF^{TIR1/AFB}$ pathway, even though it had a relatively low affinity

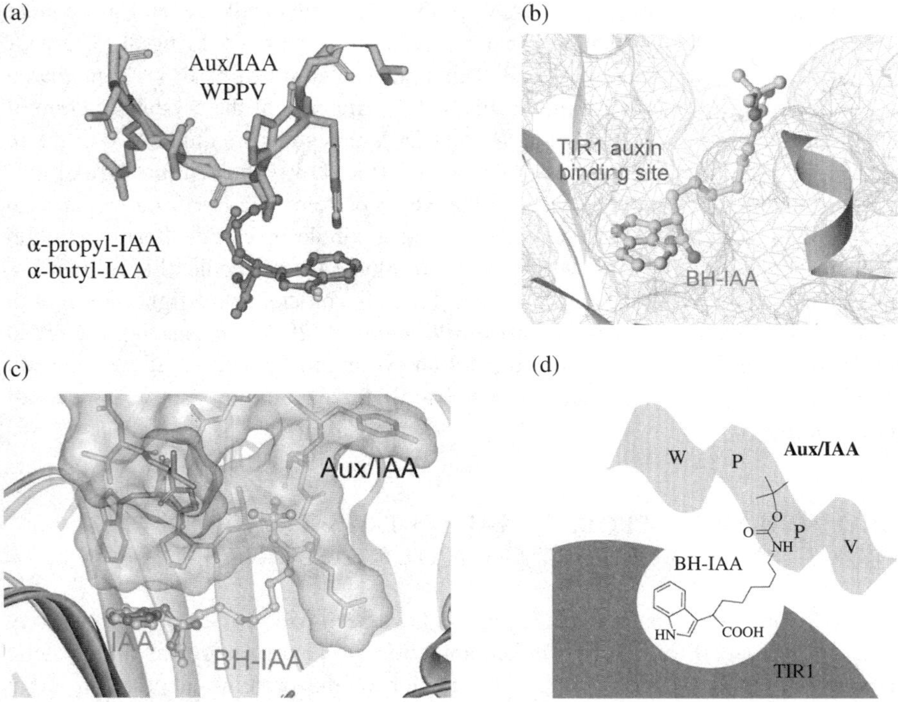

FIGURE 4.1.7 Structural constraints imposed by *a*-alkyl-IAA derivatives. (**a**) The coordinates of α-propyl-IAA (blue) and α-butyl-IAA (red) were superimposed on the TIR1-IAA7 crystal structure (PDB ID 2P1Q) to show their hypothesized interaction with the WPPV motif of IAA7 (pink). The terminal region of α-butyl-IAA clashes with the second proline in the WPPV motif and thus would block access of Aux/IAA to the TIR1 docking site. In contrast, the shorter chain of α-propyl-IAA does not prevent docking of the Aux/IAA. These observations account for why the α-butyl-IAA serves as an auxin antagonist and the α-propyl-IAA functions as an auxin agonist. (**b**)–(**d**) Images based on the crystal structure of BH-IAA interacting with TIR1 to prevent Aux/IAA protein docking. (**b**) BH-IAA (green and red) is shown interacting with the auxin-binding site of TIR1. (**c**) IAA (red) and BH-IAA (green) shown interacting with the TIR1 (shown as ribbon structure). The Aux/IAA protein coordinates (tan) were superimposed to highlight how BH-IAA would exclude the docking of domain II. (**d**) A schematic summary of how the long alkyl chain and bulky terminal groups effectively block the access of Aux/IAA proteins to the TIR1 docking site. For color detail, please see color plate section.

for TIR1 [128]. Crystal structures revealed that BH-IAA interacts with the auxin-binding site of TIR1 such that the flexible alkyl chain and bulky end-group prevent the docking of the domain II GWPPV motif of the Aux/IAA protein with TIR1 (Fig. 4.1.7). Similar observations were seen when the α-alkyl substitutions were made on either NAA or 2,4-D [128]. Based on the structural information of the (BH-IAA)-TIR1 complex, a virtual screening for candidate molecules that would have higher affinity became possible. An *in silico* docking experiment identified α-(phenylethyl-2-oxo)-IAA (PEO-IAA) as a high-scoring candidate, and subsequent structure–activity analysis showed that methylation of the phenyl ring of PEO-IAA increased its inhibitory activity [129]. This potent inhibitor of $SCR^{TIR1/AFB}$-mediated auxin signaling was named auxinole (Fig. 4.1.2). Analysis of the crystal structure of ABP1 shows that auxinole and other α-alkyl-IAA molecules are unlikely to interfere with auxin signaling through this protein. Woo et al. [130] showed that NAA interacts with a confined binding pocket of the ABP1 protein. The larger molecular size of compounds such as BH-IAA, PEO-IAA, or auxinole precludes their interaction with this ligand-binding site. As such, the α-alkyl-IAA molecules likely interfere specifically with TIR1/AFB proteins. BH-IAA also blocks auxin signaling in both *O. sativa* (rice) and the moss *Physcomitrella patens* [128] demonstrating the broad applicability of these molecules to disrupt an evolutionarily conserved mechanism. As such, these molecules represent novel tools for examining specific aspects of auxin biology.

4.1.11 CHEMICAL PROBES ON AUXIN TRANSPORT FROM CHEMICAL LIBRARY AND NATURAL SOURCES

Synthetic inhibitors of auxin polar transport have been widely used in auxin biology (Table 4.1.1 and Fig. 4.1.8). The combination of pharmacological and genetic approaches has succeeded in the identification of the proteins involved in auxin transport. Further examination of these proteins established the position of each transport activity in current models of auxin movement.

N-1-naphthylphthalamic acid (NPA) belongs to a broad class of compounds, the phytotropins, that alter plant tropic responses and represents the most widely used auxin transport inhibitor [131, 132]. NPA inhibits the polar movement of auxin, which in turn limits auxin-related plant responses, such as tropic bending or lateral root formation [133]. NPA inhibits auxin efflux in a noncompetitive manner. The number of NPA-binding proteins and their cellular roles remains under investigation. Two members of the Arabidopsis B subfamily of ABC transporters, ABCB1 and ABCB19, interact with NPA, which results in attenuated transport activity in yeast or mammalian cells [134–137]. Interestingly, NPA was reported to interact with the immunophilin-like protein, TWISTED DWARF1 (TWD1). TWD1 interacts with ABCB proteins to regulate their transport activity, and NPA binding blocks the interaction of TWD1 with ABCB1 [136–140]. These data lead to a model where TWD1 and ABCB1, or ABCB19, represent high and low affinity, respectively, NPA-binding proteins that participate in auxin efflux [140].

Plant flavonoids are considered to be endogenous modulator of local auxin distribution through their interaction with the NPA-sensitive transport complex [141, 142]. Flavonoids compete for NPA-binding sites [143], and the roots of the *Arabidopsis transparent testa4* mutant, which contains a mutation in chalcone synthase, the first enzyme in flavonoid biosynthesis, show altered gravitropic responses that can be rescued by the exogenous flavonoids [141]. The chemical structure of the active flavonoids involved and direct evidence for their cellular targets remain unknown.

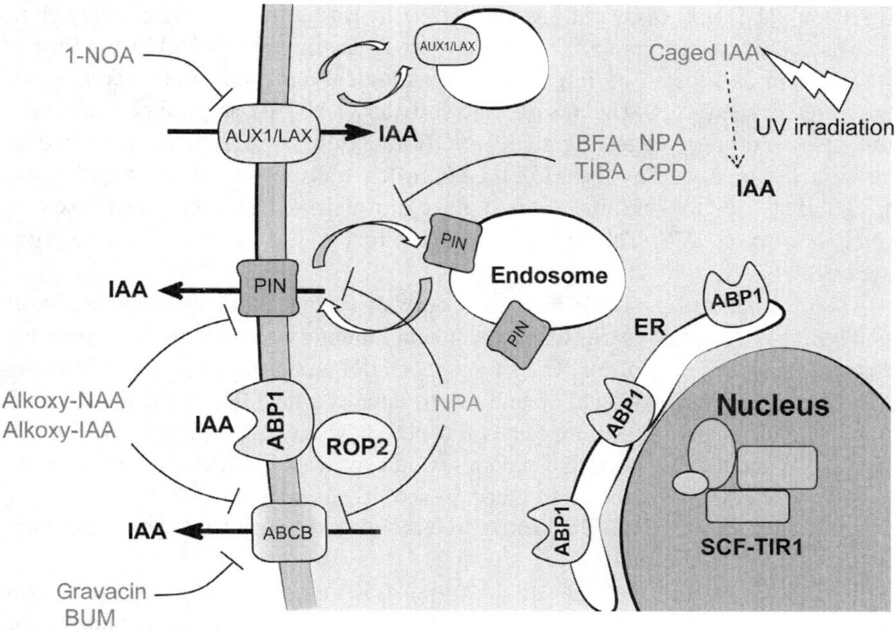

FIGURE 4.1.8 Auxin transport and hypothesized sites of action for small molecule probes. PIN and ABCB proteins are auxin efflux transporters. AUX1/LAX family proteins are auxin influx transporters. The localization of PIN and AUX1 is regulated by a protein trafficking system that involves cycling from the plasma membrane to the endomembrane system. ER represents endoplasmic reticulum, ABP1 bound on plasma membrane would perceive auxin and then auxin signal would be transduced to ROP proteins to regulate PIN protein trafficking. The small molecule probe targeting components of the auxin transport system is indicated.

Additional screens for molecules that either disrupt gravity sensing [134] or polar auxin transport [144] identified structurally distinct molecules that also target ABCB transporters. In the first screen, the molecule gravicin (3-(5-[3,4-dichlorophenyl]-2-furyl) acrylic acid) attenuates both root and shoot gravitropism in *Arabidopsis* seedlings. Map-based cloning of a gravicin-resistant mutant identified grav-r1 as an allele of ABCB19, and subsequent biochemical work showed that gravicin could compete for [^{3}H]NPA binding to microsomes and that ABCB19 is a primary site of gravicin interaction [134]. In the second screen, the search for growth and development regulators identified BUM (2-[4-(diethylamino)-2-hydroxybenzoyl] benzoic acid) based in its ability to induced dark green, epinastic leaves and to reduced primary and secondary root length. The results of biochemical and *in silico* docking experiments showed that BUM competes for NPA-binding sites, primarily at ABCB1, and that BUM disrupts the interaction between TWD1 and ABCB1 [134]. Interestingly, gravicin did not disrupt the TWD1-ABCB1 dimer formation.

A specific inhibitor of auxin influx has also been described. Imhoff et al. [145] screened a large number of aryl and aryloxyalkylcarboxylic acids for their ability to reduce 2,4-D uptake by culture tobacco cells. Based on these data, Parry et al. [146] selected the compound 1-naphthaleneoxy acetic acid (1-NOA) for further examination and tested its impact on physiological and molecular responses of *Arabidopsis* seedlings. The results showed that 1-NOA recapitulates the *aux1* mutant phenotype, suggesting that 1-NOA functions as a specific inhibitor of auxin influx mediated by the AUX1/LAX influx transporters. A later study confirmed that 1-NOA inhibits auxin uptake through AUX1 when expressed in *Xenopus* oocytes [57]. This study also shows that AUX1 is likely directly regulated by 1-NOA.

Alkoxy-auxins (Fig. 4.1.4) are auxin derivatives that work as auxin transport inhibitor [147]. Alkoxy-auxins displayed potent inhibitory activity on the gravitropic response of *Arabidopsis* plants. The structure of TIR1 auxin-binding cavity illustrated that these alkoxy-auxins would be unlikely to interact with TIR1 or the related AFB proteins, and, indeed, these molecules did not show any effect on the $SCF^{TIR1/AFB}$ pathway. When the PIN or ABCB proteins were expressed in yeast, the alkoxy-auxin analogs were effective at lowering transport activity. As such, the alkoxy-auxins perturb the transport activities of PIN and ABCB proteins, but do not alter auxin signaling through the TIR1/AFB pathway.

In addition to direct effects on IAA transport, other compounds alter auxin movement by disrupting the localization of transport proteins to the plasma membrane. Treatment of plants with the vesicle-trafficking inhibitor, brefeldin A, established the actin-dependent cycling of PIN proteins between the plasma membrane and endomembrane system as important for polar auxin transport [148]. Similarly, two other transport inhibitors, 2,3,5-triiodobenzoic acid (TIBA) and pyrenoyl benzoic acid (PBA), disrupt PIN vesicle trafficking processes by perturbing the actin network [149]. The identification of compounds that specifically target distinct components of the endomembrane system may lead to novel tools to further dissect the dynamic link between directional auxin transport and the vesicular trafficking [150].

Finally, the use of "caged" auxin molecules represents a powerful approach for investigating the impact of local, transient increases in auxin levels [151]. The chemical coupling of a bioactive molecule with a photo-removable protecting group creates the biologically inert caged compounds. The bioactive compound forms through UV-mediated photolysis, which can be accomplished using a variety of UV sources (i.e.., fluorescence microscope, hand-held UV light, and UV lasers). A diverse set of biologically active molecules (e.g., ATP, calcium, glutamate, neurotransmitters, and peptides) have been caged and widely used as valuable chemical tools for examining processes of importance for cell biologists [152]. The precise spatial and temporal control of auxin release can be performed using state-of-the-art optical devices, opening the possibility that the effect of artificial auxin gradients can be studied. Because plant esterases hydrolyze the caging groups commonly used in mammalian cell biology, Kusaka et al. [151] developed an esterase-resistant caging group to create a set of caged auxins. These molecules allowed the tight photocontrol of auxin response in *Arabidopsis* and allowed a single cell manipulation of auxin response in tobacco BY2 cells.

4.1.12 CONCLUDING REMARKS

During the past decades, many of the cellular components involved in auxin metabolism, auxin transport, and auxin perception have been identified. Progress in these areas is based on a growing number of labs working on these topics along with the development of abundant genetic resources using *A. thaliana*. One of the most exciting advances in auxin biology was the structural determination of the TIR1-Aux/IAA co-receptor. This discovery opened the door for the rational design of new auxin agonist and antagonist, which can be used in both basic and applied settings to manipulate auxin-mediated events. The suite of chemical tools available also allows the network, replete with overlapping or redundant components, of auxin synthesis, transport, and signaling, to be further defined in developmentally or spatially relevant tissues. As such, the successful application of chemical biology allows the molecular pleiotropy of auxin action to be explored. Combining these tools with advances in proteomics and metabolomics should reveal additional regulatory components of these processes. Furthermore, the incorporation of these data into mathematical models would provide a platform for further refining our understanding of the diverse processes mediated by auxin.

REFERENCES

1. Woodward, A.W. and B. Bartel, *Auxin: Regulation, action, and interaction.* Annals of Botany, 2005. **95**(5): p. 707–735.
2. Wildman, S.G., *The auxin-A, B enigma: Scientific fraud or scientific ineptitude?* Plant Growth Regulation, 1997. **22**: p. 37–68.
3. Abel, S. and A. Theologis, *Odyssey of auxin.* Cold Spring Harbor Perspectives in Biology, 2010. **2**(10): p. a004572.
4. Went, F.W., *Auxin, the plant-growth hormone. II.* Botanical Review, 1945. **11**(9): p. 487–496.
5. Kogl, F. and D. Kostermans, *Hetero-auxin als Stoffwechselprodukt niederer pflanzlicher Organismen. Isolierung aus Hefe.* XIII Mitteilung über pflanzliche Wachstumsstoffe. Hoppe-Seyler´s Zeitschrift für physiologische Chemie, 1934. **228**: p. 113–121.
6. Haagen-Smit, A.J., W.D. Leech, and W.R. Bergren, *The estimation, isolation, and identification of auxins in plant materials.* American Journal of Botany, 1942. **29**(7): p. 500–506.
7. Slovin, H., R.S. Bandurski, and J.D. Cohen, *in Biochemistry and Molecular Biology of Plant Hormones*, eds. P.J.J. Hooykaas, M.A. Hall, and K.R. Libbenga, 1999. Amsterdam; New York: Elsevier.
8. Simon, S. and J. Petrášek, *Why plants need more than one type of auxin.* Plant Science, 2011. **180**(3): p. 454–460.
9. Ludwig-Müller, J., *Indole-3-butyric acid in plant growth and development.* Plant Growth Regulation, 2000. **32**: p. 219–230.
10. Strader, L.C., et al., *Multiple facets of Arabidopsis seedling development require indole-3-butyric acid-derived auxin.* The Plant Cell, 2011. **23**(3): p. 984–999.
11. Rybel, B.D., et al., *A role for the root cap in root branching revealed by the non-auxin probe naxillin.* Nature Chemical Biology, 2012. **8**: p. 798–805.

12. Strader, L.C. and B. Bartel, *Transport and metabolism of the endogenous auxin precursor indole-3-butyric acid.* Molecular Plant, 2011. **4**(3): p. 477–486.
13. Ludwig-Muller, J. and J.D. Cohen, *Identification and quantification of three active auxins in different tissues of* Tropaeolum majus. Physiologia Plantarum, 2002. **115**(2): p. 320–329.
14. Zhao, Y., *Auxin biosynthesis and its role in plant development.* Annual Review of Plant Biology, 2010. **61**: p. 49–64.
15. Strader, L.C. and B. Bartel, *A new path to auxin.* Nature Chemical Biology, 2008. **4**(6): p. 337–339.
16. Stepanova, A.N., et al., *The Arabidopsis YUCCA1 flavin monooxygenase functions in the indole-3-pyruvic acid branch of auxin biosynthesis.* The Plant Cell, 2011. **23**(11): p. 3961–3973.
17. Mashiguchi, K., et al., *The main auxin biosynthesis pathway in Arabidopsis.* Proceedings of the National Academy of Sciences of the United States of America, 2011. **108**(45): p. 18512–18517.
18. Won, C., et al., *Conversion of tryptophan to indole-3-acetic acid by TRYPTOPHAN AMINOTRANSFERASES OF ARABIDOPSIS and YUCCAs in Arabidopsis.* Proceedings of the National Academy of Sciences of the United States of America, 2011. **108**(45): p. 18518–18523.
19. Stepanova, A.N., et al., *TAA1-mediated auxin biosynthesisis essential for hormone crosstalk and plant development.* Cell, 2008. **133**: p. 177–191.
20. Tao, Y., et al., *Rapid synthesis of auxin via a new tryptophan-dependent pathway is required for shade avoidance in plants.* Cell, 2008. **133**: p. 164–176.
21. Cheng, Y., X. Dai, and Y. Zhao, *Auxin synthesized by the YUCCA flavin monooxygenases is essential for embryogenesis and leaf formation in Arabidopsis.* Plant Cell, 2007. **19**: p. 2430–2439.
22. Cheng, Y., X. Dai, and Y. Zhao, *Auxin biosynthesis by the YUCCA flavin monooxygenases controls the formation of floral organs and vascular tissues in Arabidopsis* Genes & Development, 2006. **20**(13): 1790–1799.
23. Yamada, M., et al., *The TRANSPORT INHIBITOR RESPONSE2 gene is required for auxin synthesis and diverse aspects of plant development.* Plant Physiology, 2009. **151**(1): p. 168–179.
24. He, W., et al., *A small-molecule screen identifies* L*-kynurenine as a competitive inhibitor of TAA1/TAR activity in ethylene-directed auxin biosynthesis and root growth in Arabidopsis.* The Plant Cell, 2011. **23**(11): p. 3944–3960.
25. Schwarcz, R., W.O. Whetsell, and R.M. Mangano, *Quinolinic acid: an endogenous metabolite that produces axon-sparing lesions in rat brain.* Science, 1983. **219**: 316–318.
26. Opitz, C.A., et al. *An endogenous tumour-promoting ligand of the human aryl hydrocarbon receptor.* Nature, 2011. **478**(7368): p. 197–203.
27. Soeno, K., et al., *Auxin-biosynthesis inhibitors, identified by genomics-based approach, provide insights into auxin biosynthesis.* Plant & Cell Physiology, 2010. **51**(4): p. 524–536.
28. Boller, T., Herner, R.C., and Kende, H. *Assay for and enzymatic formation of an ethylene precursor, 1-aminocyclopropane-1-carboxylic acid.* Planta, 1979. **145**: p. 293–303.
29. Tsurusaki, K.-i., et al., *Conversion of* D*-tryptophan to indole-3-acetic acid in coleoptiles of a normal and a semi-dwarf barley* (Hordeum vulgare) *strain.* Physiologia Plantarum, 1990. **79**(2): p. 221–225.

30. Ludwig-Müller, J., et al., *An inhibitor of tryptophan-dependent biosynthesis of indole-3-acetic acid alters seedling development in Arabidopsis*. Journal of Plant Growth Regulation, 2009. **29**(2): p. 242–248.
31. Ludwig-Muller, J., *Auxin conjugates: Their role for plant development and in the evolution of land plants*. Journal of Experimental Botany, 2011. **62**(6): p. 1757–1773.
32. Staswick, P.E., I. Tiryaki, and M.L. Rowe, Jasmonate response locus JAR1 and several related Arabidopsis genes encode enzymes of the firefly luciferase superfamily that show activity on jasmonic, salicylic, and indole-3-acetic acids in an assay for adenylation. Plant Cell, 2002. **14**(6), p. 1405–1415.
33. Lee, T.T. and A.N. Starratt, *Inhibition of conjugation of indole-3-acetic acid with amino acids by 2,6-dihydroxyacetophenone in* Teucrium canadense. Phytochemistry, 1986. **25**(11): p. 2457–2461.
34. Bottcher, C., et al., *A novel tool for studying auxin-metabolism: The inhibition of grapevine indole-3-acetic acid-amido synthetases by a reaction intermediate analogue*. Public Library of Science One, 2012. **7**(5): p. e37632.
35. Chen, Q., et al., *Kinetic basis for the conjugation of auxin by a GH3 family indole-acetic acid-amido synthetase*. The Journal of Biological Chemistry, 2010. **285**(39): p. 29780–29786.
36. Bartel, B. and G.R. Fink, *ILR1, an amidohydrolase that releases active indole-3-acetic acid from conjugates*. Science, 1995. **268**(5218): p. 1745–1748.
37. Davies, R.T., et al., *IAR3 encodes an auxin conjugate hydrolase from Arabidopsis*. The Plant Cell, 1999. **11**(3): p. 365–376.
38. Campanella, J.J., J. Ludwig-Mueller, and C.D. Town, *Isolation and characterization of mutants of* Arabidopsis thaliana *with increased resistance to growth inhibition by indoleacetic acid-amino acid conjugates*. Plant Physiology, 1996. **112**(2): p. 735–745.
39. Slovin, J.P., *Phytotoxic conjugates of indole-3-acetic acid: Potential agents for biochemical selection of mutants in conjugate hydrolysis*. Plant Growth Regulation, 1997. **21**(3): p. 215–221.
40. Staswick, P.E., *The tryptophan conjugates of jasmonic and indole-3-acetic acids are endogenous auxin inhibitors*. Plant Physiology, 2009. **150**(3): p. 1310–1321.
41. Staswick, P., *Plant hormone conjugation: A signal decision*. Plant Signaling & Behavior, 2009. **4**(8): p. 757–759.
42. Went, F. and K. Thimann, *Phytohormones* 1937, New York: MacMillan Company.
43. Ferro, N., et al., *Route to novel auxin: Auxin chemical space toward biological correlation carriers*. Chemical Reviews, 2010. **110**(8): p. 4690–4708.
44. Zimmerman, P.W. and A.E. Hitchcock, *Plant hormones*. Annual Review of Biochemistry, 1948. **17**: p. 601–626.
45. Porter, W.L. and K.V. Thimann, *Molecular requirements for auxin action*. Phytochemistry, 1965. **4**: p. 229–243.
46. Farrimond, J.A., M.C. Elliott, and D.W. Clack, *Charge separation as a component of structural requirements for hormone-activity*. Nature, 1978. **274**(5669): p. 401–402.
47. McRae, D.H. and J. Bonner, *Chemical structure and antiauxin activity*. Physiologia Plantarum, 1953. **6**(3): p. 485–510.
48. Aberg, B., *The interaction of some auxin antagonists and 2,4-D in root growth*. Physiologia Plantarum, 1951. **4**(4): p. 627–640.
49. Hatano, T., et al., *A new indolic antiauxin, alpha-(5,7-dichloroindole-3-)isobutyric acid: Its chemical synthesis and biological activity*. Experientia, 1989. **45**: p. 400–402.

50. Darwin, C.R., *The Power of Movement in Plants* 1880, London: Murray.
51. Boysen-Jensen, P., *Uber die Leitung des phototropischen Reizes in der Avenakoleoptile*. Berichte der Deutschen Botanischen Gesellschaft, 1913. **31**: p. 559–566.
52. Petrasek, J. and J. Friml, *Auxin transport routes in plant development*. Development, 2009. **136**(16): p. 2675–2688.
53. Zazimalova, E., et al., *Auxin transporters—why so many?* Cold Spring Harbor Perspectives in Biology, 2010. **2**(3): p. a001552.
54. Grunewald, W., et al., *Parasitic nematodes modulate PIN-mediated auxin transport to facilitate infection*. Public Library of Science Pathogens, 2009. **5**(1): p. e1000266.
55. Marchant, A., et al., *AUX1 regulates root gravitropism in Arabidopsis by facilitating auxin uptake within root apical tissues*. European Molecular Biology Organization Journal, 1999. **18**(8): p. 2066–2073.
56. Bennett, M.J., et al., *Arabidopsis AUX1 gene: A permease-like regulator of root gravitropism*. Science, 1996. **273**(5277): p. 948–950.
57. Yang, Y.D., et al., *High-affinity auxin transport by the AUX1 influx carrier protein*. Current Biology, 2006. **16**(11): p. 1123–1127.
58. Titapiwatanakun, B., et al., *ABCB19/PGP19 stabilises PIN1 in membrane microdomains in Arabidopsis*. The Plant Journal, 2009. **57**(1): p. 27–44.
59. Titapiwatanakun, B. and A.S. Murphy, *Post-transcriptional regulation of auxin transport proteins: cellular trafficking, protein phosphorylation, protein maturation, ubiquitination, and membrane composition*. Journal of Experimental Botany, 2009. **60**(4): p. 1093–1107.
60. Chapman, E.J. and M. Estelle, *Mechanism of auxin-regulated gene expression in plants*. Annual Review of Genetics, 2009. **43**: p. 265–285.
61. Nemhauser, J., F. Hong, and J. Chory, *Different plant hormones regulate similar processes through largely nonoverlapping transcriptional responses*. Cell, 2006. **126**(3): p. 467–475.
62. Hagen, G., A. Kleinschmidt, and T. Guilfoyle, *Auxin-regulated gene-expression in intact soybean hypocotyl and excised hypocotyl sections*. Planta, 1984. **162**(2): p. 147–153.
63. Chae, K., et al., *Arabidopsis SMALL AUXIN UP RNA63 promotes hypocotyl and stamen filament elongation*. The Plant journal, 2012. **71**(4): p. 684–697.
64. Spartz, A.K., et al., *The SAUR19 subfamily of SMALL AUXIN UP RNA genes promote cell expansion*. The Plant Journal: For Cell and Molecular Biology, 2012. **70**(6): p. 978–990.
65. Szemenyei, H., M. Hannon, and J.A. Long, *TOPLESS mediates auxin-dependent transcriptional repression during Arabidopsis embryogenesis*. Science, 2008. **319**(5868): p. 1384–1386.
66. Gray, W.M., et al., *Auxin regulates SCF(TIR1)-dependent degradation of AUX/IAA proteins*. Nature, 2001. **414**(6861): p. 271–276.
67. Ouellet, F., P.J. Overvoorde, and A. Theologis, *IAA17/AXR3: Biochemical insight into an auxin mutant phenotype*. The Plant Cell, 2001. **13**(4): p. 829–841.
68. Kim, J., K. Harter, and A. Theologis, *Protein–protein interactions among the Aux/IAA proteins*. Proceedings of the National Academy of Sciences of the United States of America, 1997. **94**(22): p. 11786–11791.
69. Guilfoyle, T.J. and G. Hagen, *Auxin response factors*. Current Opinion in Plant Biology, 2007. **10**(5): p. 453–460.
70. Dharmasiri, N., S. Dharmasiri, and M. Estelle, *The F-box protein TIR1 is an auxin receptor*. Nature, 2005. **435**(7041): p. 441–445.

71. Kepinski, S. and O. Leyser, *The Arabidopsis F-box protein TIR1 is an auxin receptor.* Nature, 2005. **435**(7041): p. 446–451.

72. Mockaitis, K. and M. Estelle, *Auxin receptors and plant development: A new signaling paradigm.* Annual Review of Cell and Developmental Biology, 2008. **24**: p. 55–80.

73. del Pozo, J.C., et al., *AXR1-ECR1-dependent conjugation of RUB1 to the Arabidopsis Cullin AtCUL1 is required for auxin response.* The Plant Cell, 2002. **14**(2): p. 421–433.

74. Cheng, Y., X. Dai, and Y. Zhao, *AtCAND1, a HEAT-repeat protein that participates in auxin signaling in Arabidopsis.* Plant Physiology, 2004. **135**(2): p. 1020–1026.

75. Gray, W.M., et al., *Arabidopsis SGT1b is required for SCF(TIR1)-mediated auxin response.* The Plant Cell, 2003. **15**(6): p. 1310–1319.

76. Calderon-Villalobos, L.I., et al., *Auxin perception—structural insights.* Cold Spring Harbor Perspectives in Biology, 2010. **2**(7): p. a005546.

77. Tan, X., et al., *Mechanism of auxin perception by the TIR1 ubiquitin ligase.* Nature, 2007. **446**(7136): p. 640–645.

78. Dharmasiri, N., et al., *Plant development is regulated by a family of auxin receptor F box proteins.* Developmental Cell, 2005. **9**(1): p. 109–119.

79. Parry, G., et al., *Complex regulation of the TIR1/AFB family of auxin receptors.* Proceedings of the National Academy of Sciences of the United States of America, 2009. **106**(52): p. 22540–22545.

80. Greenham, K., et al., *The AFB4 auxin receptor is a negative regulator of auxin signaling in seedlings.* Current Biology, 2011. **21**(6): p. 520–525.

81. Jurado, S., et al., *The Arabidopsis cell cycle F-box protein SKP2A binds to auxin.* The Plant Cell, 2010. **22**(12): p. 3891–3904.

82. Jurado, S., et al., *SKP2A protein, an F-box that regulates cell division, is degraded via the ubiquitin pathway.* Plant Signaling & Behavior, 2008. **3**(10): p. 810–812.

83. Napier, R.M., K.M. David, and C. Perrot-Rechenmann, *A short history of auxin-binding proteins.* Plant Molecular Biology, 2002. **49**(3–4): p. 339–348.

84. Tromas, A., I. Paponov, and C. Perrot-Rechenmann, *AUXIN BINDING PROTEIN 1: Functional and evolutionary aspects.* Trends in Plant Science, 2010. **15**(8): p. 436–446.

85. Badescu, G.O. and R.M. Napier, *Receptors for auxin: Will it all end in TIRs?* Trends in Plant Science, 2006. **11**(5): p. 217–223.

86. Timpte, C., *Auxin binding protein: Curiouser and curiouser.* Trends in Plant Science, 2001. **6**(12): p. 586–590.

87. Steffens, B., et al., *The auxin signal for protoplast swelling is perceived by extracellular ABP1.* The Plant Journal, 2001. **27**(6): p. 591–599.

88. Yamagami, M., et al., *Two distinct signaling pathways participate in auxin-induced swelling of pea epidermal protoplasts.* Plant Physiology, 2004. **134**(2): p. 735–747.

89. Jones, A.M., et al., *Auxin-dependent cell expansion mediated by overexpressed auxin-binding protein 1.* Science, 1998. **282**(5391): p. 1114–1117.

90. Chen, J.G., et al., *The role of auxin-binding protein 1 in the expansion of tobacco leaf cells.* The Plant Journal, 2001. **28**(6): p. 607–617.

91. David, K.M., et al., *The auxin-binding protein 1 is essential for the control of cell cycle.* The Plant Journal, 2007. **50**(2): p. 197–206.

92. Chen, J.G., et al., *ABP1 is required for organized cell elongation and division in Arabidopsis embryogenesis.* Genes & Development, 2001. **15**(7): p. 902–911.

93. Braun, N., et al., *Conditional repression of AUXIN BINDING PROTEIN1 reveals that it coordinates cell division and cell expansion during postembryonic shoot development in Arabidopsis and tobacco*. The Plant Cell, 2008. **20**(10): p. 2746–2762.
94. Tromas, A., et al., *The AUXIN BINDING PROTEIN 1 is required for differential auxin responses mediating root growth*. Public Library of Science One, 2009. **4**(9): p. e6648.
95. Effendi, Y., et al., *The heterozygous abp1/ABP1 insertional mutant has defects in functions requiring polar auxin transport and in regulation of early auxin-regulated genes*. The Plant Journal, 2011. **65**(2): p. 282–294.
96. Kleine-Vehn, J., et al., *Recycling, clustering, and endocytosis jointly maintain PIN auxin carrier polarity at the plasma membrane*. Molecular Systems Biology, 2011. **7**: p. 500.
97. Robert, S., et al., *ABP1 mediates auxin inhibition of clathrin-dependent endocytosis in Arabidopsis*. Cell, 2010. **143**(1): p. 111–121.
98. Xu, T., et al., *Cell surface- and rho GTPase-based auxin signaling controls cellular interdigitation in Arabidopsis*. Cell, 2010. **143**(1): p. 99–110.
99. Jones, M.A., et al., *The Arabidopsis Rop2 GTPase is a positive regulator of both root hair initiation and tip growth*. The Plant Cell, 2002. **14**(4): p. 763–776.
100. Lincoln, C., J.H. Britton, and M. Estelle, *Growth and development of the axr1 mutants of Arabidopsis*. The Plant Cell, 1990. **2**(11): p. 1071–1080.
101. Leyser, H.M., et al., *Arabidopsis auxin-resistance gene AXR1 encodes a protein related to ubiquitin-activating enzyme E1*. Nature, 1993. **364**(6433): p. 161–164.
102. Liscum, E. and J.W. Reed, *Genetics of Aux/IAA and ARF action in plant growth and development*. Plant Molecular Biology, 2002. **49**(3–4): p. 387–400.
103. Ruegger, M., et al., *The TIR1 protein of Arabidopsis functions in auxin response and is related to human SKP2 and yeast Grr1p*. Genes & Development, 1998. **12**(2): p. 198–207.
104. Walsh, T.A., et al., *Mutations in an auxin receptor homolog AFB5 and in SGT1b confer resistance to synthetic picolinate auxins and not to 2,4-dichlorophenoxyacetic acid or indole-3-acetic acid in arabidopsis*. Plant Physiology, 2006. **142**(2): p. 542–552.
105. Gleason, C., R.C. Foley, and K.B. Singh, *Mutant analysis in Arabidopsis provides insight into the molecular mode of action of the auxinic herbicide dicamba*. Public Library of Science One, 2011. **6**(3): p. e17245.
106. Calderon Villalobos, L.I., et al., *A combinatorial TIR1/AFB-Aux/IAA co-receptor system for differential sensing of auxin*. Nature Chemical Biology, 2012. **8**(5): p. 477–485.
107. McRae, D.H., R.J. Foster, and J. Bonner, *Kinetics of auxin interaction*. Plant Physiology, 1953. **28**(3): p. 343–355.
108. Oono, Y., et al., *p-Chlorophenoxyisobutyric acid impairs auxin response in Arabidopsis root*. Plant Physiology, 2003. **133**(3): p. 1135–1147.
109. Rahman, A., et al., *A small acidic protein 1 (SMAP1) mediates responses of the Arabidopsis root to the synthetic auxin 2,4-dichlorophenoxyacetic acid*. The Plant Journal, 2006. **47**(5): p. 788–801.
110. Biswas, K.K., et al., *Genetic characterization of mutants resistant to the antiauxin p-chlorophenoxyisobutyric acid reveals that AAR3, a gene encoding a DCN1-like protein, regulates responses to the synthetic auxin 2,4-dichlorophenoxyacetic acid in Arabidopsis roots*. Plant Physiology, 2007. **145**(3): p. 773–785.
111. Blackwell, H.E. and Y. Zhao, *Chemical genetic approaches to plant biology*. Plant Physiology, 2003. **133**(2): p. 448–455.

112. Oono, Y., et al., *age Mutants of Arabidopsis exhibit altered auxin-regulated gene expression.* The Plant Cell, 1998. **10**(10): p. 1649–1662.

113. Ulmasov, T., G. Hagen, and T.J. Guilfoyle, *ARF1, a transcription factor that binds to auxin response elements*. Science, 1997. **276**(5320): p. 1865–1868.

114. Ballas, N., L.M. Wong, and A. Theologis, *Identification of the auxin-responsive element, AuxRE, in the primary indoleacetic acid-inducible gene, PS-IAA4/5, of pea* (Pisum sativum). Journal of Molecular Biology, 1993. **233**(4): p. 580–596.

115. Hayashi, K.i., *Yokonolide B, a novel inhibitor of auxin action, blocks degradation of AUX/IAA factors*. Journal of Biological Chemistry, 2003. **278**(26): p. 23797–23806.

116. Hayashi, K., et al., *Yokonolide A, a new inhibitor of auxin signal transduction, from Streptomyces diastatochromogenes B59*. The Journal of Antibiotics (Tokyo), 2001. **54**(7): p. 573–581.

117. Hayashi, K., et al., *Toyocamycin specifically inhibits auxin signaling mediated by SCFTIR1 pathway*. Phytochemistry, 2009. **70**(2): p. 190–197.

118. Yamazoe, A., et al., *Characterization of terfestatin A, a new specific inhibitor for auxin signaling*. Plant Physiology, 2005. **139**(2): p. 779–789.

119. Lipinski, C.A., et al., *Experimental and computational approaches to estimate solubility and permeability in drug discovery and development settings*. Advanced Drug Delivery Reviews, 1997. **23**(1–3): p. 3–25.

120. Armstrong, J.I., et al., *Identification of inhibitors of auxin transcriptional activation by means of chemical genetics in Arabidopsis*. Proceedings of the National Academy of Sciences of the United States of America, 2004. **101**(41): p. 14978–14983.

121. Sungur, C., et al., *The small molecule 2-furylacrylic acid inhibits auxin-mediated responses in* Arabidopsis thaliana. The Plant Cell Physiology, 2007. **48**(12): p. 1693–1701.

122. Christian, M., et al., *Identification of auxins by a chemical genomics approach*. Journal of Experimental Botany, 2008. **59**(10): p. 2757–2767.

123. Savaldi-Goldstein, S., et al., *New auxin analogs with growth-promoting effects in intact plants reveal a chemical strategy to improve hormone delivery*. Proceedings of the National Academy of Sciences of the United States of America, 2008. **105**(39): p. 15190–15195.

124. Grozinger, C.M., et al., *Identification of a class of small molecule inhibitors of the sirtuin family of NAD-dependent deacetylases by phenotypic screening*. The Journal of Biological Chemistry, 2001. **276**(42): p. 38837–38843.

125. Chuang, H.W., W. Zhang, and W.M. Gray, *Arabidopsis ETA2, an apparent ortholog of the human cullin-interacting protein CAND1, is required for auxin responses mediated by the SCF(TIR1) ubiquitin ligase*. The Plant Cell, 2004. **16**(7): p. 1883–1897.

126. Zhao, Y., et al., *SIR1, an upstream component in auxin signaling identified by chemical genetics*. Science, 2003. **301**(5636): p. 1107–1110.

127. Dai, X., et al., *Genetic and chemical analyses of the action mechanisms of sirtinol in Arabidopsis*. Proceedings of the National Academy of Sciences of the United States of America, 2005. **102**(8): p. 3129–3134.

128. Hayashi, K., et al., *Small-molecule agonists and antagonists of F-box protein-substrate interactions in auxin perception and signaling*. Proceedings of the National Academy of Sciences of the United States of America, 2008. **105**(14): p. 5632–5637.

129. Hayashi, K., et al., *Rational design of an auxin antagonist of the SCF(TIR1) auxin receptor complex*. ACS Chemical Biology, 2012. **7**(3): p. 590–598.

130. Woo, E.J., et al., *Crystal structure of auxin-binding protein 1 in complex with auxin.* European Molecular Biology Organization Journal, 2002. **21**(12): p. 2877–2885.
131. Beyer, E.M., A.L. Johnson, and P.B. Sweetser, *A new class of synthetic auxin transport inhibitors.* Plant Physiology, 1976. **57**(6): p. 839–841.
132. Katekar, G.F. and A.E. Geissler, *Auxin transport inhibitors: IV.* Evidence of a common mode of action for a proposed class of auxin transport inhibitors: the phytotropins. Plant Physiology, 1980. **66**(6): p. 1190–1195.
133. Reed, R.C., S.R. Brady, and G.K. Muday, *Inhibition of auxin movement from the shoot into the root inhibits lateral root development in Arabidopsis.* Plant Physiology, 1998. **118**(4): p. 1369–1378.
134. Rojas-Pierce, M., et al., *Arabidopsis P-glycoprotein19 participates in the inhibition of gravitropism by gravacin.* Chemistry & Biology, 2007. **14**(12): p. 1366–1376.
135. Noh, B., A.S. Murphy, and E.P. Spalding, *Multidrug resistance-like genes of Arabidopsis required for auxin transport and auxin-mediated development.* The Plant Cell, 2001. **13**(11): p. 2441–2454.
136. Geisler, M., et al., *Cellular efflux of auxin catalyzed by the Arabidopsis MDR/PGP transporter AtPGP1.* The Plant Journal, 2005. **44**(2): p. 179–194.
137. Blakeslee, J.J., et al., *Interactions among PIN-FORMED and P-glycoprotein auxin transporters in Arabidopsis.* The Plant Cell, 2007. **19**(1): p. 131–147.
138. Bouchard, R., et al., *Immunophilin-like TWISTED DWARF1 modulates auxin efflux activities of Arabidopsis P-glycoproteins.* The Journal of Biological Chemistry, 2006. **281**(41): p. 30603–30612.
139. Murphy, A.S., et al., *Identification, purification, and molecular cloning of N-1-naphthylphthalmic acid-binding plasma membrane-associated aminopeptidases from Arabidopsis.* Plant Physiology, 2002. **128**(3): p. 935–950.
140. Bailly, A., et al., *Modulation of P-glycoproteins by auxin transport inhibitors is mediated by interaction with immunophilins.* The Journal of Biological Chemistry, 2008. **283**(31): p. 21817–21826.
141. Brown, D.E., et al., *Flavonoids act as negative regulators of auxin transport in vivo in arabidopsis.* Plant Physiology, 2001. **126**(2): p. 524–535.
142. Peer, W.A. and A.S. Murphy, *Flavonoids and auxin transport: Modulators or regulators?* Trends in Plant Science, 2007. **12**(12): p. 556–563.
143. Jacobs, M. and P.H. Rubery, *Naturally occurring auxin transport regulators.* Science, 1988. **241**(4863): p. 346–349.
144. Kim, J.Y., et al., *Identification of an ABCB/P-glycoprotein-specific inhibitor of auxin transport by chemical genomics.* The Journal of Biological Chemistry, 2010. **285**(30): p. 23309–23317.
145. Imhoff, V., et al., *Inhibitors of the carrier-mediated influx of auxin in suspension-cultured tobacco cells.* Planta, 2000. **210**(4): p. 580–588.
146. Parry, G., et al., *Novel auxin transport inhibitors phenocopy the auxin influx carrier mutation aux1.* The Plant Journal, 2001. **25**(4): p. 399–406.
147. Tsuda, E., et al., *Alkoxy-auxins are selective inhibitors of auxin transport mediated by PIN, ABCB, and AUX1 transporters.* The Journal of Biological Chemistry, 2011. **286**(3): p. 2354–2364.
148. Geldner, N., et al., *Auxin transport inhibitors block PIN1 cycling and vesicle trafficking.* Nature, 2001. **413**(6854): p. 425–428.

149. Dhonukshe, P., et al., *Auxin transport inhibitors impair vesicle motility and actin cytoskeleton dynamics in diverse eukaryotes*. Proceedings of the National Academy of Sciences of the United States of America, 2008. **105**(11): p. 4489–4494.

150. Norambuena, L., G.R. Hicks, and N.V. Raikhel, *The use of chemical genomics to investigate pathways intersecting auxin-dependent responses and endomembrane trafficking in* Arabidopsis thaliana. Methods in Molecular Biology, 2009. **495**: p. 133–143.

151. Kusaka, N., et al., *Manipulation of intracellular auxin in a single cell by light with esterase-resistant caged auxins*. Chembiochem, 2009. **10**(13): p. 2195–2202.

152. Yu, H., et al., *Chemistry and biological applications of photo-labile organic molecules*. Chemical Society Reviews, 2010. **39**(2): p. 464–473.

153. Ferro, N., et al., *Unrevealed structural requirements for auxin-like molecules by theoretical and experimental evidences*. Phytochemistry, 2007. **68**(2): p. 237–250.

154. Hayashi, M., et al., *2,4-Dichlorophenoxybutyric acid-resistant mutants of Arabidopsis have defects in glyoxysomal fatty acid beta-oxidation*. The Plant Cell, 1998. **10**(2): p. 183–195.

155. Oono, Y., *p-Chlorophenoxyisobutyric acid impairs auxin response in Arabidopsis root*. Plant Physiology, 2003. **133**(3): p. 1135–1147.

156. Hayashi, K., et al., *Active core structure of terfestatin A, a new specific inhibitor of auxin signaling*. Bioorganic & Medicinal Chemistry, 2008. **16**(9): p. 5331–5344.

157. Katayama, M. and R.K. Gautam, *Synthesis and biological activities of fluorinated plant growth regulators, 4,4,4-trifluoro-3-(3-indolyl)butyric acids and 4,4,4-trifluoro-3-(2-indolyl) butyric acid bearing a methyl group on the indole nucleus*. Journal of Pesticide Science, 1997. **22**(4): p. 331–337.

158. Hatano, T., M. Katayama, and S. Marumo, *5,6-Dichloroindole-3-acetic acid as a potent auxin: Its synthesis and biological activity*. Experientia, 1987. **43**(11): p. 1237–1239.

159. Leopold, A.C. and W.H. Klein, *Maleic hydrazide as an antiauxin in plants*. Science, 1951. **114**(2949): p. 9–10.

160. Overbeek, J.v., R. Blondeau, and V. Horne, *Trans-cinnamic acid as an anti-auxin*. American Journal of Botany, 1951. **38**(8): p. 589–595.

161. Audus, L.J. and M.E. Shipton, *2,4-Dichloranisole—auxin interactions in root-growth*. Physiologia Plantarum, 1952. **5**(3): p. 430–455.

4.2

BRASSINOSTEROIDS SIGNALING AND BIOSYNTHESIS

TAKESHI NAKANO AND TADAO ASAMI

4.2.1 BRASSINOSTEROID RESEARCH MOVES INTO THE ERA OF CHEMICAL BIOLOGY

Brassinosteroids are bioactive compounds involved in the basic regulation of plant growth, for example, regulation of cell elongation/division. At the organ level, brassinosteroids play an important role in photomorphogenesis and regulation of chloroplast function, vascular bundle development, and plant disease resistance. Brassinosteroids belong to a class of chemical compounds called plant hormones. A unique property of brassinosteroids, in comparison with other plant hormones, is that brassinosteroid-like compounds with a similar chemical structure are found in other species. In higher organisms, several compounds are known to have a similar chemical structure or steroid skeleton; these include testosterone (a male sex hormone controlling differentiation of reproductive organs), estrogens and progesterone (female sex hormones regulating pregnancy), and corticosterone (adrenal cortex hormone) in mammals and ecdysone (a hormone that controls molting) in insects and crustaceans. Thus, steroid hormones are bioactive substances conserved widely across species. These steroid hormones have bioactivity in specific organs in an individual species, for example, functional regulation of chloroplasts in plants, regulation of pregnancy in mammals, and regulation of molting in insects. Additionally, these hormones have universal roles, such as stimulation of growth, that are common across species. This parallel existence of universality and specificity of steroid hormone bioactivity is of interest in steroid hormone research (Fig. 4.2.1).

Plant Chemical Biology, First Edition. Edited by Dominique Audenaert and Paul Overvoorde.

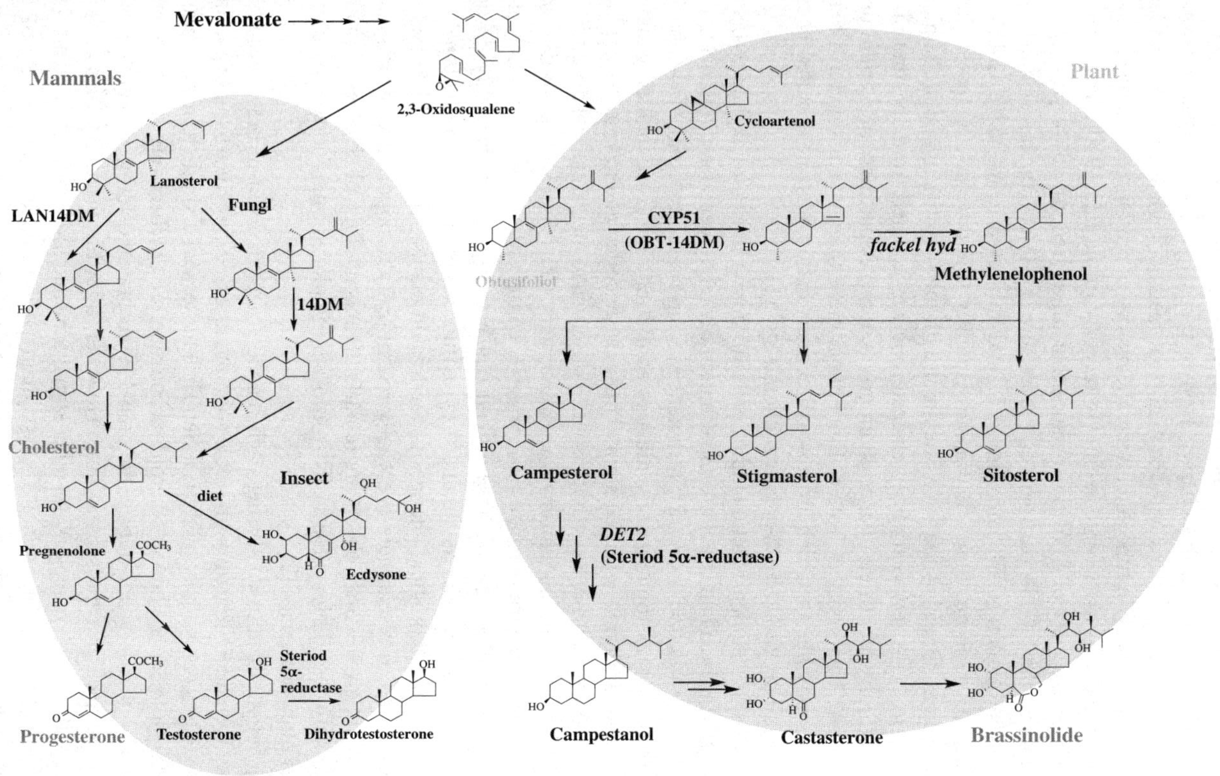

FIGURE 4.2.1 Comparison of biosynthesis pathways between plant brassinosteroid and animal or insect steroid hormone. For color detail, please see color plate section.

Recently, the scientific discipline of chemical biology has increased in popularity. The goal of chemical biology is to elucidate biological mechanisms by using small bioactive organic compounds. Brassinosteroid research has seen the increasing use of chemical biology-based research methods in a very effective manner. Some of these methods will be described in this chapter. However, because the history of brassinosteroid research largely overlaps with the history of methodological advances in plant research, we will go back in time and review the history of brassinosteroid research.

Brassinosteroid was discovered relatively recently in the late 1970s. The history begins with the discovery of a novel substance that stimulated growth of young leguminous plants by the U.S. Department of Agriculture (USDA). This novel substance was found in the pollen of *Brassica napus* L. and subjected to intensive isolation/structure identification processes. It was subsequently found to have a steroid skeleton and named as "brassinolide" to associate it with the scientific name of the plant "*Brassica napus* L." A few years later, castasterone was isolated from chestnut inset gall and structurally identified by a research group in the Department of Agricultural Chemistry at the University of Tokyo. Since then, more than 50 analogs have been isolated from many plants, which are now collectively called brassinosteroids. These discoveries were propelled by "analytical chemistry" [1].

In the 1990s, molecular genetic studies on *Arabidopsis thaliana* (Japanese name: Shiroironazuna), which had become an internationally recognized lab plant, led to the discovery of *de-etiolation 2* (*det 2*), a mutant lacking brassinosteroid biosynthesis. This mutant strain was found among mutants exhibiting dwarfism, including shortened leaves and short and widened stem. This gene encodes an enzyme resembling human steroid 5α-reductase. Its activity is well conserved to the extent that the enzyme can be replaced with the human enzyme counterpart. This discovery proved at the gene level that steroid hormones are evolutionarily conserved not only in animals and insects but also in plants. Some have argued that brassinosteroids probably do not have as prominent an activity as auxin and cytokinin; auxin or cytokinin treatment is known to cause morphological changes in the treated plants. However, the botanical society and botanical researchers eventually recognized the importance of brassinosteroids in plant growth after observing remarkable morphological changes in plants lacking brassinosteroid biosynthesis. This information placed brassinosteroids in the category of plant hormones [2].

Subsequently, enzymes catalyzing each reaction step involved in brassinosteroid biosynthesis were identified through studies of mutants exhibiting dwarfism. In parallel with these studies on synthases, BRI1, a Ser/Thr kinase with a single transmembrane domain, was isolated as a brassinosteroid receptor before the discovery of other plant hormones. For a few years during the mid-1990s, brassinosteroid research made remarkable advances for the second time. However, after the discovery of brassinosteroid biosynthesis and receptors, no other signal transduction factors were identified in the subsequent stagnant period. This stagnant period ended with the arrival of chemical biology and through the use of brassinazole (Brz), a brassinosteroid biosynthesis inhibitor. We isolated the transcription factor BZR1/BIL1 [3] in 2002. This transcription factor served as the foundation for a subsequent global increase in the number of discoveries of brassinosteroid-related transcription factor groups [4].

Moreover, even when chemical compounds are not used in mutant isolation and gene identification, inhibitors have been widely used. For example, when a candidate gene for a brassinosteroid signal transduction factor is isolated using biochemical or molecular biological methods, chemical biology-based functional analysis is performed to evaluate the protein encoded by the gene using a brassinosteroid regulator, such as Brz, to clarify the relationship between the gene and brassinosteroid. This principle has become increasingly popular. In the following sections, we review the biosynthesis and signal transduction factors of brassinosteroids and introduce the latest botanical chemical biology research on brassinosteroids from the viewpoint of utilization of brassinosteroid regulators.

4.2.2 BRASSINOSTEROID REGULATORS

Many methods are employed to discover/produce a novel biologically active compound. Typical examples include the following: screening of naturally occurring compounds, a method typically used in the discovery of antibiotics; screening using a chemical library that has been upgraded, particularly in recent years; design and synthesis of bioactive compounds utilizing researcher experience and intuition; and *in silico* design and *in silico* screening to exploit the recent advances in computer technology. Nevertheless, studying the structure–activity relationship through derivation by chemical synthesis is essential. Moreover, when using chemical derivation to enhance the activity of a compound having a targeted bioactivity, for example, even when it is impossible to control the targeted protein in an intended manner, it is important to synthesize a wide variety of compounds and build a library in order to develop a compound that can inhibit protein functions. If a biological phenomenon related to a previously studied targeted protein is studied using chemical biology-based methods, a previously created library can be used as a target-oriented library without any modification, thereby making it easier to search for regulator compounds and increasing efficiency. It is beneficial for both botanical and biological researchers to collect chemically synthesized compounds for further developing a chemical library. Moreover, taking into account the technical possibility and the past contributions of agricultural chemicals in agriculture, practical application of the collected bioactive compounds, as well as utilization of useful genes identified in gene research, is very likely to occur in the future.

Previously reported brassinosteroid activity regulators are described next (Table 4.2.1). The first reported brassinosteroid inhibitor is KM-01, which was isolated from a microbial culture medium. KM-01 was observed to inhibit brassinosteroid activity in a rice lamina inclination test [5]. Although the site of action is not clear, KM-01 exhibits highly potent activity, and thus is an interesting test subject for application to *Arabidopsis* and other plants. However, a major limitation is that no KM-01 supply system has been established; thus, this compound cannot be used.

Compounds described in References [5–19] are azole-type biosynthesis inhibitors. Among these, Brz91, Brz2001, and Brz220 were discovered by Asami et al. The azole ring contained in azole-type compounds is known to have affinity to heme contained

TABLE 4.2.1 Structure and action site of brassinosteroid regulators. ? indicates that targeting of DET2 by spironolactone is tentative.

Chemical	Structure	Target	Reference
KM-01	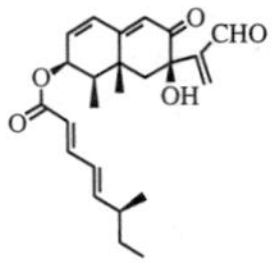	Unknown	[5]
Brassinazole (Brz91)	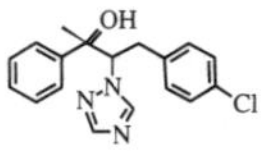	DWF4 (cytochrome P450)	[6–9]
Brassinazole (Brz2001)	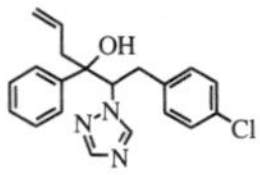	DWF4 (cytochrome P450)	[10]
Brassinazole (Brz220)	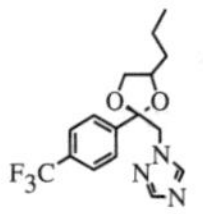	DWF4 (cytochrome P450)	[11,12]
DPPM4	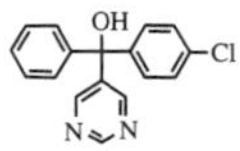	DWF4 (cytochrome P450)	[13]
Triadimefon	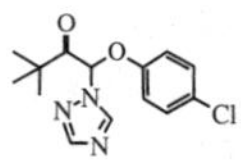	DWF4 (cytochrome P450)	[14]
Imazalil	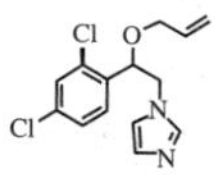	DWF4 (cytochrome P450)	[15]
Spironolactone	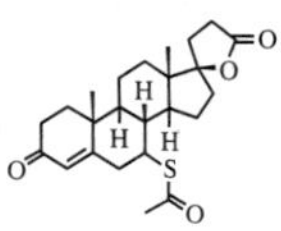	DET2?	[16]
Brassinopride	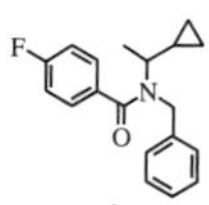	Upstream of CPD	[17]
Endosidin1	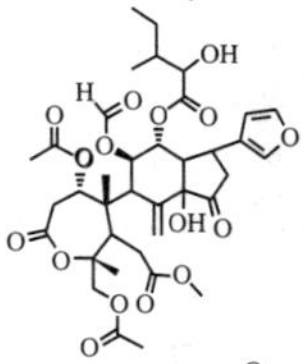	BRI1 recycling	[18]
Bikinin	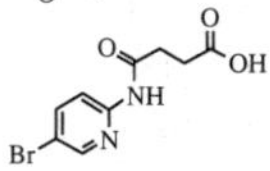	BIN2 (GSK3)	[19]

in P450 oxidase. Using this property, various azole derivatives with different chemical structures were synthesized through chemical modification of structural components other than the azole ring. *A. thaliana* was allowed to grow on culture media containing these compounds to perform the first screening for brassinosteroid biosynthesis inhibitors. This was done using the same dwarf mutant plants lacking brassinosteroid biosynthesis as an indicator. Next, plant morphological changes were observed in the presence of a dwarfism-inducing compound and brassinosteroid to screen for compounds capable of rescuing the dwarfism phenotype by addition of brassinosteroid. In the early stages of the study, the lead compound showed insufficient morphological recovery after brassinosteroid treatment. This was probably due to the compound's inhibitory activity toward P450 enzymes that exist in the biosynthesis pathways of other plant hormones. Thus, to reduce these side effects, more compounds were synthesized based on the lead compound skeleton. Further screening was performed using the degree of morphological recovery from dwarfism induced by treatment with a synthesized compound. The result was synthesis of the highly specific Brz compound group. Subsequent studies showed that the site of action of Brz 91, Brz 2001, and Brz 220 is DWF4, a cytochrome P450 involved in hydroxylation of the C-22 position of the brassinosteroid side chain [9–14]. In particular, Brz 91 is currently marketed as a reagent, which can be regarded as success in terms of its availability to many researchers. Imazalil's site of action has not been reported [15]. A series of physiological experiments using Brz showed that dwarfism can be induced not only in the experimental plant *Arabidopsis* but also in the dicots *Nicotiana tabacum*, *Solanum lycopersicum*, *Pisum sativum* L., and *Ipomoea nil*, and in the monocots *Oryza sativa*, and *Triticum aestivum*. Using Brz, a state of brassinosteroid deficit can be induced in various plant species at any time point of the growth process and at any intensity. A unique Brz application was reported in a study of *Gossypium* spp., where cotton floral buds were treated with Brz during the growing period. Brassinosteroid deficit was shown to inhibit cotton fiber elongation. Brz will likely be used widely in various applications.

Spironolactone is a brassinosteroid biosynthesis inhibitor. Previous studies showed that the spironolactone site of action is different from that of the aforementioned azole-type compounds [16]. This difference is useful in characterizing inhibitor-resistant mutants; using spironolactone and azole-type inhibitor concomitantly, it can be easily distinguished whether the inhibition is due to a mutation on the action site or a mutation in the signal transduction pathway. Brassinopride inhibits brassinosteroid biosynthesis; however, the action site has not been determined [17]. This compound inhibits biosynthesis of both brassinosteroid and ethylene. This unique property can be regarded as a side effect; however, brassinopride can be used to observe the morphological changes caused by double inhibition. The inhibitor must be used for a particular purpose depending on the study objective. Endosidin1 inhibits a process related to brassinosteroid receptor BRI 1 recycling [18]. Because this area of research is largely unexplored, analysis of a new factor is necessary; however, endosidin1 is currently not readily available. Bikinin exerts its effect on the brassinosteroid signal transduction factor BIN2 protein and may be useful in signal transduction studies [19]. Among the aforementioned inhibitors,

KM-01 was isolated using the same screening method used to identify antibiotics. Azole-type compounds were discovered through chemical syntheses. Brassinopride, endosidin1, and bikinin were discovered through chemical library screening. In the future, elucidation of the receptor tertiary structure will help discover receptor inhibitors by using *in silico* methods.

4.2.3 BRASSINOSTEROID RECEPTORS

Some mutants grow equally well in the dark as in light or exhibit dwarfism in the dark. If these mutants are brassinosteroid biosynthesis mutants, brassinosteroid addition should restore their wild-type morphology. However, dwarfism was not rescued by brassinosteroid addition. In the early 1990s, some research groups searched for a mutant that, unlike the wild type, did not show root elongation inhibition by brassinosteroid. As a result, the gene responsible for the *Arabidopsis* mutant *bri1* was identified by Chory et al. The mutant was a Leu-rich repeat (LPR) receptor-like Ser/Thr kinase containing repeated units (25 repeats) [20]. GFP was used to determine that *BRI1* localizes to the plasma membrane [21]. Moreover, the BRI1 protein that was synthesized *in vivo* using a human cell culture system and purified using antibody autophosphorylated serine and threonine residues confirmed that BRI1 functions as a plasma membrane kinase [21]. Moreover, in a transformant that contains about 20 times more BRI1 protein than the wild-type strain, the morphological traits similar to those caused by excessive brassinosteroid, such as excessive petiole elongation and true leaf epinasty, were observed [22].

Next, using a chemical probe, researchers attempted to prove that BRI1 is truly the brassinosteroid receptor [23]. Chemical probes are used to label bioactive compounds; for example, a chemical binding reaction can label biotin with a fluorescent compound or antibody so that it can be observed; the labeled compound or antibody is then used to observe the affinity between a target protein and the probe or the target protein *in vivo* activity. Currently, chemical probes are attracting much interest in the field of chemical biology research on mammals. During preparation, chemical probes must be attached to the side chain of the compound such that the compound bioactivity is not affected. Thus, prior to probe preparation, various analogues are synthesized to conduct a structure–activity relationship study to determine the side chain that is important for bioactivity and the side chain that can be used to attach the probe without affecting compound bioactivity. A detailed structure–activity relationship study was carried out for brassinosteroids. Following this study, Seto et al. successfully added a photoaffinity functional group and a biotin group to castasterone through chemical synthesis to use for analyses. As a result, affinity was confirmed between the probe and the purified plant BRI1 complex [22]. In studies, binding was observed between biotin-labeled castasterone and the characteristic domain consisting of 70 amino acids in the LPR domain, but not included in the LPR, of the *E. coli* BRI1 protein. Based on this observation, BRI1 was shown to be the brassinosteroid receptor [23].

4.2.4 CHEMICAL BIOLOGY RESEARCH BY GENETIC TECHNIQUES UTILIZING BRASSINOSTEROID BIOSYNTHESIS INHIBITORS: CHEMICAL GENETICS

As stated previously, the triazole derivative Brz was developed as a brassinosteroid biosynthesis enzyme DWF4-specific inhibitor. With Brz, researchers can create a state of brassinosteroid deficiency in specific plants and under specific conditions. Thus, a molecular genetic study was conducted using Brz to select mutants. The first step of this molecular genetics approach was to screen for mutants exhibiting plant hormone deficiency or those that show resistance to various plant hormones. This approach has been successful. The second step was to conduct a suppressor screen for a mutation that complements deletion mutants. However, suppressor screening has several requirements, for example, suppressor screening requires a variant causing a particular mutation in at least two ecotypes if the mutation is induced by a chemical or radiation. Thus, compared with the first screening, in suppressor screening, the study material preparation was expected to be more difficult. The biosynthesis inhibitor Brz is a very effective tool in conducting such screening. Today, this field is called chemical genetics. Based on the previous observation that is similar to the brassinosteroid deficit mutant, wild-type *A. thaliana* grows in the dark in the presence of Brz and exhibits the same morphological traits as the plant growing under light. Thus, we hypothesized that if a revertant that exhibits hypocotyl elongation and closed cotyledon under the presence of Brz is isolated, the isolated revertant will be a mutant in which brassinosteroid signal transduction or biosynthesis is activated in the downstream domain. Based on this hypothesis, we conducted a study to identify the *bil* (Brz-insensitive-long hypocotyl) mutant (Fig. 4.2.2).

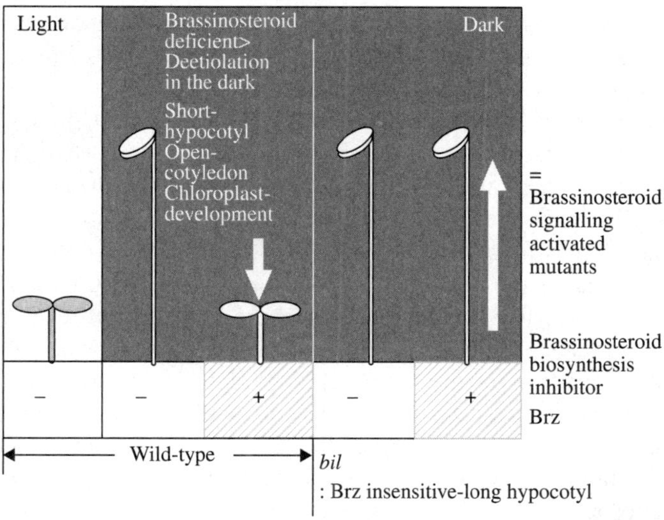

FIGURE 4.2.2 Strategy of chemical genetics research combined with the use of brassinosteroid biosynthesis inhibitor Brz, and Brz regulatory activity on plant hypocotyl elongation as an indicator.

Wild-type plants growing in dark on solid culture medium containing 3 mM of Brz exhibit dwarfism, and the length of the hypocotyl is approximately 20% of the wild-type plant grown under light on culture medium not containing Brz. The authors conducted a screening study under these conditions to identify *bil1* from ethyl methane sulphonate (EMS)-induced mutant lines of *A. thaliana*. During this period, *bzr1 (Brz-resistant1)* was selected by Professor Joan Chory's research lab at the Salk Institute for Biological Studies (USA). Subsequent collaborative studies revealed that *bil1* and *bzr1* are the same gene, and that in *bzr1/bil1*, the protein was stabilized by amino acid substitution caused by a point mutation, which is expressed as a Brz-resistant gain-of-function mutant [3]. Moreover, the *bes1* (*bri1-EMS-suppressor1*) mutant was isolated from the EMS-treated receptor mutant *bri1-5*. However, this *bri1-5* mutant was a mutant of *Arabidopsis* Ws2, for which little genetic data were available. Moreover, the gene responsible for mutation must be mapped under the continuous presence of the *bri1* mutation. Although the mutant was obtained, it took several years to isolate the gene responsible. However, the *bzr1/bil1* mutant was isolated, and the mutant phenotype was similar to that of the *bes1* mutation; thus, the base *BZR1/BIL1* gene sequence and the six related genes in the *bes1* mutant were analyzed. These results demonstrated that at the amino acid level, amino acid substitution occurs at the same position as in the *bil1/bzr1* mutation in a gene family gene that is highly homologous (88% identical) to *bil1/bzr1* [24]. The bes1 mutant may be the first beneficiary of chemical biology-based *bzr1/bil1* mutation research in brassinosteroids. Subsequently, several new discoveries were made regarding bes1, for example, bes1 was shown to be a transcriptional factor, along with BZR1/BIL1 and BES1 proteins, and was found to be involved in the transcriptional regulation of genes related to brassinosteroid-induced morphogenesis as well as the gene responsible for brassinosteroid biosynthesis regulated by a feedback mechanism. The protein bes1 migrates from the cytoplasm to the nucleus upon brassinosteroid stimulation. Today, bes1 has achieved unwavering recognition as a major transcriptional factor in the brassinosteroid signaling pathway [25, 26].

Significant progress has been made in mutually independent studies to find the complex comprising factors involved in these processes. Moreover, many complex comprising factors were found near the plasma membrane-localized receptor BRI1. The authors will continue the search for *bil* mutants through chemical biology-based studies using the brassinosteroid biosynthesis inhibitor Brz, in order to isolate the gene involved in signal transduction in the cell plasma between the plasma membrane and the nucleus since few associated factors have been isolated from this region. Currently, several candidate genes have been obtained. Among them, *BIL4* is considered a positive regulator of the brassinosteroid signaling pathway because *BIL4* overexpression induces Brz-resistance [27]. Structurally, BIL4 was hypothesized to be a membrane protein containing seven transmembrane domains, which is speculated to be evolutionarily conserved in many higher plants, mammals (e.g., humans and mice), Drosophilidae, and nematodes; however, no research had been conducted on this novel gene in any of these groups. Entire plant genomes have been sequenced and analyzed for many species, leading to the post-genome era and the use of reverse genetics technique. Reverse genetics is a computer-based analytical method of identifying functional gene candidates in a plant by searching previously obtained data from other species. Reverse genetics enables a fast search for a gene that can be used as a clue. However, novel genes such as *BIL4* tend to be overlooked. A significant advantage of classical genetics and chemical genetics is that novel genes can be identified using these disciplines.

4.2.5 CHEMICAL BIOLOGY RESEARCH BY PLANT PHYSIOLOGICAL TECHNIQUES UTILIZING BRASSINOSTEROID BIOSYNTHESIS INHIBITORS

In addition to the gene identification methods using chemical genetics techniques described in the previous section, other molecular biological methods are used to identify brassinosteroid signaling factors. Many of these methods use the brassinosteroid biosynthesis inhibitor Brz to analyze the function of identified genes, and progress has been made in chemical biology research of the major genes mentioned (Fig. 4.2.3).

The authors isolated *bzr1/bil1* using the following traits as indicators: (i) the wild-type *Arabidopsis* grown in the dark in the presence of Brz exhibits morphological traits, such as short hypocotyl and open cotyledons, as if it were growing under light; and (ii) under the same condition, a mutant with an activated brassinosteroid signaling pathway exhibits hypocotyl elongation, reflecting Brz-resistance. These morphological changes are widely considered to be an effective tool in demonstrating the function of brassinosteroid signaling factors. The presence of BES1-interacting factors around the basic helix-loop-helix (bHLH) transcription factor BES1 has been investigated using various methods. The following factors were obtained by yeast two-hybrid screening using BES1: the bHLH transcriptional factor BIM [25]; IWS1 [28] that interacts with BES1 and is the gene responsible for *seb1*, the morphologically complementing mutant of the *bes1* mutant; and MYB30 that also interacts with BES1 as the *myb* transcription factor that is highly expressed in the *bes1* mutant. When genes corresponding to these factors are highly expressed in the plant body, the plant exhibits hypocotyl elongation, reflecting Brz-resistance. When these genes are expressed at a low level, the plant exhibits a short hypocotyl, reflecting Brz hypersensitivity. This suggests that these factors are positive regulators of the brassinosteroid signaling pathway.

The plasma membrane-localized Ser/Thr kinase BIN2 is known to interact with and phosphorylate BZR1/BIL1 and BES1. The auxin-induced transcription factor ARF2 was isolated by yeast two-hybrid screening using BIN2. This factor has been studied in auxin research. Based on the observation that the *arf2*-knockout mutant exhibits hypocotyl elongation, it has been hypothesized that ARF2 is under the control of both brassinosteroid and auxin.

Furthermore, some suggest that several HLH transcription factors lacking the typical basic region present in transcription factors are involved in brassinosteroid signaling. First, the HLH obtained from the activation-tagging mutant *atbs1* from the *Arabidopsis* brassinosteroid receptor mutant *bri1-9* exhibits hypocotyl elongation,

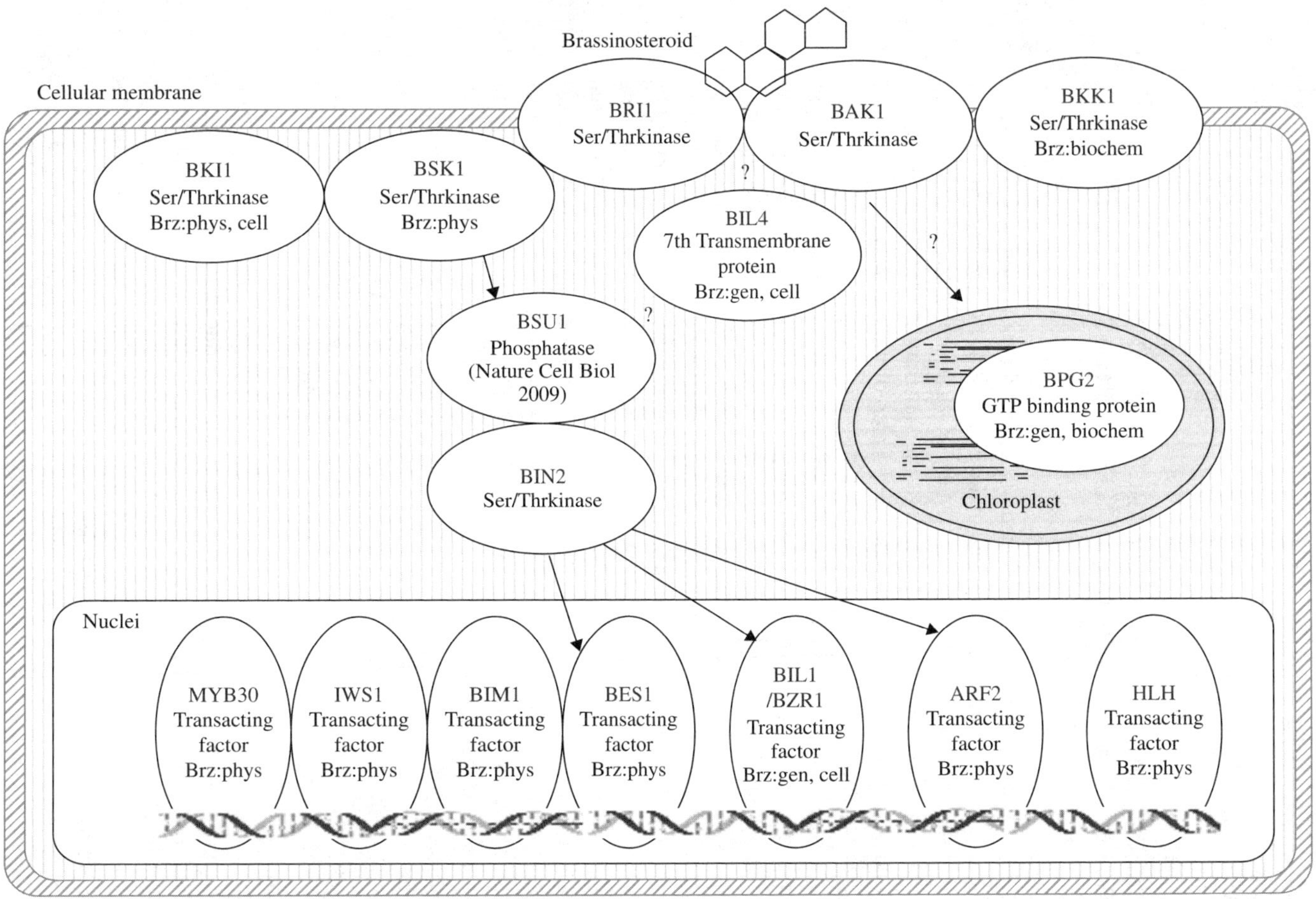

FIGURE 4.2.3 Applications of brassinosteroid biosynthesis inhibitor Brz for analysis of brassinosteroid signal transduction-related factors and their function. Brz: genetics: applications in molecular genetics; physiology: applications in plant physiology; cell: applications in cell imaging; biochemistry: applications in biochemistry. ? indicates potential signaling events.

providing proof of the function. BU1, obtained from rice plants, is a more interesting brassinosteroid-induced HLH factor [29]. In rice plants, the internode stem length where brassinosteroid exerts its effect was studied. The results showed that Brz decreased the internode length in the wild-type rice plant; however, rice plants overexpressing BU1 exhibited internode elongation and Brz-resistance. These observations indicate that BU1 is a positive regulator of the brassinosteroid signaling pathway. Chemical biology research conducted using Brz is an important study example because of the following reasons. Firstly, Brz was proven to be effective not only in the lab plant *Arabidopsis* but also in rice plants, which are actual crops, thereby demonstrating its usability in agriculture. Secondly, its biological importance has been demonstrated by proving its effectiveness in a different plant species, that is, monocots such as rice plants, in addition to dicots such as *Arabidopsis*.

4.2.6 CHEMICAL BIOLOGY RESEARCH BY UTILIZATION OF BRASSINOSTEROID BIOSYNTHESIS INHIBITOR IN CELL IMAGING AND BIOCHEMICAL STUDIES

In brassinosteroid signaling research, neighboring regions of transcription factors or the receptor BRI1 have been investigated. The kinase BSK interacts with the receptor BRI1 that was isolated in a search for brassinosteroid-induced proteins. BSK is thought to act as a positive regulator of brassinosteroid signaling near the plasma membrane intracellular surface. Plant strains overexpressing BSK1 exhibit hypocotyl elongation, reflecting Brz-resistance. This is used to demonstrate function of BSK1. However, other two factors, BKK1 and BKI1, have distinctive new usages.

The brassinosteroid receptor BRI1 is a single-transmembrane Ser/Thr kinase containing extracellular 20 LRPs (Leu-rich repeats). Other similar single-transmembrane Ser/Thr kinases are known to have shorter extracellular LPRs of approximately five repeats. Among these kinases, BKK1 forms a heterodimer with BRI in a similar manner as in the heterodimerization of BRI1 and BAK1 [30]. Furthermore, using immunoprecipitation, BKK1 and BRI1 were partially purified from plants grown in the presence of Brz. The purified fraction was assayed using a phosphorylated Thr-specific antibody. The results showed that the autophosphorylation level of BKK1 and BRI1 decreased with Brz treatment, and that phosphorylation level increased when brassinosteroid was administered to plants grown in the presence of Brz. This result is the first case in which Brz was shown to be an effective tool in biochemical functional analysis.

BKI1 is another kinase identified to interact with the receptor BRI1 and exert its effect near the plasma membrane intercellular surface. Plants overexpressing *BKI1* exhibit short hypocotyl, suggesting high Brz sensitivity; thus, BKI1 is a negative regulator of the brassinosteroid signaling pathway. Intracellular localization of BKI1 was studied using a transformant in which BKI1 was labeled with yellow fluorescent protein (YFP). This study showed that the intensity of the BKI-YFP fluorescent signal near the plasma membrane was stronger in Brz-treated plants than in nontreated plants. When these plants were treated with brassinosteroid, a fluorescent signal was dissociated from the plasma membrane and diffused throughout the cytoplasm. The bHLH transcription factor BZR1/BIL1 is typically localized in the cytoplasm and nucleus. However, the fluorescent signal of the BZR1/BIL1 transformant grown in the presence of Brz exhibited decreased nuclear signal intensity. The intensity was approximately 50% of that observed in nontreated plants. When plants were treated with brassinosteroid in the presence of Brz, fluorescent signal intensity in the nucleus 1 h after treatment increased to approximately 400% of the intensity observed in nontreated plants. The migration mechanism of BRZ1/BIL1 from the cytoplasm to the nucleus has gradually been elucidated. Cell imaging, combined with the use of fluorescent protein, has become a powerful tool in studying protein functions in living cells. If this tool is combined with chemical biology to study how various compounds affect cellular processes, it will likely enable more detailed functional analyses.

4.2.7 CHEMICAL GENETICS WITH A FOCUS ON THE REGULATORY ACTIVITY OF BRASSINOSTEROID BIOSYNTHESIS INHIBITORS ON THE CHLOROPLAST FUNCTION

Various Brz applications related to the control mechanism of brassinosteroid-induced plant morphogenesis were discussed earlier. In addition to regulating plant morphogenesis, brassinosteroids have another important regulatory activity, which is chloroplast regulation. This was elucidated mainly through the analysis of mutants lacking brassinosteroid biosynthesis. It was confirmed using Brz. Researchers previously elucidated that plants grown in the dark normally have elongated hypocotyls; however, wild-type plants grown in the dark in the presence of Brz undergo photomorphogenesis, which is typically characterized by short hypocotyls and open cotyledons. At the same time, expression of genes encoding enzymes that carry out photosynthesis in the chloroplast was induced. Additionally, these enzymes are typically expressed at a very low level in the dark. Subsequently, parallel to research on morphological regulation of hypocotyl, chloroplast regulation was studied using chemical genetics. From these investigations, *BPG2* was isolated as a gene candidate and was thought to have an important novel role in chloroplast regulation [31].

When *Arabidopsis* plants grown in the dark in the presence of Brz for a few days are moved and placed under light, cotyledon greening is observed to be more rapid than that in plants grown on standard culture medium. This greening acceleration is also visible in plants grown under light beginning at germination. Thus, it was hypothesized that if a mutant in which greening acceleration does not occur when grown in the presence of Brz in the light is isolated, that mutant can be said to lack a functional factor involved in the signal pathway leading to chloroplast regulation by brassinosteroid. Based on this hypothesis, the *bpg* (Brz-insensitive-pale green) was identified (Fig. 4.2.4).

Even mature *bpg2* mutants showed low greening levels. Chlorophyll assay data showed that Brz increases the chlorophyll content in wild-type plants to approximately 130%; however, in *bpg* mutants, the endogenous chlorophyll content is approximately 50% of that in wild-type plants and Brz does not increase the chlorophyll content in *bpg* mutants.

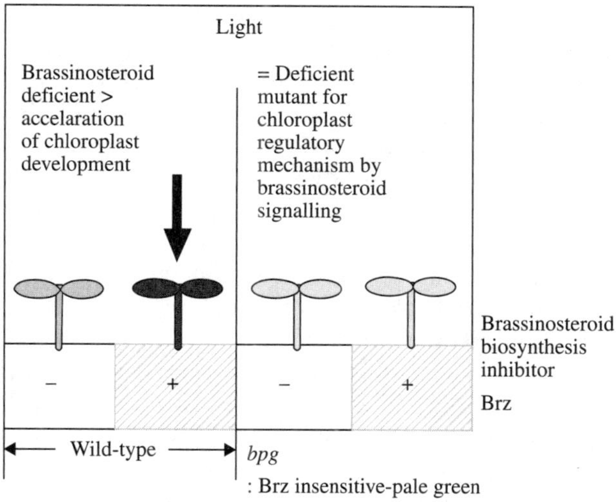

FIGURE 4.2.4 Chemical genetics research strategies combined with the use of brassinosteroid biosynthesis inhibitor Brz, and Brz regulatory activity on chloroplast function as an indicator.

The gene responsible for the mutation has a single zinc finger domain and four GTP-binding domains, which are widely conserved in many higher plants, green algae, and prokaryotes including *Bacillus subtilis*. *BPG2* is a novel gene that had never been reported in higher plants. However, the following observations suggested that *BPG2* is important in the chloroplast. Analysis using a BPG2-GFP fusion protein showed that *BPG2* is localized in the chloroplast. Additionally, *BPG2* gene expression in response to light and Brz is consistent with the gene expression pattern of major photosynthesis genes. Electron microscopy of the chloroplast in *bpg2* mutants showed that the thylakoid membrane, where photosynthesis occurs, was insufficiently developed, and excess starch grains had accumulated. From an evolutionary perspective, the origin of chloroplasts in plants is thought to be caused by the coevolution of cyanobacteria with a primitive eukaryote through an endosymbiotic relationship. However, cyanobacteria do not have a gene homologous to *BPG2*. This phenomenon is still a mystery. However, *BPG2* may be a rare case of a chloroplast gene reflecting the history of gene evolution in which the gene moved from the genome of nonphotosynthetic primitive prokaryote to the nuclear genome of a primitive plant, resulting in the primitive plant adding the chloroplast localization signal to regulate the chloroplast.

The molecular mechanism of BPG2 activity in chloroplast function was analyzed in detail. No difference in expression level of photosynthesis genes encoded by the chloroplast genome was observed between the wild type and *bpg2* mutants and the mRNA expression level of photosynthesis genes encoded by the nuclear genome. Further detailed analyses showed that in bpg2 mutants, abnormal chloroplast rRNA splicing, which is encoded by the chloroplast genome, occurs. As a result, a large amount of unspliced long chloroplast rRNAs accumulate in those mutants. Such unspliced rRNAs are not found in wild-type plants. It is known that rRNAs play an important role in ribosome assembly, the site of protein synthesis. Thus, protein synthesis in the chloroplast was analyzed. Among the proteins encoded by the chloroplast genome, two proteins with major roles in photosynthesis were studied, namely, the RubisCo L protein, which catalyzes a key step in the fixation of atmospheric CO_2 in the Calvin–Benson cycle that carries out carbon fixation by photosynthesis, and the D1 protein, a core protein in the photochemical system that produces NADPH and ATP. In wild-type plants, Brz treatment increases the accumulation of both RubisCo L and D1 proteins. In *bpg2* mutants, synthesis of these proteins decreases to approximately 10% of wild-type values, and Brz does not cause increased accumulation. This indicates that *BPG2* plays an important role in protein synthesis in the chloroplast. With regard to the involvement of *BPG2* in the splicing of chloroplast rRNA, it is still not clear whether *BPG2* directly exerts its effect or whether it exerts its effect via the splicing enzymes. However, other studies indicate that in *bpg2* mutants, the level of accumulation of LHCP (light-harvesting chlorophyll-binding protein), which binds to chlorophyll and absorbs light energy, decreases to the same level as RubisCo L and D1. LHCP is encoded by the nuclear genome, transcribed/translated there, and migrates to the chloroplast. Additionally, Brz does not cause increased LHCP accumulation. These findings suggest that *BPG2* is involved not only in protein synthesis of the chloroplast genome but also in stabilization of proteins migrating from the nucleus to the chloroplast; in other words, *BPG2* is broadly involved in chloroplast functional regulation.

4.2.8 OUTLOOK FOR THE FUTURE

In the aforementioned sections, the molecular mechanism of the brassinosteroid signal transduction pathway and various applications of the brassinosteroid biosynthesis inhibitor Brz were reviewed. Many of the applications use the lab plant *Arabidopsis*; this does not mean that brassinosteroid research is important only in basic studies. In 2008, SELES, a grain-trading company in the United States, created a rice transformant in which the brassinosteroid biosynthesis gene DWF4 is attached to the green tissue-specific promoter *AS* [32]. When this rice plant, which overexpresses *DWF4*, was planted directly on cultivated land, an increase was observed in the number of panicles per plant, the number of seed grains per plant, and the grain weight per seed. Compared with nontransformant rice plants, the rice grain yield per plant increased to 144% in rice transformant, and the amount of rice grain yield per field area also increased to 158% of wild-type values. Additionally, green leaf growth was stimulated in the transformant and biomass was increased. Compared with the nontransformant, in the rice transformant, the leaf weight (not including seed grains) was increased to 144%. In agricultural terms, these results can be expressed as a 58% increase in yield per cultivated area. This suggests that the addition of *DWF4* resulted in a surprisingly beneficial trait to the rice plant.

These innovative research studies used the brassinosteroid biosynthesis gene. Past research on brassinosteroids has relied on chemical biology-based functional analyses combined with the use of Brz. Future research in this field should aim to discover genes, other than those discovered in the past, that can be used to increase yield and biomass by studying brassinosteroid signal transduction-related genes.

REFERENCES

1. G. Adam & V. Marquardt: Brassinosteroid, *Phytochemistry*, **25**, 1787–1799 (1986).
2. T. Asami, T. Nakano, S. Fujioka: Plant brassinosteroid hormones, *Vitam. Horm.*, **72**, 479–504 (2005).
3. Z. Wang et al.: Nuclear-localized BZR1 mediates brassinosteroid-induced growth and feedback suppression of brassinosteroid biosynthesis, *Dev. Cell*, **2**, 505–513 (2002).
4. S. Y. Park et al.: Abscisic acid inhibits type 2C protein phosphatases via the PYR/PYL family of START proteins, *Science*, **324**, 1068–1071 (2009).
5. S. K. Kim, T. Asano, S. Marumo: Biological activity of brassinosteroid inhibitor KM-01 produced by a fungus *Drechslera avenae*, *Biosci. Biotechnol. Biochem.*, **59**, 1394–1397 (1995).
6. T. Asami & S. Yoshida: Brassinosteroid biosynthesis inhibitors, *Trends Plant Sci.*, **4**, 348–353 (1999).
7. Y. K. Min et al.: New lead compounds for brassinosteroid biosynthesis inhibitors, *Bioorg. Med. Chem. Lett.*, **9**, 425–430 (1999).
8. T. Asami et al.: Characterization of brassinazole, a triazole-type brassinosteroid biosynthesis inhibitor, *Plant Physiol.*, **123**, 93–99 (2000).
9. T. Asami et al.: Selective interaction of triazole derivatives with DWF4, a cytochrome P450 monooxygenase of the brassinosteroid biosynthetic pathway, correlates with brassinosteroid deficiency *in planta*, *J. Biol. Chem.*, **276**, 25687–25691 (2001).
10. K. Sekimata et al.: A specific brassinosteroid biosynthesis inhibitor, Brz2001: evaluation of its effects on Arabidopsis, cress, tobacco, and rice, *Planta*, **213**, 716–721 (2001).
11. K. Sekimata et al.: A specific and potent inhibitor of brassinosteroid biosynthesis possessing a dioxolane ring, *J. Agr. Food Chem.*, **50**, 3486–3490 (2002).
12. K. Sekimata et al.: Brz220 a novel brassinosteroid biosynthesis inhibitor: stereochemical structure–activity relationship, *Tetrahedron: Asym.*, **13**, 1875–1878 (2002).
13. J. M. Wang et al.: Biological evaluation of 5-substituted pyrimidine derivatives as inhibitors of brassinosteroid biosynthesis, *Biosci. Biotechnol. Biochem.*, **65**, 817–822 (2001).
14. T. Asami et al.: Triadimefon, a fungicidal triazole-type P450 inhibitor, induces brassinosteroid deficiency-like phenotypes in plants and binds to DWF4 protein in the brassinosteroid biosynthesis pathway, *Biochem. J.*, **369**, 71–76 (2003).
15. S. P. O. Werbrouck et al.: Physiological and morphological evidence of brassinosteroid-biosynthesis inhibition by the fungicide imazalil, *Physiol. Plant.*, **119**, 69–77 (2003).
16. T. Asami et al.: A mammalian steroid action inhibitor spironolactone retards plant growth by inhibition of brassinosteroid action and induces light-induced gene expression in the dark, *J. Steroid Biochem. Mol. Biol.*, **91**, 41–47 (2004).
17. J. M. Gendron et al.: Chemical genetic dissection of brassinosteroid–ethylene interaction, *Mol. Plant*, **1**, 368–379 (2008).
18. S. Robert et al.: Endosidin1 defines a compartment involved in endocytosis of the brassinosteroid receptor BRI1 and the auxin transporters PIN2 and AUX1, *Proc. Natl. Acad. Sci. USA*, **105**, 8464–8469 (2008).
19. B. D. Rybel et al.: Chemical inhibition of a subset of *Arabidopsis thaliana* GSK3-like kinases activates brassinosteroid signaling, *Chem. Biol.*, **16**, 594–604 (2008).
20. J. M. Li & J. Chory: A putative leucine-rich repeat receptor kinase involved in brassinosteroid signal transduction, *Cell*, **90**, 929–938 (1997).

21. D. M. Friedrichsen et al.: Brassinosteroid-insensitive-1 is a ubiquitously expressed leucine-rich repeat receptor serine/threonine kinase, *Plant Physiol.*, **123**, 1247–1255 (2000).
22. Z. Y. Wang et al.: BRI1 is a critical component of a plasma-membrane receptor for plant steroids, *Nature*, **410**, 380–383 (2001).
23. T. Kinoshita et al.: Binding of brassinosteroids to the extracellular domain of plant receptor kinase BRI1, *Nature*, **433**, 167–171 (2005).
24. Y. Yin et al.: BES1 accumulates in the nucleus in response to brassinosteroids to regulate gene expression and promote stem elongation, *Cell*, **109**, 181–191 (2002).
25. Y. Yin et al.: A new class of transcription factors mediates brassinosteroid-regulated gene expression in Arabidopsis, *Cell*, **120**, 249–259 (2005).
26. J. He et al.: BZR1 is a transcriptional repressor with dual roles in brassinosteroid homeostasis and growth responses, *Science*, **307**, 1634–1638 (2005).
27. A. Yamagami et al.: Chemical genetics reveal the novel transmembrane protein BIL4, which mediates plant cell elongation in brassinosteroid signaling, *Biosci. Biotechnol. Biochem.*, **73**, 415–421 (2009).
28. L. Li et al.: Arabidopsis IWS1 interacts with transcription factor BES1 and is involved in plant steroid hormone brassinosteroid regulated gene expression, *Proc. Natl. Acad. Sci. USA*, **107**, 3918–3923 (2010).
29. A. Tanaka et al.: BRASSINOSTEROID UPREGULATED1, encoding a helix–loop–helix protein, is a novel gene involved in brassinosteroid signaling and controls bending of the lamina joint in rice, *Plant Physiol.*, **151**, 669–680 (2009).
30. K. He et al.: BAK1 and BKK1 regulate brassinosteroid-dependent growth and brassinosteroid-independent cell-death pathways, *Curr. Biol.*, **17**, 1109–1115 (2007).
31. T. Komatsu et al.: The chloroplast protein BPG2 functions in brassinosteroid-mediated post-transcriptional accumulation of chloroplast rRNA, *Plant J.*, **61**, 409–422 (2010).
32. C. Wu et al.: Brassinosteroids regulate grain filling in rice, *Plant Cell*, **20**, 2130–2145 (2008).

4.3

CHEMICAL GENETIC APPROACHES ON ABA SIGNAL TRANSDUCTION

EUNJOO PARK AND TAE-HOUN KIM

4.3.1 ABSCISIC ACID SIGNAL TRANSDUCTION: LESSONS FROM EXOGENOUS ABA TREATMENT

Just like other multicellular eukaryotes, plants have their own system to maintain their lives from seeds to adult stages by adapting to ever-changing environmental conditions [1–3]. These systems have distinct pathways in perception of abiotic or biotic signals, but these two independent signals do not necessarily exclude each other to generate proper responses [4–6]. Efficient use of limited resources such as light or water and avoidance of pathogenic attacks and excessive environmental stresses such as heat, cold, and ozone are the key mechanisms for plants to survive. Plants perceive various kinds of environmental stimuli and that is translated into generation of signal cascades for resistant responses. During this process, the phytohormone ABA is known to play a major role in the control of plants' stress responses against abiotic harsh conditions. The initial identification of the components that comprise of ABA signal transduction and the key regulators that affect either negatively or positively ABA signaling have relied on genetic screening of mutants showing aberrant responses to ABA treatments [7, 8] (Fig. 4.3.1).

For the screening of ABA response mutants, delayed seed germination and root growth inhibition phenotypes by ABA treatment have been successfully used to isolate novel mutants. From the screening of positive ABA signaling components, a series of *ABA insensitive* (*abi*) mutations was identified [9] (Fig. 4.3.2). Positional cloning of corresponding dominant mutations of *abi1-1* and *abi2-1* discovered that

Plant Chemical Biology, First Edition. Edited by Dominique Audenaert and Paul Overvoorde.

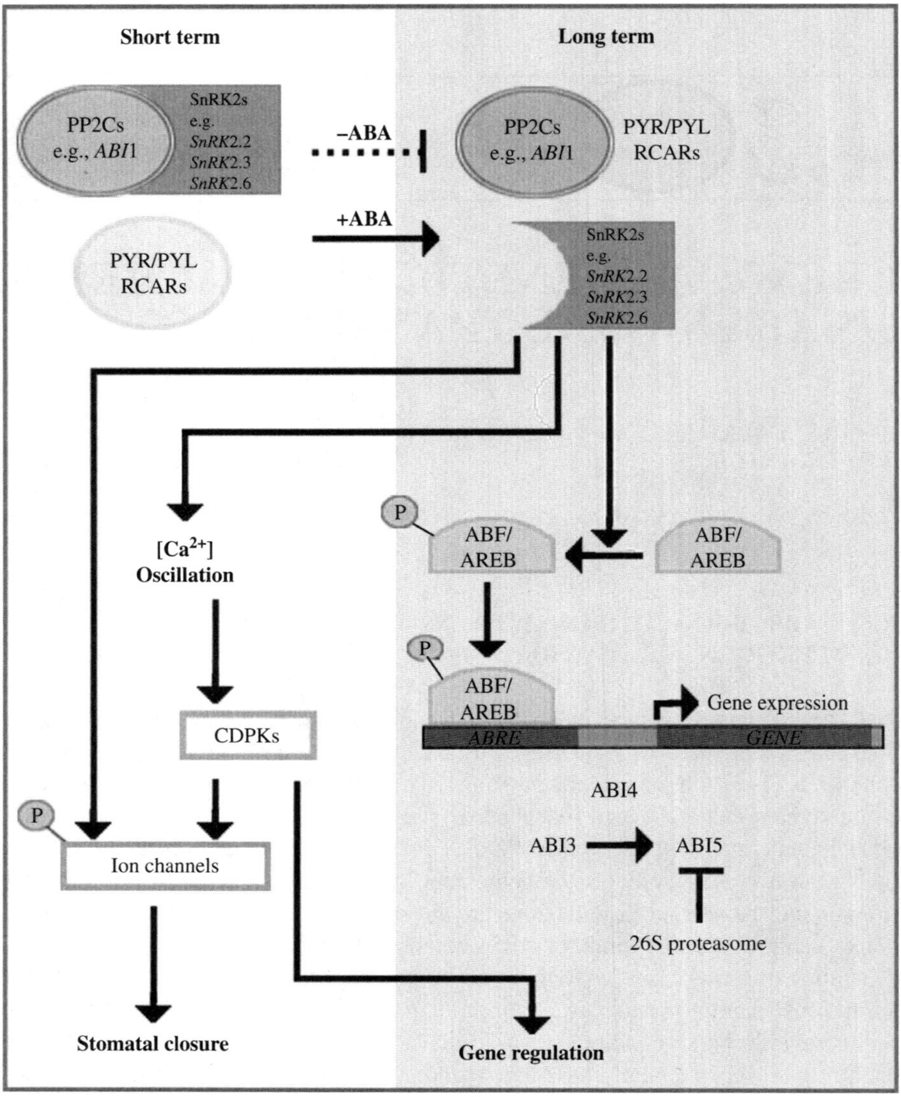

FIGURE 4.3.1 ABA signaling pathway from ABA perception to downstream responses. Short term stomatal closure responses and long term gene expression responses are presented here as ABA responses. PP2Cs negatively regulate SnRK2s in normal condition. Upon ABA perception under abiotic stress condition, such as drought, ABA activates the signaling pathway by promoting PP2Cs and PYR/PYL/RCARs interaction. As a result, unbound SnRK2s stimulate phosphorylation of ABR/AREBs and downstream genes. At the same time, activated SnRK2s regulate stomatal closure through targeting Ca^{2+} oscillation and ion channels. For color detail, please see color plate section.

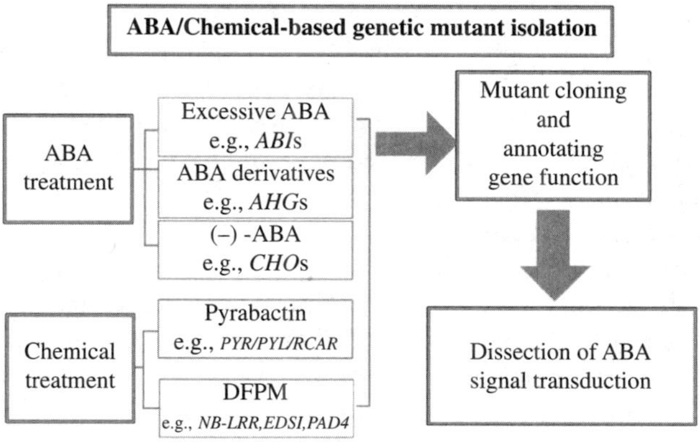

FIGURE 4.3.2 ABA- and chemical-based genetic mutant screening and characterization.

ABI1 and *ABI2* both encode similar proteins of Ser/Thr protein phosphatase 2C groups (A PP2Cs) [10–12]. Genetic analyses have revealed that *ABI1* and *ABI2* function as negative regulators of ABA signal transduction and position at the early steps of ABA signaling [13].

Mutations in the *ABI3*, *ABI4*, and *ABI5* genes produced ABA-insensitive seed germination phenotypes. *ABI3* encodes a transcription factor of B3 domain and acts as an upstream regulator of *ABI5* [14, 15]. The gene product of *ABI4* belongs to a transcription factor that contains an APETALA2 (AP2) putative DNA-binding domain [16]. *ABI4* was strongly expressed in seeds and slightly in the tissues of vegetative stages. The function of *ABI4* is required in seedling growth under the influence of ABA signaling. Like other *ABI*s, mutations in *ABI4* caused pleiotropic effects in plant growth and development, and it may not function alone in a regulatory hierarchy of ABA signaling but in a combinatorial network of *ABI3*, *ABI4*, and *ABI5* [17, 18].

ABI5 represents a member of bZIP transcription factors [19]. The mRNA level and protein expression of *ABI5* are kept in tight control under normal conditions, yet stressful conditions such as drought and high salt treatments make the *ABI5* gene expression increased *via* ABA signal transduction. The increased level of ABI5 leads to phenotypic changes in early stages of development; delay of seed germination and seedling growth arrest. In addition to transcriptional control of the *ABI5* gene expression through *ABI3*, protein stability of ABI5 is under control of the 26S proteasome machinery (Fig. 4.3.1). So far several E3 ligases responsible for the ubiquitination of ABI5 have been identified. By regulation of ABI5 ubiquitination, these E3 ligases play a role in maintaining the ABI5 protein levels for proper ABA responses [20]. For example, *KEG* (*Keep on Going*) E3 ligase regulates ABI5 degradation by attaching polyubiquitins to ABI5 for proteasome degradation [21]. Additionally, SUMO (Small Ubiquitin-related MOdifier) E3 ligase, SIZ1, was identified as a modifier of ABI5 producing negative regulation of ABA signal transduction [22]. As a bZIP transcription factor, ABI5 induces the expression of *AtEm LEA* (*Late Embryogenesis Abundant*) in vegetative tissues by binding to *ABRE* (*ABA responsive element*) in the *AtEm* promoter directly [23]. Interestingly, yeast-1-hybrid screening using an *ABRE* as bait identified a series of bZIP transcription factors named ABFs (ABRE binding factors), which are homologous to ABI5 [24]. ABFs are also known as AREBs (ABA-Responsive Element Binding proteins) and their functions in ABA signal transduction were characterized biochemically and genetically [25, 26]. Especially a triple mutant of *areb1 areb2 abf3* showed ABA-insensitive phenotypes, suggesting these transcription factors function as downstream positive components for ABA-dependent transcriptional regulation [26].

Another essential positive regulator of ABA signaling, OST1 (OPEN STOMATA1), was identified from a genetic screening for mutants with aberrant leaf temperature phenotypes in response to drought stress [27]. The mutant *ost1* produced a colder leaf temperature phenotype compared to wild-type because ABA signaling in response to drought stress was disrupted and guard cells remained open as a result. *OST1* encodes one of SNF1-related protein kinase 2s (SnRK2s), SnRK2.6, and acts as an upstream positive signaling component during ABA signal transduction (Fig. 4.3.1). The major roles of SnRK2s in ABA signaling have been genetically investigated using triple mutants of *snrk2.2 snrk2.3 snrk2.6*, as well as a decuple mutant that includes 10 *SnRK2* genes [28–30]. Consistent with genetic data, biochemical functions of SnRK2.2/2.3/2.6 were shown to phosphorylate the AREB1 and ABF3 proteins, which regulate the downstream ABA-induced transcription [31].

4.3.2 ABSCISIC ACID SIGNAL TRANSDUCTION: LESSONS FROM ABA DERIVATIVES TREATMENT

In addition to ABA response-based mutant screening, use of ABA stereoisomers also led to the identification of specific novel components that affect certain parts of ABA signal transduction (Fig. 4.3.2 and Fig. 4.3.3). With this approach, Nambara et al. tried to find novel genetic factors that are involved in a specific pathway of ABA signal transduction [32]. They hypothesized that it might be possible to screen genetic mutations sensing of the (+)-ABA or (-)-ABA differently. They screened for mutants germinating differently from wild-type in the presence of (-)-ABA and further categorized some of the selected mutants based on (+)-ABA sensitivity. Among 26 independent mutant lines, new loci involved in ABA signal transduction were discovered; *chotto1* (*cho1*) and *chotto2* (*cho2*) including 6 new alleles of *abi3* [32].

Moreover, structurally modified versions of ABA have been used for dissecting ABA signal transduction in a similar way of chemical genetics approaches (Fig. 4.3.3). Nishimura et al. used an acetylenic ABA analog, PBI-51 ((-)-2,3-dihydroacetylenic ABA alcohol), aiming for isolation of ABA-hypersensitive mutants, *ahg* (ABA-hypersensitive germination) (Fig. 4.3.2) [33]. It was based on the observation that PBI-51 showed an antagonistic effect on ABA-induced genes expression in *Brassica napus*, as well as an agonistic effect on seed germination and post-germination growth assays in *Arabidopsis*. This bidirectional function of PBI-51 might be due to differential regulation of various ABA recognizing protein complexes. Genetic analyses of the isolated *ahg* mutants categorized them into four of novel ABA-hypersensitive loci (*AHG1*, *AHG2*, *AHG3*, and *AHG4*). *AHG1* and *AHG3* both encode

Name	Structure	Function	Related genes
(+)-ABA	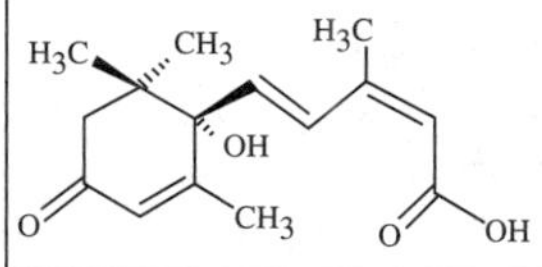	Phytohormone	e.g., *ABI*s [9–26]
(–)-ABA	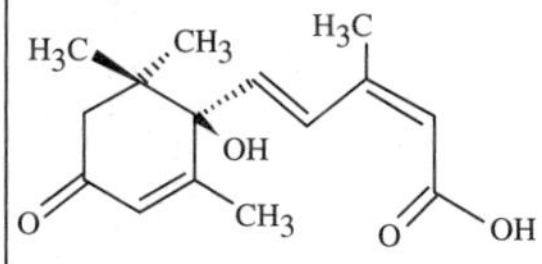	(+)-ABA enantiomer	e.g., *CHO*1, *CHO*2 [32]
Tetralone ABA analog	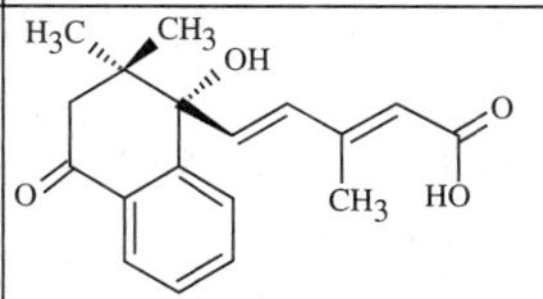	(ABA analog (1)	[38, 39]
ABA bicyclic analog	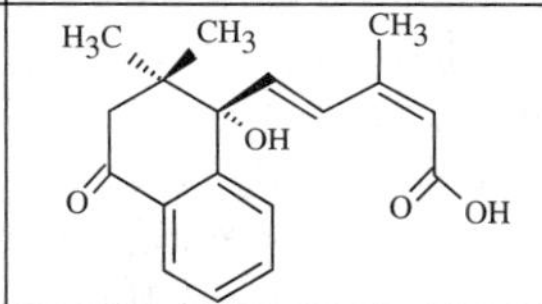	ABA analog (induction of *FAE* mRNA)	[40]
Biotinylated analog	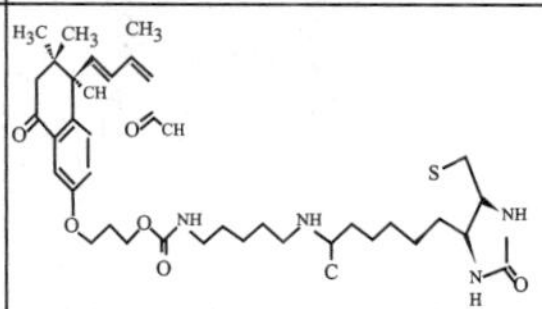	Probe (separation of ABA interacting proteins)	e.g., ABA8′-hydroxylase [40]
PBI 425	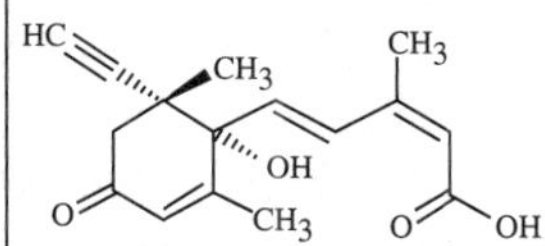	ABA analog (2)	e.g., *MYBR* [38, 39]
PBI 51	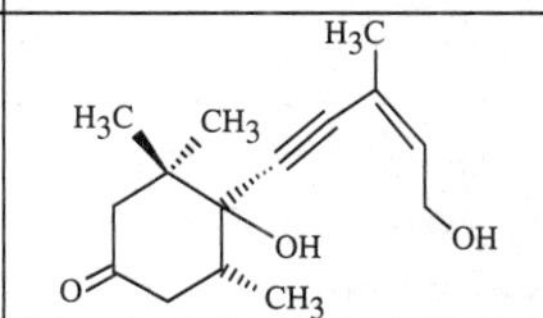	**ABA antagonist** in *Brassica napus* **ABA agonist** in *Arabidopsis thaliana*	e.g., *AHG1, AHG2, AHG3, AHG4* [33–37]
Pyrabactin	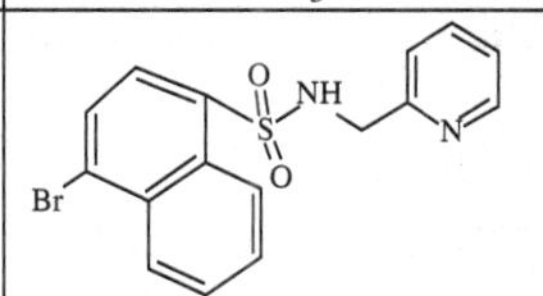	Selective ABA agonist	e.g., *PYR/PYL/RCAR* [7, 43]
DFPM	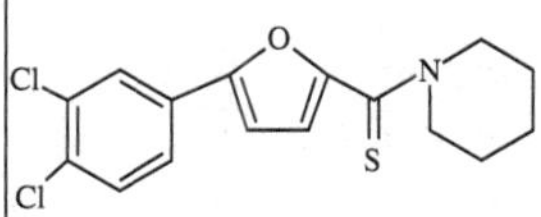	Selective ABA antagonist	e.g., *NB-LRR* [6, 54]

FIGURE 4.3.3 Structures and its functions of ABA signaling-related small compounds. (1) Growth inhibitor in suspension-cultured cells of maize and seeds germination inhibitor in Arabidopsis, (2) 8′-hydroxylase irreversible inhibitor, also used for identifying MYBR1 as an ABA responsible gene.

similar ABI1-related PP2C protein phosphatases that function more specifically in the seeds during germination in contrast to *ABI1* and *ABI2* [34, 35]. *AHG2* was identified as a gene encoding a poly(A)-specific ribonuclease, AtPARN [36]. Together with the finding of AtPARN-dependent mRNA degradation, regulation of RNA metabolism by ABA is now recognized as an important part of ABA signal transduction [37]. Although the signaling mechanism by PBI-51 in these mutants remains unexplained, this approach proved the concept that the use of ABA analogs could identify novel genes functioning in specific pathways or at different times of ABA signal transduction.

Besides the mutant screening-based genetic approaches, physiological and gene expression studies of ABA derivatives such as tetralone ABA8 and PBI-425 ((+)-8′-acetylene ABA) revealed that structural modification of ABA could specifically activate certain pathways of ABA signal transduction in the cells (Fig. 4.3.3) [38, 39]. In the test of maize cell growth and *Arabidopsis* seed germination, tetralone ABA8 was shown to induce stronger effects than ABA [38]. Microarray analyses of PBI-425-treated samples identified novel ABA-regulated genes perhaps by inducing selectively a part of the ABA-regulated transcriptional network [39].

One of the major advantages of using ABA derivatives is that modified ABA analogues can be used as an affinity probe for identifying ABA-binding proteins including ABA receptors (Fig. 4.3.3). For this approach, it is important to design a probe that retains all the functional groups of ABA and a motif for tagging such as biotinylation for easier purification of chemical–protein complexes. A bicyclic ABA analogue, ABA probe (+)-17, contains all the important C-1 carboxylic acid group, vinyl carbon, and C-4′ carbonyl group but, with an aromatic ring at side, has a biotin group for affinity purification [40]. Effectiveness of ABA probe (+)-17 was tested to determine whether it was biologically functional as ABA in *B. napus*. Expression level of *FAE (3-ketoacyl-CoA synthase)* mRNA was much higher when treated with (+)-17 than (+)-ABA at 72 h. Moreover, the next affinity column purification experiments with ABA probe (+)-17 as bait revealed that ABA 8′-hydroxylase could bind to ABA probe (+)-17. Therefore, it supports the idea that modified ABA analogues with a proper tagging may possibly be used for identifying novel ABA-binding proteins [40]. Notably, various ABA biosynthesis enzymes have been identified as ABA-binding proteins [41]. For example, UDP-glycosyltransferase, UGT71B6 (71B6) can catalyze glucosylation of naturally present ABAs [42]. Different affinities of UGT71B6 to structurally different ABA derivatives indicate that glucosylation of ABAs can affect ABA signaling through regulation of ABA homeostasis. In addition, it strongly suggests that metabolic conversion of ABA derivatives needs to be considered for possible use of ABA-like chemicals in agriculture.

However, despite the numerous advantages of these approaches, it was not possible to identify ABA receptors and the corresponding genes due to issues of genetic redundancy and/or lethality of the ABA perception machinery. Recent advances with chemical genetics approaches have successfully revealed the novel ABA signaling components that build up more detailed signaling pathways (Fig. 4.3.1) and indicate network interactions with other cellular signal transduction.

4.3.3 ABSCISIC ACID SIGNAL TRANSDUCTION: NEW INSIGHTS FROM CHEMICAL GENETICS

Although classical forward and reverse genetics approaches contributed extensively to understand the key features of the ABA and abiotic stress signaling network, especially the perception machinery of endogenous ABA and the following upstream signaling components of ABA signal transduction had been enigmatic for decades. The recent findings of soluble ABA receptors, PYR/PYL/RCARs, ignited the serial discovery of the most upstream signaling events staring from ABA perception to release of negative regulation of SnRK2 protein kinases by PP2C groups (Fig. 4.3.1) [43–45]. Among the independent approaches for finding of *PYR/PYL/RCARs*, identification of the ABA agonist pyrabactin and the corresponding genetic loci *pyr1* (*pyrabactin resistance 1*) is particularly notable to the eyes of chemical genetics researchers (Fig. 4.3.2 and Fig. 4.3.3) [43]. Map-based cloning of *pyr1* led attention to the bet V group of proteins, which was hypothesized to contain a biding pocket for a steroid hormone. The following genetic and biochemical analyses of PYR1 confirmed that PYR/PYL/RCARs are soluble ABA receptors and they interact with the negative regulators of ABA signaling, PP2C-type protein phosphatases generally in an ABA-dependent manner. As in the case of auxin perception where the natural auxin IAA (indole-3-acetic acid) functions as "molecular glue" to form a receptor complex with the TIR F-box receptor and the AUX/IAA transcriptional repressor [46, 47], ABA-bound PYR/PYL/RCARs form a protein complex with a PP2C group phosphatases. The binding of ABA-PYR/PYL/RCAR to PP2C interrupts the catalytic activity of PP2C [48–50]. The binding pocket of PP2C is competitively recognized by a SnRK2 protein kinase whose activation is under the negative control of PP2C in normal nonstress conditions [51, 52]. When ABA activates abiotic stress signal transduction, subsequent release of PP2Cs from the negative regulation of SnRK2s induces phosphorylation of downstream factors such as ABF/AREBs and SLAC1 (Slow Anion Channel-Associated1) perhaps through CDPKs (calcium-dependent protein kinases) and generates changes of transcriptional or ion transport activities (Fig. 4.3.1) [31, 53].

Whereas pyrabactin was isolated as an ABA agonist, the ABA signaling inhibitor DFPM was isolated as an ABA antagonist that interfering with the ABA-induced gene expression response (Fig. 4.3.2 and Fig. 4.3.3) [54]. The mechanism by which DFPM negatively affects ABA signaling lead to identification of signaling interactions between abiotic ABA signal transduction and biotic disease signaling pathways.

These examples emphasized the power of chemical genetics for dissection of complicated signaling networks, and particularly the chemical genetic approaches on ABA signaling will be addressed more in detail in the following section.

4.3.4 PYRABACTIN AND ITS CELLULAR TARGET PYR1: DISCOVERY OF ABA RECEPTORS, *PYR/PYL/RCAR*S

As mentioned earlier, the discovery of the ABA receptors *PYR/PYL/RCARs* highlighted the efficiency of chemical genetics approaches on investigation of plant signal transduction (Fig. 4.3.4 and Fig. 4.3.5). *PYR/PYL/RCAR* was also discovered by several

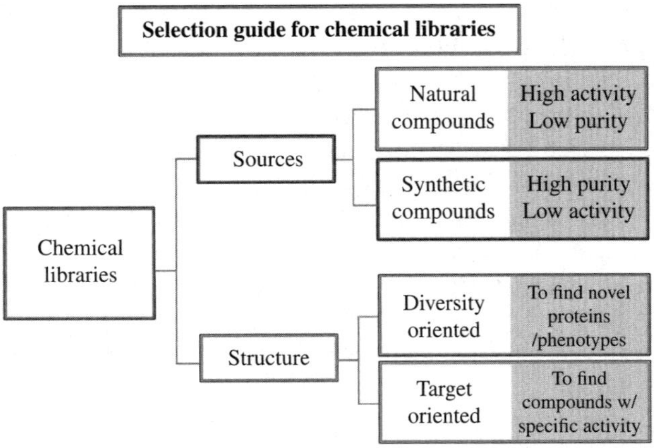

FIGURE 4.3.4 A selection guide for chemical libraries. Chemical libraries can be classified based on its source; synthetic or natural, and its structure; diversity focused or target focused.

Theoretical	Experimental
1. Library selection	Subject: *Arabidopsis thaliana* Applied chemicals: ChemBridge E library : bioactive compounds collection Physiological marker: seeds germination
2. Chemical screening Agonist antagonist a. Primary screening b. Secondary screening Mutant test Marker gene expression Transcriptome, etc.	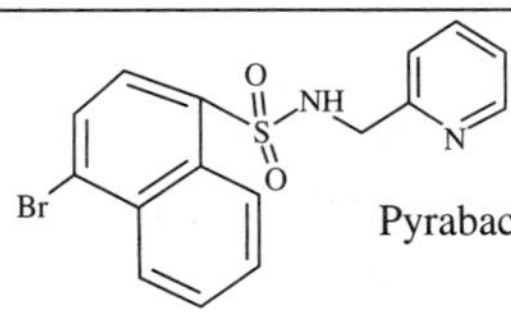 Pyrabactin Small compounds test to already known mutants: *abi*1-1,*aba*1-1,*rgl*2-1 Microarray analyses: chemical treated seeds and seedlings
3. Genetic screening Mutant selection Map-based cloning Protein function test	Mutant map based cloning: *PYR*1 Reverse genetics: *pyr*1/*pyl*1/*pyl*2/*pyl*4 Y2H, Co-IP of PYR1

FIGURE 4.3.5 Theoretical scheme of chemical genetics and its experimental application. Left panel is about the theoretical chemical genetics approach and right panel shows an example of application, pyrabactin and PYR1 case.

other approaches, including proteomics and yeast-two-hybrid (Y2H) screening of the PP2C interactive partners [44, 45]. Here, we discuss into more detail the discovery and characterization of *PYR1* through a chemical genetic approach. The chemical called pyrabactin (4-bromo-*N*-[pyridin-2-yl methyl] naphthalene-1-sulfonamide) was originally identified as an *Arabidopsis* germination inhibiting compound from chemical library screening [7, 43]. This chemical inhibited seed germination in a similar way like ABA, yet the microarray analyses of pyrabactin effects indicated a certain difference to those of ABA effects. In addition, the ABA-insensitive mutant *abi1-1* showed an insensitive seed germination phenotype to pyrabactin. Specificity of ABA signaling components for recognizing pyrabactin was confirmed by the fact that the similar ABA effects did not occur with apyrabactin (nonfunctional analogue to pyrabactin) treatment. However, there were still several possibilities that the changes of endogenous ABA concentration would intervene in the pyrabactin pathway or that other phytohormones got involved in this mechanism. Hence, pyrabactin sensitivity assays were done with the ABA-insensitive mutant that also showed reduced sensitivity toward pyrabactin. Also, as the mutants defective in ABA biosynthesis and giberellic acid perception showed normal sensitivities to pyrabactin treatment, pyrabactin represented a potent ABA agonist for ABA signaling. Forward chemical genetic screening led to the identification of insensitive mutants to pyrabactin during seed germination, named as *pyrabactin resistance1* (*pyr1*). Through map-based cloning, the mutation of *pyr1* was identified, and there were 13 more of its homologues in the *Arabidopsis* genome called from *PYL1*(*PYR1-LIKE 1*) to *PYL13*. The sequence analysis of *PYR1* suggested it belongs to START (steroidogenic acute regulatory-related lipid transfer) superfamily, which is also known as bet V I-fold superfamily [43], inferring *PYR1* as a direct putative pyrabactin interacting protein. Isolation of pyrabactin as an ABA agonist and identification of the pyrabactin target gene *PYR1* through forward chemical genetics represent a near perfect example of how chemical genetics can offer to circumvent the difficulties of traditional methods and lead to the discovery of not yet identified novel signaling components (Fig. 4.3.4 and Fig. 4.3.5).

However, forward chemical genetics alone was not enough to understand the complex genetics network for pleiotropic hormone signal transduction. Because of high expression levels of *PYR/PYL/RCARs* mRNAs in seeds and guard cells, functional overlap of PYR/PYL/RCARs was expected. In fact, in the case of the *pyr1* mutant, reduced pyrabactin sensitivity but with normal ABA sensitivity was observed. Subsequently, the generation and characterization of triple and quadruple mutant lines, *pyr1 pyl1 pyl4* and *pyr1 pyl1 pyl2 pyl4*, respectively, through reverse genetics, indeed demonstrated expected (+)-ABA insensitivity phenotypes [43]. These data support the idea that PYR/PYL/RCARs are involved in multiple ABA signal transduction in parallel, and that pyrabactin activates specifically the PYR1-dependent linear ABA signal transduction. It proves that a synthetic agonist could express simpler synthetic phenotypes, whereas the natural agonist can generate complex traits.

Furthermore, Y2H analyses of PYR1 showed a direct interaction with the group A type 2 C protein phosphatases. Interestingly, the interaction between PYR1 and the PP2Cs was ABA-dependent, showing reduced interaction in the absence of ABA. In other words, PP2C could bind to PYR1 only with ABA. Moreover, YFP-PP2C and HA-PYR1 constructs were transiently expressed in *Nicotiana benthamiana* for co-immunoprecipitation analyses, and also similar results as Y2H were displayed. Another piece of evidence for substantiating the premise was presented based on the HSQC (heteronuclear single quantum coherence) NMR studies using ^{15}N-labeled PYR1 and the pyrabactin-insensitive allele PYR1^{P88S}. Addition of ABA induced conformational changes of both PYR1 and PYR^{P88S}, indicating that ABA binding to PYR^{P88S} is not defected. However, the addition of HAB1 PP2C in the same NMR study demonstrated that HAB1 interacts with PYR1-ABA but not with PYR^{P88S}-ABA, suggesting that protein complex formation of PYR1-ABA-HAB1 is important to generate ABA signal transduction [43]. In fact, whereas the *in vitro* measurements of dissociation constant (K_d) of ABA for PYR/PYLs showed micromolar-level high values, K_d became lower nanomolar ranges when the compatible PP2Cs were added to the tests [55]. Taken together, it turned out to be probable that ABA could facilitate an interaction of PYR1 with the clade A PP2Cs as similarly shown in the perception of another phytohormone auxin; auxin promotes interaction of IAA with SCF^{TIR1} while it acts as a molecular glue for these two perception components [46, 47]. After the first report of PYR1, X-ray crystallographic investigation of PYR/PYL/RCAR groups determined the structure of ABA receptors and interaction with PP2Cs [48–50]. Since the genes involved in this early steps of ABA signaling are members of gene families, combinatorial matching of *PYR/PYL/RCARs* with *PP2C* partners produces diversity of ABA receptor complexes [55, 56]. The various receptor complexes with particular components in it showed differential affinities to ABA and possibly to other ABA derivatives, suggesting a hypothesis of tissue-specific or certain abiotic stress-specific regulatory mechanisms for ABA signal transduction.

In conclusion, the *PYR/PYL/RCAR* gene family encodes a soluble ABA receptor. Overlapping functions of *PYR/PYL/RCARs* have been investigated through chemical genetics approaches showing ABA-dependent regulation of PP2Cs and then activation of SnRK2s (Fig. 4.3.1). Isolation of the novel ABA agonist pyrabactin and identification of the genetic targets represent a successful application of chemical genetics to dissect the complex signaling networks (Fig. 4.3.5).

4.3.5 ABA ANTAGONIST DFPM: MECHANISMS FOR CROSSTALKS BETWEEN ABIOTIC AND BIOTIC SIGNALING PATHWAYS

Another successful use of chemical genetics approach on ABA signal transduction was the isolation of an ABA antagonist DFPM ([5-(3,4-dichlorophenyl)furan-2-yl]-piperidine-1-ylmethanethione) and its genetic characterization [54]. DFPM was screened from a combinatorial synthetic chemical library composed of 9600 chemicals as an inhibitor of ABA-induced *RAB18* reporter gene expression. DFPM not only inhibits the ABA-induction of gene expression but also interferes with the ABA-induced stomatal closure response. However, the antagonistic effects of DFPM were limited for certain phenotypes as DFPM did not affect much on ABA-delay of seed germination. Consistently, microarray analyses of DFPM and ABA cotreated samples indicated that DFPM inhibited a part of ABA-regulated genes but not all. DFPM did not affect the early ABA perception steps by PYR/RCAR receptors or the ABA-activation of SnRK2 kinases. It rather interferes with formation of cytosolic Ca^{2+} signaling and then the downstream SLAC1 anion channel activation step. During the transcriptome analyses, DFPM was also found to stimulate the expression of genes involved in biotic disease-resistant responses. In addition to the transcriptional activation of plant immune signaling, ABA-inhibition activity of DFPM requires the functions of major early regulators of pathogen resistance responses, including *EDS1 (Enhanced Disease Susceptibility1), PAD4 (Phytoalexin Deficient4), RAR1 (Required for Mla12 resistance)*, and *SGT1b (suppressor of G-two allele of Skp1b)* [54]. It is notable that the requirement of these genetic elements for DFPM inhibition of ABA signaling is also needed for the natural pathogen *Pseudomonas*-interference with ABA signal transduction. On the contrary to the previously suggested antagonistic relationship between ABA and plant immune signal transduction [57, 58], DFPM inhibition and *Pseudomonas* inhibition of ABA signaling did not require *EDS16* (*Enhanced Disease Susceptibility1*) and *NPR1* (*Nonexpressor of PR genes1*), which are critical genes for biosynthesis and signaling steps of salicylic acids (SA), respectively. Instead, the signals for the ABA signaling inhibition come from an activation of NB-LRR (nucleotide binding-leucine rich repeat) immune receptors such as *SNC1* (*Suppressor of npr1-1, Constitutive1*), indicating that DFPM- or pathogen-stimulation of upstream signaling components of the effector-triggered immune signaling pathways may regulate ABA signal transduction antagonistically [6].

Because the putative signals generated by NB-LRR and EDS1/PAD4 seem to disrupt cytosolic Ca^{2+} oscillation and Ca^{2+} signaling steps of ABA signal transduction, it is hypothesized that abiotic ABA signal transduction and biotic disease signal transduction might crosstalk through regulation of Ca^{2+} signals [54]. Considering the precise and controlled regulation of stress–response mechanisms against abiotic and biotic stresses is critical for plant survival, it is natural to predict that the signaling networks for individual stress responses interact with each other in a complex pattern [6]. It is again chemical genetics approaches to dissect complex signaling network into a simpler level to understand and identify the participating components through either specific activation of certain pathways or general activation of parallel pathways.

4.3.6 SUMMARY

We have discussed a list of small compounds that function to affect ABA signal transduction (Fig. 4.3.3). At the beginning, we look over what the researchers have found on natural ABA signal transduction using synthetic ABA and its analogues. Complex nature of plant hormone signal transduction and the functional genetic redundancy due to frequent gene duplication events in plants hamper the effort of researchers to overcome further the limit of traditional genetics approaches. With existing genetic approaches on hand, new chemical genetics approaches with novel compounds have been providing successful tools to unravel the long-sought questions on ABA hormone biology, hunting for a receptor and the mechanism through which regulates crosstalk to other cellular signaling pathways (Fig. 4.3.5). General workflow of this approach includes the following with some modification depending on the purpose of experiments: (i) library selection among the structure diversity groups or from the different source of chemicals (Fig. 4.3.4). At any event, selected compounds from any library should lead to identify putative target proteins responsible for the phenotype of interests. (ii) Chemical screening. Primary screening would narrow down the handful compounds into the manageable number of these, and then secondary screening gives you more detailed clues to link observed phenotype changes to the function of compounds. (iii) Genetic screening. If newly found compounds characterize a specific phenotype, then identifying the targets will be the next step *via* either forward or reverse chemical genetics. Once the target proteins are discovered, further experiments will guide the function of novel proteins and interactions with already known signaling components. The ABA biology field has been particularly well benefited by the power of chemical genetics approaches. Discoveries of ABA agonists and antagonists shed light on more detailed signaling mechanism from ABA perception to downstream controls and guide to understand how ABA signaling can be communicated with other cellular stress responses to maximize growth and survival (Fig. 4.3.1). Hence, future chemical genetics methods with more diverse sources of chemical libraries and the more targeted phenotypes with specificity would produce better understanding of complex signaling networks. Eventually, our aims challenge to apply this chemical genetics approach to nonmodeled crops and have crops resistant to drought condition or other abiotic stress condition.

REFERENCES

1. Schroeder JI, et al.: Guard cell signal transduction. *Annu Rev Plant Physiol Plant Mol Biol* 2001, **52**:627–658.
2. Giraudat J: Abscisic acid signaling. *Curr Opin Cell Biol* 1995, **7**:232–238.
3. Shinozaki K, Yamaguchi-Shinozaki K: Molecular responses to dehydration and low temperature: differences and cross-talk between two stress signaling pathways. *Curr Opin Plant Biol* 2000, **3**:217–223.
4. Mauch-Mani B, Mauch F: The role of abscisic acid in plant–pathogen interactions. *Curr Opin Plant Biol* 2005, **8**:409–414.
5. Cao FY, Yoshioka K, Desveaux D: The roles of ABA in plant–pathogen interactions. *J Plant Res* 2011, **124**:489–499.
6. Kim TH: Plant stress surveillance monitored by ABA and disease signaling interactions. *Mol Cells* 2012, **33**:1–7.
7. Cutler SR, et al.: Abscisic acid: emergence of a core signaling network. *Annu Rev Plant Biol* 2010, **61**:651–679.
8. Kim TH, et al.: Guard cell signal transduction network: advances in understanding abscisic acid, CO_2, and Ca^{2+} signaling. *Annu Rev Plant Biol* 2010, **61**:561–591.
9. Finkelstein RR, Somerville CR: Three classes of abscisic acid (ABA)-insensitive mutations of Arabidopsis define genes that control overlapping subsets of ABA responses. *Plant Physiol* 1990, **94**:1172–1179.
10. Leung J, et al.:Arabidopsis ABA response gene ABI1: features of a calcium-modulated protein phosphatase. *Science* 1994, **264**:1448–1452.
11. Meyer K, Leube MP, Grill E: A protein phosphatase 2C involved in ABA signal transduction in *Arabidopsis thaliana*. *Science* 1994, **264**:1452–1455.

12. Leung J, Merlot S, Giraudat J: The Arabidopsis ABSCISIC ACID-INSENSITIVE2 (ABI2) and ABI1 genes encode homologous protein phosphatases 2C involved in abscisic acid signal transduction. *Plant Cell* 1997, **9**:759–771.

13. Merlot S et al.: The ABI1 and ABI2 protein phosphatases 2C act in a negative feedback regulatory loop of the abscisic acid signalling pathway. *Plant J* 2001, **25**:295–303.

14. Giraudat J, et al.: Isolation of the Arabidopsis ABI3 gene by positional cloning. *Plant Cell* 1992, **4**:1251–1261.

15. Lopez-Molina L, et al.: ABI5 acts downstream of ABI3 to execute an ABA-dependent growth arrest during germination. *Plant J* 2002, **32**:317–328.

16. Finkelstein RR, et al.: The Arabidopsis abscisic acid response locus ABI4 encodes an APETALA 2 domain protein. *Plant Cell* 1998, **10**:1043–1054.

17. Soderman EM, et al.: Regulation and function of the Arabidopsis ABA-insensitive4 gene in seed and abscisic acid response signaling networks. *Plant Physiol* 2000, **124**:1752–1765.

18. Reeves WM, et al.: Direct targets of the transcription factors ABA-Insensitive(ABI)4 and ABI5 reveal synergistic action by ABI4 and several bZIP ABA response factors. *Plant Mol Biol* 2011, **75**:347–363.

19. Finkelstein RR, Lynch TJ: The Arabidopsis abscisic acid response gene ABI5 encodes a basic leucine zipper transcription factor. *Plant Cell* 2000, **12**:599–609.

20. Lopez-Molina L, Mongrand S, Chua NH: A postgermination developmental arrest checkpoint is mediated by abscisic acid and requires the ABI5 transcription factor in Arabidopsis. *Proc Natl Acad Sci U S A* 2001, **98**:4782–4787.

21. Stone SL, et al.: KEEP ON GOING, a RING E3 ligase essential for Arabidopsis growth and development, is involved in abscisic acid signaling. *Plant Cell* 2006, **18**:3415–3428.

22. Miura K, et al.: Sumoylation of ABI5 by the Arabidopsis SUMO E3 ligase SIZ1 negatively regulates abscisic acid signaling. *Proc Natl Acad Sci U S A* 2009, **106**:5418–5423.

23. Bensmihen S, et al.: Analysis of an activated ABI5 allele using a new selection method for transgenic Arabidopsis seeds. *FEBS Lett* 2004, **561**:127–131.

24. Choi H, et al.: ABFs, a family of ABA-responsive element binding factors. *J Biol Chem* 2000, **275**:1723–1730.

25. Lee SJ, et al.: DREB2C interacts with ABF2, a bZIP protein regulating abscisic acid-responsive gene expression, and its overexpression affects abscisic acid sensitivity. *Plant Physiol* 2010, **153**:716–727.

26. Yoshida T, et al.: AREB1, AREB2, and ABF3 are master transcription factors that cooperatively regulate ABRE-dependent ABA signaling involved in drought stress tolerance and require ABA for full activation. *Plant J* 2010, **61**:672–685.

27. Mustilli AC, et al.: Arabidopsis OST1 protein kinase mediates the regulation of stomatal aperture by abscisic acid and acts upstream of reactive oxygen species production. *Plant Cell* 2002, **14**:3089–3099.

28. Fujita Y, et al.: Three SnRK2 protein kinases are the main positive regulators of abscisic acid signaling in response to water stress in Arabidopsis. *Plant Cell Physiol* 2009, **50**:2123–2132.

29. Fujii H, Zhu JK: Arabidopsis mutant deficient in 3 abscisic acid-activated protein kinases reveals critical roles in growth, reproduction, and stress. *Proc Natl Acad Sci U S A* 2009, **106**:8380–8385.

30. Fujii H, Verslues PE, Zhu JK: Arabidopsis decuple mutant reveals the importance of SnRK2 kinases in osmotic stress responses in vivo. *Proc Natl Acad Sci U S A* 2011, **108**:1717–1722.
31. Sirichandra C, et al.: The Arabidopsis ABA-activated kinase OST1 phosphorylates the bZIP transcription factor ABF3 and creates a 14-3-3 binding site involved in its turnover. *PLoS One* 2010, **5**:e13935.
32. Nambara E, et al.: A screen for genes that function in abscisic acid signaling in *Arabidopsis thaliana*. *Genetics* 2002, **161**:1247–1255.
33. Nishimura N, et al.: Isolation and characterization of novel mutants affecting the abscisic acid sensitivity of Arabidopsis germination and seedling growth. *Plant Cell Physiol* 2004, **45**:1485–1499.
34. Nishimura N, et al.: ABA-hypersensitive germination1 encodes a protein phosphatase 2C, an essential component of abscisic acid signaling in Arabidopsis seed. *Plant J* 2007, **50**:935–949.
35. Yoshida T, et al.: ABA-hypersensitive germination3 encodes a protein phosphatase 2C (AtPP2CA) that strongly regulates abscisic acid signaling during germination among Arabidopsis protein phosphatase 2Cs. *Plant Physiol* 2006, **140**:115–126.
36. Nishimura N, et al.: Analysis of ABA hypersensitive germination2 revealed the pivotal functions of PARN in stress response in Arabidopsis. *Plant J* 2005, **44**:972–984.
37. Kuhn JM, Schroeder JI: Impacts of altered RNA metabolism on abscisic acid signaling. *Curr Opin Plant Biol* 2003, **6**:463–469.
38. Nyangulu JM, et al.: Synthesis and biological activity of tetralone abscisic acid analogues. *Org Biomol Chem* 2006, **4**:1400–1412.
39. Huang D, et al.: Structural analogs of ABA reveal novel features of ABA perception and signaling in Arabidopsis. *Plant J* 2007, **50**:414–428.
40. Nyangulu JM, et al.: An affinity probe for isolation of abscisic acid-binding proteins. *J Am Chem Soc* 2005, **127**:1662–1664.
41. Cutler AJ, et al.: Inhibitors of abscisic acid 8′-hydroxylase. *Biochemistry* 2000, **39**:13614–13624.
42. Priest DM, et al.: The use of abscisic acid analogues to analyse the substrate selectivity of UGT71B6, a UDP-glycosyltransferase of *Arabidopsis thaliana*. *FEBS Lett* 2005, **579**:4454–4458.
43. Park SY, et al.: Abscisic acid inhibits type 2C protein phosphatases via the PYR/PYL family of START proteins. *Science* 2009, **324**:1068–1071.
44. Ma Y, et al.: Regulators of PP2C phosphatase activity function as abscisic acid sensors. *Science* 2009, **324**:1064–1068.
45. Nishimura N, et al.: PYR/PYL/RCAR family members are major in-vivo ABI1 protein phosphatase 2C-interacting proteins in Arabidopsis. *Plant J* 2010, **61**:290–299.
46. Tan X, et al.: Mechanism of auxin perception by the TIR1 ubiquitin ligase. *Nature* 2007, **446**:640–645.
47. Calderon-Villalobos LI, et al.: Auxin perception—structural insights. *Cold Spring Harb Perspect Biol* 2010, **2**:a005546.
48. Nishimura N, et al.: Structural mechanism of abscisic acid binding and signaling by dimeric PYR1. *Science* 2009, **326**:1373–1379.
49. Santiago J, et al.: The abscisic acid receptor PYR1 in complex with abscisic acid. *Nature* 2009, **462**:665–668.

50. Miyazono K, et al.: Structural basis of abscisic acid signalling. *Nature* 2009, **462**:609–614.

51. Ng LM, et al.: Structural basis for basal activity and autoactivation of abscisic acid (ABA) signaling SnRK2 kinases. *Proc Natl Acad Sci U S A* 2011, **108**:21259–21264.

52. Soon FF, et al.: Molecular mimicry regulates ABA signaling by SnRK2 kinases and PP2C phosphatases. *Science* 2012, **335**:85–88.

53. Geiger D, et al.: Guard cell anion channel SLAC1 is regulated by CDPK protein kinases with distinct Ca^{2+} affinities. *Proc Natl Acad Sci U S A* 2010, **107**:8023–8028.

54. Kim TH, et al.: Chemical genetics reveals negative regulation of abscisic acid signaling by a plant immune response pathway. *Curr Biol* 2011, **21**:990–997.

55. Joshi-Saha A, Valon C, Leung J: A brand new START: abscisic acid perception and transduction in the guard cell. *Sci Signal* 2011, **4**:re4.

56. Szostkiewicz I, et al.: Closely related receptor complexes differ in their ABA selectivity and sensitivity. *Plant J* 2010, **61**:25–35.

57. Yasuda M, et al.: Antagonistic interaction between systemic acquired resistance and the abscisic acid-mediated abiotic stress response in Arabidopsis. *Plant Cell* 2008, **20**:1678–1692.

58. de Torres Zabala M, et al.: Antagonism between salicylic and abscisic acid reflects early host–pathogen conflict and moulds plant defence responses. *Plant J* 2009, **59**:375–386.

4.4

JASMONIC ACID

CHRISTIAN MEESTERS AND ERICH KOMBRINK

4.4.1 INTRODUCTION: JASMONIC ACID AND RELATED BIOACTIVE MOLECULES

Jasmonic acid (JA) and its derivatives, collectively referred to as jasmonates, comprise a group of oxylipin-signaling molecules that share a high degree of structural and functional similarity with prostaglandins found in animals. Both types of compound are derived from fatty acid oxidation pathways that are activated in response to stress. Correspondingly, jasmonate levels increase drastically and rapidly upon wounding, UV irradiation, ozone treatment, and other abiotic stresses [1]. In addition, responses to biotic stress such as herbivore attack or infection by microbial pathogens also depend on jasmonates and in healthy, nonstressed plants jasmonates mediate developmental processes such as root growth, seed germination, tuber formation, tendril coiling, trichome initiation, flower development, and senescence [1, 2]. Jasmonates exert their function by large-scale reprogramming of gene expression, which is in part mediated by the transcription factor MYC2 [3–5]. Elucidation of the mode of action of jasmonates has long been hampered by the fact that plants contain many JA derivatives, biosynthetic precursors, and JA metabolites, which differ considerably in their biological activities [6–11]. In addition, the identification of the JA receptor has been impaired by its modular structure comprising two components, the F-box protein COI1 and a JAZ protein, which cooperatively interact with one particular stereoisomer of the bioactive JA–isoleucine conjugate (JA-Ile) forming a ternary complex and thereby initiating the JA response [12–14].

Most of our current knowledge about synthesis and function of jasmonates has been derived from mutants that are defective in JA biosynthesis or the physiological

Plant Chemical Biology, First Edition. Edited by Dominique Audenaert and Paul Overvoorde.

response to JA treatment [15–17]. However, in addition to such biological approaches, chemical strategies have also always been integral part of jasmonate research. During the first years after discovery of jasmonates as plant constituents and recognition of their strong influence on plant growth and development [18, 19], extensive studies on structure–activity relationships were carried out, which included the synthesis of numerous JA derivatives and determining their impact on various plant responses [1, 20]. This type of work not only provided first insight into structural requirements for jasmonate bioactivity, it also aimed at developing tools for target identification. This approach culminated in the synthesis of the highly active jasmonate analog, coronalon, which was inspired by the extremely high bioactivity of the bacterial phytotoxin coronatine (COR) [21, 22].

Because of its exceptional structure and bioactivity, COR has been essential for two important discoveries in jasmonate research: (i) application of COR in a genetic screen yielded the *Arabidopsis thaliana* mutant *coronatine insensitive 1* (*coi1*), which is defective in a central regulatory unit of jasmonate signaling [23]; and (ii) its defined stereochemical structure provided the blueprint for identification and synthesis of the plant endogenous bioactive jasmonate, (+)-7-*iso*-JA-L-Ile [13].

In this chapter, we will outline aspects of biosynthesis and action of jasmonates originating from combined chemical and biological research strategies, which is trying to explore how chemistry contributed to our understanding of jasmonate-related biology and, *vice versa*, how biological questions contributed to advance jasmonate-related chemistry. Although the recent discovery of JAZ proteins as transcriptional repressors, which together with COI1 function as JA receptor, provides a coherent and conclusive model of JA recognition and signaling, many questions remain open. These are also addressed in recent review articles, which cover different facets of this rapidly evolving research field of jasmonate signaling [1, 20, 24–27].

4.4.2 JASMONATE BIOSYNTHESIS PATHWAY

The plant-specific pathway of JA biosynthesis has been unraveled; it comprises at least 10 mostly well-characterized enzymatic reactions that are outlined in Figure 4.4.1 [1, 28]. The initial step is the release of α-linolenic acid (C18:3) (**1**) from plastidial galactolipids by phospholipases such as DAD1 or GDL [29, 30], followed by oxidation of α-linolenic acid by 13-lipoxygenase (LOX) to 13(*S*)-hydroperoxy-octadecatrienoic acid [13(*S*)-HPOT] (**2**). This can be metabolized to different classes of oxylipins [31, 32], but its conversion to 12,13(*S*)-epoxy-octadecatrienoic acid [12,13(*S*)-EOT] (**3**) by allene oxide synthase (AOS) is the first committed step of JA biosynthesis. In *Arabidopsis* AOS is a single copy gene and, correspondingly, the *aos* loss-of-function mutant is deficient in JA [33, 34]. The allylic epoxide product of AOS is acted on by allene oxide cyclase (AOC) giving rise to optically pure (9*S*,13*S*)-12-oxo-phytodienoic acid [*cis*-(+)-OPDA] (**4**), which is the endproduct of the plastid-localized part of the JA biosynthesis pathway and represents the first cyclic compound with pronounced bioactivity [1, 35]. The exclusive accumulation of a specific OPDA stereoisomer indicates that AOC enforces strong steric restrictions on the chemical reaction path of

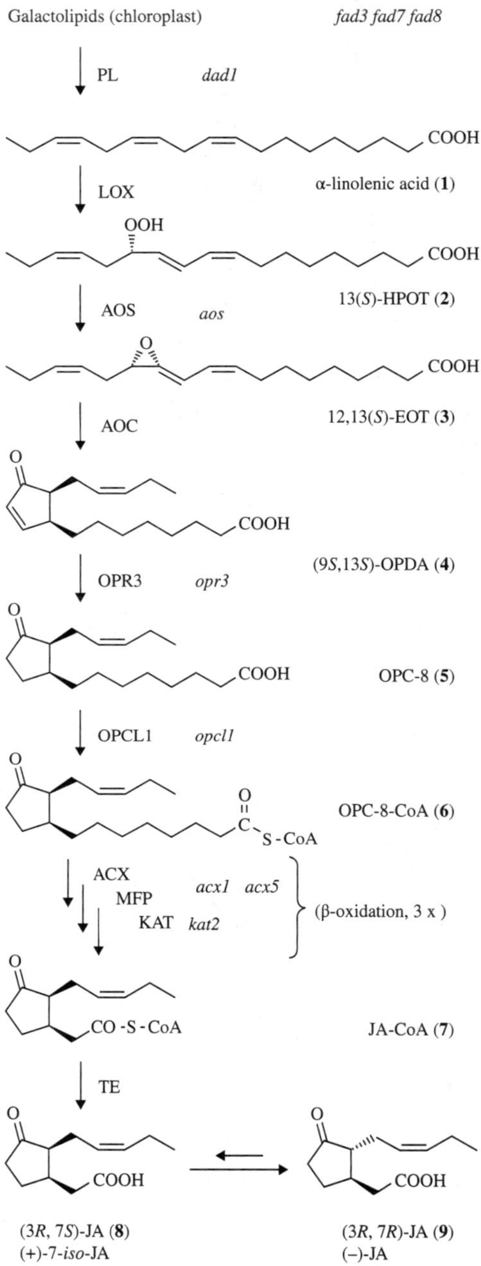

FIGURE 4.4.1 Pathway of jasmonic acid biosynthesis. Upon release of α-linolenic acid (C18:3) (**1**) from galactolipids by phospholipases (PL) in the chloroplast, the hydroperoxy compound 13(*S*)-HPOT (**2**) is formed by 13-lipoxygenase (LOX). The unstable allene oxide 12,13(*S*)-EOT (**3**) is generated by allene oxide synthase (AOS) and further converted to (9*S*,13*S*)-OPDA (**4**) by allene oxide cyclase (AOC). In the peroxisome OPDA (**4**) is reduced to OPC-8 (**5**) by OPDA reductase 3 (OPR3). Subsequent activation to the corresponding CoA ester (**6**) by OPC-8:CoA ligase (OPCL1) allows shortening of the carboxylic acid side chain *via* the fatty acid β-oxidation machinery, comprising acyl-CoA oxidase (ACX), multifunctional protein (MFP) and L-3-ketoacyl CoA thiolase (KAT). The endproduct, jasmonoyl-CoA (**7**), is cleaved by a putative thioesterase (TE) yielding (+)-7-*iso*-JA (**8**), which equilibrates with the more stable (–)-JA (**9**). *Arabidopsis* mutants that have been important for understanding JA biosynthesis are printed in *italics*. Reaction scheme modified after [20].

cyclization, which is reflected by the tight structure of the enzyme's substrate-binding pocket [36]. Note that the same set of enzymes acting on hexadecatrienoic acid (C16:3) generate dinor-OPDA [37]. Interestingly, in *Arabidopsis* chloroplasts, OPDA and dnOPDA have also been identified as constituents of membrane galactolipids and upon selective release by lipases may act as genuine signals or contribute to JA biosynthesis [38, 39].

For further conversion, OPDA is translocated from chloroplasts to peroxisomes, where the second half of JA biosynthesis occurs. While the plastidic export mechanism of OPDA is unknown, peroxisomal import is mediated in part by the ATP-binding cassette transporter CTS1 (identical to PXA1/PED3) [40]. OPDA reductase 3 (OPR3) catalyzes the reduction of OPDA to 3-oxo-2-(2′[*Z*]-pentenyl)-cyclopentan-1-octanoic acid (OPC-8) (**5**) (Fig. 4.4.1). OPR3 is encoded by gene families in all plants studied so far [28]. However, in *Arabidopsis* and tomato, only the isoform OPR3 is found in the peroxisome and the JA-deficient phenotype of the *opr3* mutant indicates that other isoforms do not contribute to JA biosynthesis [41, 42]. Shortening of the octanoic side chains of OPC-8 produced by OPR3 requires three rounds of β-oxidation [20, 28], which is initiated by activation of the carboxylic acid moiety to the corresponding CoA ester (**6**) by OPC-8:CoA ligase 1 (OPCL1), one member of a large superfamily of acyl-coenzyme A synthetase in *Arabidopsis* [43–45]. The *Arabidopsis opcl1* mutant is only partially impaired in JA biosynthesis, indicating that other acyl-CoA synthetases contribute to JA production [43, 44]. The core enzymes of peroxisomal β-oxidation, acyl-CoA oxidase (ACX), multifunctional protein (MFP), and L-3-ketoacyl-CoA thiolase (KAT) are also encoded by gene families, and it has been demonstrated that specific isoforms participate in wound-induced JA biosynthesis, that is, ACX1, ACX5, and KAT2/PED1/PKT3 in *Arabidopsis* and ACX1A in tomato [46–49]. The endproduct of β-oxidation, jasmonoyl-CoA (**7**), is cleaved by a yet unknown thioesterase (TE) giving rise to (+)-7-*iso*-JA [(3*R*,7*S*)-JA] (**8**) that equilibrates to the more stable (−)-JA [(3*R*,7*R*)-JA] (**9**). For key enzymes of JA biosynthesis, the crystal structures have been determined, providing insights into the mechanisms of catalysis and modes of substrate binding that give rise to stereospecific products [20, 28].

4.4.3 METABOLIC CONVERSION AND STRUCTURE–ACTIVITY RELATIONSHIP OF JASMONATES

In addition to epimerization, JA is also subject to enzymatic modifications as evident from numerous JA metabolites that were shown to be constituents in various plant species and tissues or to accumulate in response to stress [1, 50]. Among these metabolites are the methyl ester (JAMe) (**20**), glucose ester (JA-Glc) (**16**), and amino acid conjugates of JA, for example, JA-Ile (**10**), JA-Leu, and JA-Val, the hydroxylation products 12-OH-JA (tuberonic acid) (**12**) and 11-OH-JA (**15**), as well as their sulfated, glucosylated, and amino acid conjugated derivatives (**11**, **13**, **14**), the decarboxylation product *cis*-jasmone (**18**), and the reduced derivatives 9,10-dihydro-JA (**17**) and cucurbic acid (**19**) (Fig. 4.4.2). Even the conjugates with the ethylene precursor aminocyclopropane carboxylic acid (ACC),

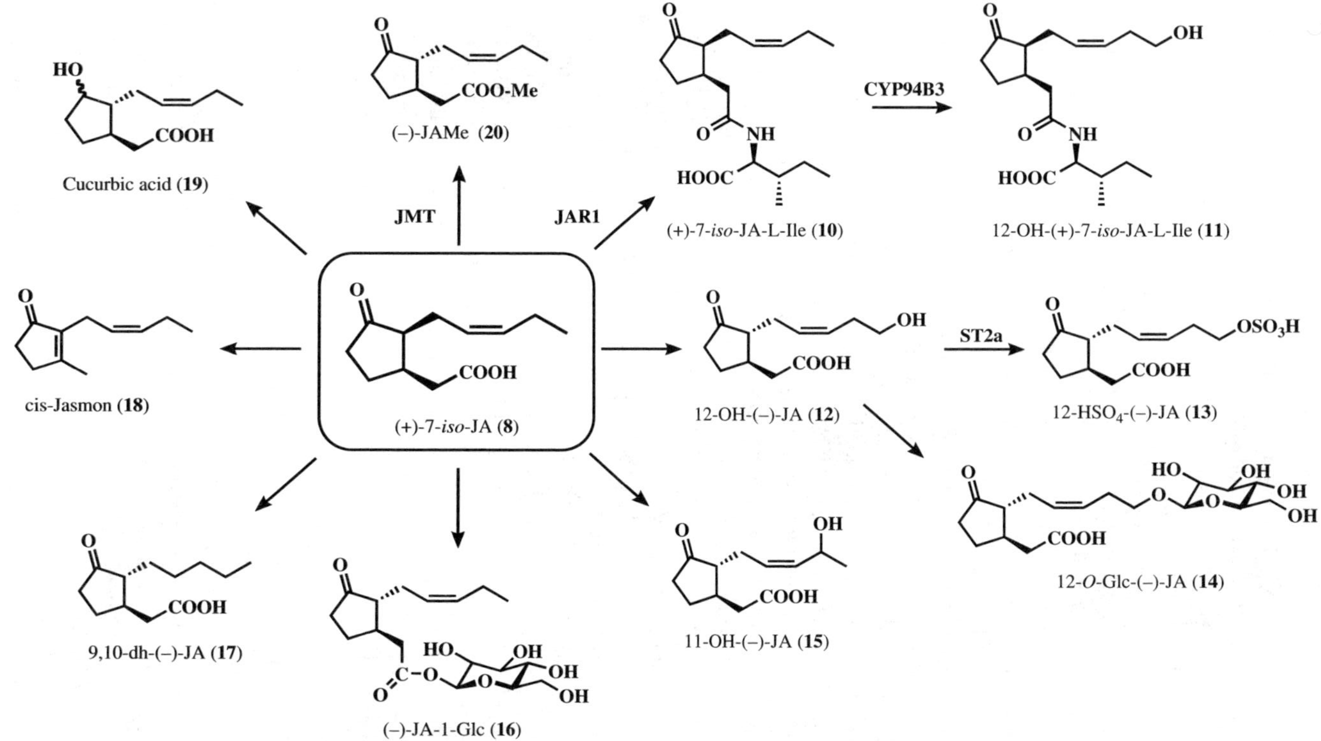

FIGURE 4.4.2 Metabolism of jasmonic acid. Conjugation of (+)-7-*iso*-JA (**8**) to isoleucine by jasmonoyl-isoleucine synthetase (JAR1) generates the natural bioactive jasmonate (+)-7-*iso*-JA-L-Ile (**10**), which is converted to inactive 12-OH-(+)-7-*iso*-JA-L-Ile (**11**) by JA-Ile hydroxylase (CYP94B3). JAR1 presumably also catalyzes the conjugation of JA to other amino acid such as valine, leucine, phenylalanine, tryptophan or the ethylene precursor aminocyclopropane carboxylic acid (not shown). Other common JA transformations include hydroxylation to 12-OH-JA (**12**), which can be sulfated by a sulfotransferase (ST2a) to 12-HSO_4-JA (**13**) or glucosylated to 12-*O*-glucosyl-JA (**14**) or conjugated with isoleucine to **11** (presumably by JAR1), hydroxylation to 11-OH-JA (**15**), and methylation at the carboxyl group by JA methyltransferase (JMT) to JAMe (**20**). Enzymes catalyzing formation of JA-1-glucoside (**16**) or reduction to 9,10-dehydro-JA (**17**), *cis*-jasmone (**18**) or cucurbic acid (**19**) are not yet known. Note that only a selection of JA metabolites is shown and their absolute stereoconfiguration is mostly not established or represented by racemic mixtures.

glucose esters of JA precursors and other variants were found. The development of new, sensitive analytical techniques and isolation methods has lead to an ever-increasing number of identified JA derivatives [50–52]. The abundance and bioactivity of different metabolites may vary considerably [8], which immediately raises two important questions: (i) What is the biological significance of such large number of JA derivatives? (ii) What is the impact of structural modifications on different biological activities?

After the first reported JA-induced plant responses, for example, growth inhibition [18, 19], the increasing structural and functional diversity of identified, natural jasmonates stimulated a strong interest in structure–activity relationships and extensive synthesis of novel JA derivates, standard approaches executed in drug research that precede target identification steps. From this early chemical–biological exercise, distinct structural requirements for JA activity were defined based on various biological responses, such as tuber formation, root growth inhibition, tendril coiling, alkaloid formation, or expression of JA-responsive genes [53–57]. They can be summarized as follows [1, 20]: (i) a cyclopentanone ring carrying a keto group at C-6 is essential. (ii) The intact pentenyl side chain is required for activity. (iii) The carboxylic acid side chain retains activity only with an even number of C-atoms. (iv) Formation of methyl ester or amino acids conjugates generally increases bioactivity. (v) Hydroxylation at C-11 or C-12 or reduction between C-11 and C-12 impair biological activity. (vi) (−)-JA [(3*R*,7*R*)-JA] and its derivatives are more active than (+)-JA [(3*S*,7*S*)-JA] and its derivatives.

Initially, bioactivity of jasmonates was exclusively determined in direct application experiments, and the results were obviously dependent on the concentration and lipophilicity of the compound, duration of treatment, and the potential metabolic conversion to a more or less active derivative. Therefore, such data provided only first hints of the functional significance of particular derivatives or distinct structural units. Some of the JA metabolites, such as 12-OH-JA (**12**), 12-HSO4-JA (**13**), and 12-O-Glc-JA (**14**), may even occur in distinct tissues or organs of some plants (e.g., *Zea mays* and *Glycine max*) at concentrations that are orders of magnitude higher than those of JA, suggesting a function as inactivated storage form of JA [8]. Indeed, it has been demonstrated that formation of 12-OH-JA and 12-HSO_4-JA was accompanied with a partial suppression of JA signaling [8, 58]. Likewise, accumulation of several glucosides of JA and JA derivatives has been observed upon wound-induced JA biosynthesis, and collectively, these data indicate a role of JA metabolism in turning off JA signaling by conversion of active jasmonates into metabolites that are at least partially inactive [8, 51, 52].

COR (**22**), a phytotoxin produced by several strains of *Pseudomonas syringae*, is a structural and functional JA mimic with extremely high biological activity (Fig. 4.4.3). It was successfully used to screen for JA-insensitive mutants of *Arabidopsis thaliana*, with *coi1–1* being the most informative [23, 59, 60]. Similar high activity was found for structurally related, synthetic compounds, such as coronalon (**23**) and other 6-substituted 4-oxo-indanoyl-isoleucine conjugates [21, 22]. The defined stereochemical structures of these highly active compounds and the differential bioactivity of defined stereoisomers of JA or JA methyl ester observed in earlier studies revealed the importance of the chiral centers at C-3 and C-7 [55, 61].

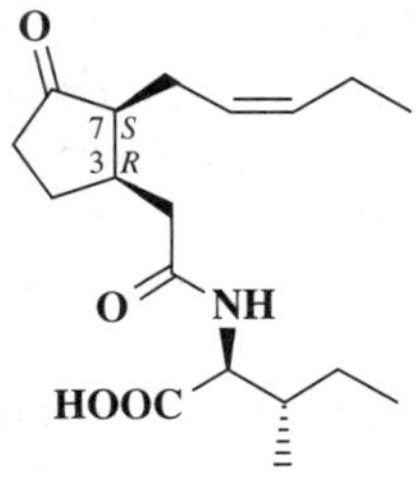

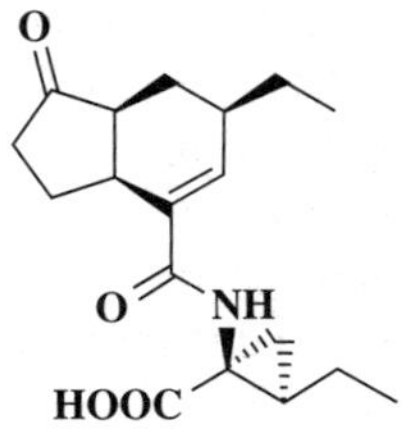

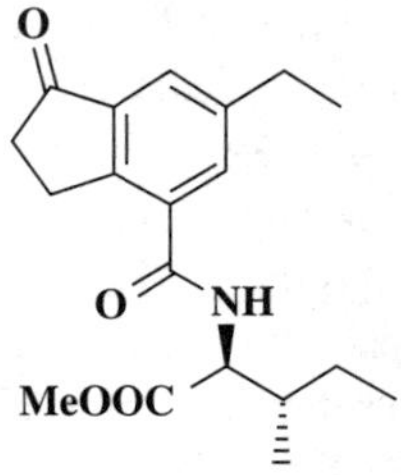

(3*R*, 7*S*)-JA-L-lle (**10**)
(+)-7-*iso*-JA-L-lle

Coronatine (**22**)

Coronalon (**23**)

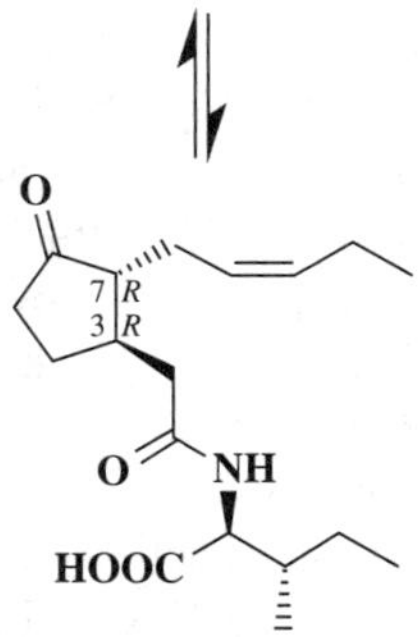

(3*R*, 7*R*)-JA-L-lle (**21**)
(–)-JA-L-lle

FIGURE 4.4.3 Structures of bioactive jasmonates. The absolute (*R,S*) configuration at the chiral centers C-3 and C-7 of the JA-Ile stereoisomers (+)-7-*iso*-JA-L-Ile [(3*R*,7*S*)-JA-L-Ile] (**10**) and (–)-JA-L-Ile [(3*R*,7*R*)-JA-L-Ile] (**21**) is indicated. JA stereoisomers with the (3*S*) configuration do not naturally occur. In coronatine (**22**) epimerization is not possible, because the configuration is fixed by the second ring structure and coronalon (**23**) contains a planar, aromatic structure without chiral centers.

Similarly, stereochemically restricted JA analogs carrying additional substitutions at C-7 showed lower activity than the parent compounds [61, 62]. The presence of two chiral centers allows four possible stereoisomers of jasmonates; however, only the (3*R*,7*S*) and (3*R*,7*R*) configurations occur in natural compounds, for example, JA or JA-Ile (**8**, **9**, **10**, **21**), since isomerization at C-3 apparently does not occur. Furthermore, compounds with a (3*R*,7*S*) configuration were generally more active than (3*R*,7*R*)-isomers [13, 55, 61].

4.4.4 (+)-7-*ISO*-JASMONOYL-L-ISOLEUCINE IS THE NATURAL BIOACTIVE JASMONATE

The identification of the COI1–JAZ module as site for interaction with JA-Ile (see later text) allowed the development of a cell-free test system for analyzing the bioactivity of jasmonates more precisely. Essentially, jasmonate-dependent COI1–JAZ interaction is detected *in vitro* by co-immunoprecipitation (pull-down) experiments [13], and obviously it was of interest to test JA derivatives for recognition by the jasmonate perception module. Initially, only JA-Ile and no other jasmonate was found to promote COI1–JAZ interactions and therefore was suggested to represent the bioactive JA derivative [14, 59]. In view of the fact that the ratio of the stereoisomers, (+)-7-*iso*-JA [(3*R*,7*S*)-JA] (**8**) versus (−)-JA [(3*R*,7*R*)-JA] (**9**), in plant tissues increases drastically in response to wounding [63], and because the jasmonate mimics COR (**22**) and coronalon (**23**), in which epimerization is not possible, show extremely high bioactivity, a search for the native, most active stereoisomer of different JA amino acid conjugates and other jasmonates was initiated [13]. This approach included the stereospecific synthesis of amino acid conjugates from different JA stereoisomers [13]. Among 40 different JA derivatives tested, only (+)-7-*iso*-JA-L-Ile [(3*R*,7*S*)-JA-L-Ile] (**10**) was exceptionally active (as well as the non-natural (+)-JA-L-Ile [(3*S*,7*S*)-JA-L-Ile]), whereas all other compounds showed only weak or no activity, including JA, JAMe, OPDA, and other JA amino acid conjugates [13]. The previously reported activity of (−)-JA-L-Ile (**21**) [14, 59] could be attributed to low, residual contamination by (+)-7-*iso*-JA-L-Ile (**10**) resulting from epimerization [13].

The pronounced bioactivity of (+)-7-*iso*-JA-L-Ile (**10**) in comparison to (−)-JA-L-Ile (**21**) suggests that its cellular level is tightly controlled by stereospecific synthesis and degradation. For the biosynthetic enzyme jasmonoyl-isoleucine synthetase (JAR1) from tomato, a strong preference for (+)-7-*iso*-JA over (−)-JA and isoleucine over other amino acids has been recently demonstrated [64, 65], suggesting that JAR1 activity is tightly coupled to (+)-7-*iso*-JA formation before its epimerization can occur. In this context, it is a puzzling observation that the recently resolved JAR1 crystal structure harbored only (−)-JA-L-Ile, although the racemic mixture was used for reaction [66]. Inactivation of jasmonate signaling originating from (+)-7-*iso*-JA-L-Ile was recently shown to involve hydroxylation of the jasmonate residue by a cytochrome P450 monooxygenase (CYP94B3) leading to 12-hydroxy-JA-Ile (**11**) and attenuation of jasmonate responses, but the stereospecificity of this catabolic reaction was not addressed [67, 68]. In addition, jasmonate signaling may also be turned off by other mechanisms, such as epimerization at C7 converting active (+)-7-*iso*-JA-L-Ile to inactive (−)-JA-L-Ile or methyl esterification to (+)-7-*iso*-JA-L-Ile-Me, which was also shown to be inactive [13]. The natural occurrence of JA-L-Ile methyl esters indicates that this mechanism of hormone inactivation may indeed operate *in planta* [69], and the unexpected stability of (+)-7-*iso*-JA-L-Ile over time in acidic and neutral pH conditions suggests that epimerization in the cell is a regulated process [13]. However, enzymes involved in any of these steps are currently unknown.

4.4.5 DISSECTING THE JA SIGNALING PATHWAY

Genetic and molecular genetic analyses have been pivotal for our current understanding of JA biosynthesis and function in signaling processes. In contrast to many other alternative approaches, mutant analysis offers the potential to provide clear, unequivocal information about how gene products affect plant function.

Some JA-related mutants were initially identified based on recognizable phenotypic characteristics, such as defective flower development in the mutants *delayed dehiscence 1* (*dde1/opr3*) and *delayed dehiscence 2* (*dde2/aos*), *defective in anther dehiscence 1* (*dad1*), or deficiency in unsaturated fatty acids of the *fatty acid desaturase* triple mutant *fad3 fad7 fad8*, but subsequently turned out to be affected in JA biosynthesis [30, 34, 41, 70, 71]. Other JA-related mutants were identified in screens by applying jasmonates (JAMe, COR, etc.) to chemically enforce a phenotype and selecting resistant plants [15–17]. Among the large number of these mutants, three have turned out to be of particular importance: *jasmonate resistant 1* (*jar1*), *coi1*, and *jasmonate insensitive 1* (*jai1/jin1/myc2*).

JAMe and other jasmonates strongly impair root growth, and the *jar1* mutant of *Arabidopsis* was found to be insensitive to this inhibition [57]. Subsequent biochemical analyses showed that the enzyme encoded by the JAR1 gene catalyzes the conjugation of JA with isoleucine to JA-Ile [9, 65, 72]. In the *jar1* mutant, JA-Ile levels are drastically reduced and most, but not all, JA responses impaired (e.g., fertility is retained), indicating the important and central signaling function of JA-Ile, as outlined earlier [9, 24, 73].

The *coi1* mutant of *Arabidopsis* was also isolated in a screen for plants that were resistant to root growth inhibition by COR [23]. Subsequent work showed that *coi1* is also insensitive to JAMe, male sterile and impaired in virtually all jasmonate responses [1, 16]. The identification of COI1 as F-box protein was seminal for subsequent work on JA signaling [74], but the suggested function of COI1 acting as an E3 ubiquitine ligase in an Skp/Cullin/F-box complex (SCF^{COI1}) and as such being involved in marking other regulators for degradation was only recently confirmed by the discovery of the so-called jasmonate-ZIM-domain (JAZ) proteins [12, 14, 75]. Although extensive genetic screens provided numerous mutants affected in JA signaling, the genetic approach failed to identify SCF^{COI1} target candidates [16].

Another informative mutant originating from a screen for jasmonate insensitive root growth is the *Arabidopsis* mutant *jai1/jin1*, which is defective in the basic helix–loop–helix–leucine zipper (bHLHzip)-type transcription factor MYC2 [5]. Importantly, MYC2 is involved in differential regulation of two branches of the JA signaling pathway. One of these branches, positively regulated by MYC2, activates expression of genes involved in the wound response, for example, *VSP2*, *LOX3*, *TAT*, the other branch, negatively regulated by MYC2, is required for the expression of pathogen defense genes, for example, *PDF1.2*, *PR1*, and *PR4*. In comparison to *coi1*, the *jai1/jin1/myc2* mutant is defective in fewer JA responses and shows a weaker phenotype, suggesting that MYC2 acts downstream of COI1 and may be involved in mediating crosstalk between different signaling pathways [3, 5].

Since genetic screens did not provide additional signaling candidate components acting downstream of COI1 or targets of SCF^{COI1}, alternative experimental approaches were initiated, including yeast-two-hybrid screening, transcriptional profiling, and biochemical purification. An important step toward understanding JA signaling was the discovery of the JAZ proteins, which function in repression of JA-responsive genes and are encoded by a gene family of 12 members in *Arabidopsis* [76]. They were identified by three different experimental approaches, their rapid JA-induced expression in stamens of *Arabidopsis* flowers [14], by cloning of the dominant JA-resistant *Arabidopsis* mutant *jai3–1* [12], and identification and cloning of a gene acting as repressor of JA-inhibited root growth in *Arabidopsis*, which turned out to represent a splice variant of JAZ10 [75]. All JAZ proteins contain a ZIM domain of 27 amino acids in their central part and a Jas domain near the C-terminus; they are translocated into the nucleus *via* a nuclear localization signal present in the Jas domain [24, 76, 77]. Besides the large number of JAZ proteins, their functional diversity may be further increased by the occurrence of different splice variants, for example, as demonstrated for JAZ10, which may attribute to the sensitivity and specificity of JA signaling [75, 78]. Another regulatory role seems to be provided by homo- and heterodimerization of particular JAZ proteins, and the demonstration of 38 interactions among the 66 possible heterodimeric combinations clearly underscores the potential regulatory flexibility in JA signaling [78, 79]. The output of JAZ protein action is dependent on the specific transcription factors affected by JAZ proteins [5, 80–82].

Finally, the last pieces for understanding the JA signaling puzzle were recently supplied by biochemical analysis [83]. Tandem affinity purification (TAP) was applied in order to identify new JAZ interactors and to monitor the dynamics of JAZ complex assembly. First, the half-life of a JAZ1–firefly luciferase fusion protein upon treatment of cultured cells with JA was determined, and subsequently this time scale was used for purifying proteins interacting with the JAZ-TAP-tagged bait. In addition to several known JAZ-interactors, such as JAZ12 reflecting the reported dimerization of JAZ proteins, MYC3 as close relative to MYC2, and COI1, a novel protein called NINJA (novel interactor of JAZ) was identified as an adaptor protein that interacts with the ZIM domain of most JAZ proteins, with the exception of JAZ7 and JAZ8 [83]. In addition, NINJA interacts *via* its EAR motif with the corepressor TOPLESS (TPL), which was previously identified to mediate auxin-dependent transcriptional repression in *Arabidopsis* [83, 84]. By contrast, JAZ8-mediated repression of JA responses is not mediated by the adaptor protein NINJA but rather relies on an endogenous EAR motif that directly interacts with TPL [85]. These exiting findings extend the mechanistic similarity between auxin and jasmonate signaling beyond the common SCF-ubiquitin-ligase-catalyzed degradation of repressor proteins Aux/IAA and JAZ, respectively [20, 26, 86, 87].

With all the identified components, the scenario of JA signaling *via* the SCF^{COI1} complex can be summarized as shown in Figure 4.4.4. In cells containing low levels of the bioactive jasmonate, (+)-7-*iso*-JA-L-Ile, JAZ proteins bind to the positive regulators of JA signaling, MYC2 and related transcription factors, which reside on promoters of JA-responsive genes [12, 14, 75]. By additional interaction with the adaptor protein NINJA, the general corepressor TPL is recruited to the MYC2–JAZ–NINJA complex, which is thought to effectively prevent downstream signaling [83]. This repression is relieved under conditions that activate JA biosynthesis and lead to concomitant increase of endogenous (+)-7-*iso*-JA-L-Ile levels *via* JAR1 activity, which may occur very rapidly, for example, 5 min after wounding [52, 88–90]. High (+)-7-*iso*-JA-L-Ile levels will promote the physical interaction between COI1 and JAZ protein, which allows ubiquitination of JAZ protein by the SCF^{COI1} complex [12, 14, 59]. The functional output is

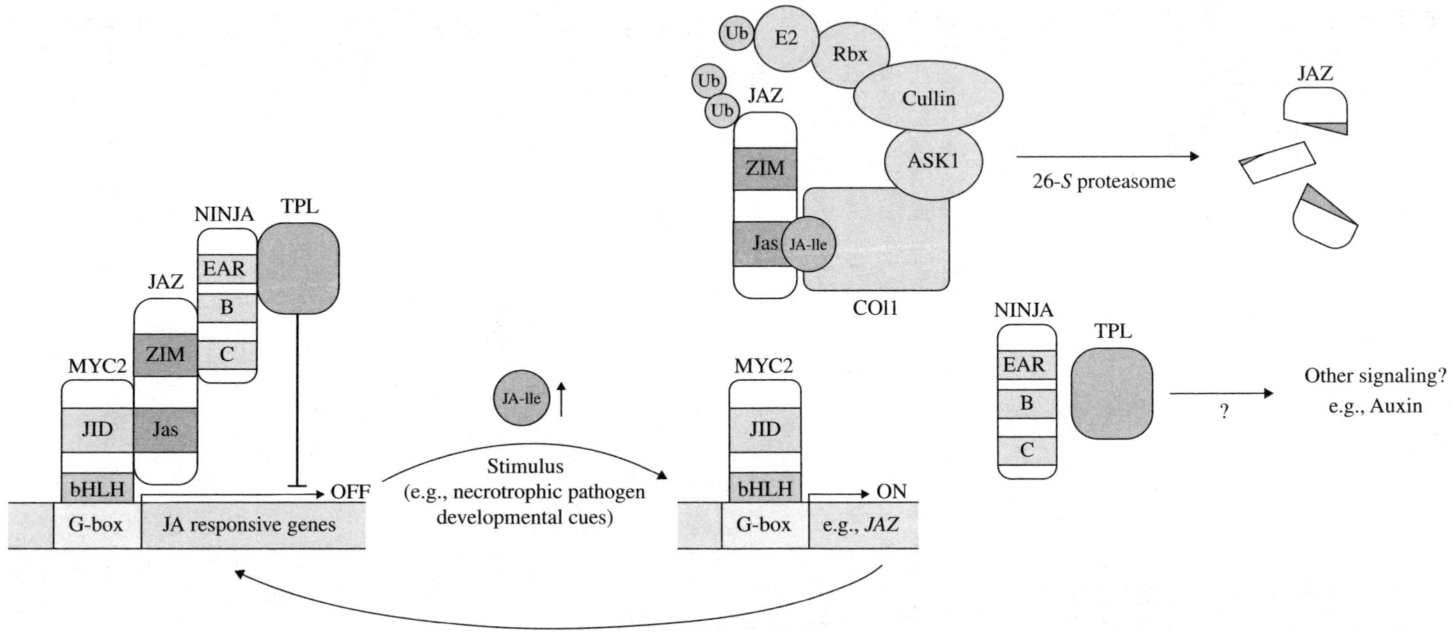

FIGURE 4.4.4 Model of jasmonate action in regulation of gene expression. In the resting state, at low level of JA-Ile, transcription factors (e.g., MYC2) are bound to their target sequence at the promoter of jasmonate-responsive gene (e.g., G-box), but their activity is repressed by interaction with JAZ proteins, the adaptor protein NINJA, and the co-repressor TOPLESS (TPL). Upon stimulation by stress (e.g., wounding or infection by necrotrophic pathogens) or developmental cues, (+)-7-*iso*-JA-L-Ile (**10**) rapidly accumulates and binds to the jasmonate receptor COI1 (F-box protein), which is part of the SCF^{COI1} complex, comprising the protein components ASK1, Cullin, Rbx, and the E2 ubiquitin-conjugating enzyme. JAZ proteins are recruited from their initial binding site to the SCF^{COI1} complex, which acts as an E3 ubiquitin ligase transferring ubiquitin (Ub) from the E2 ubiquitin-conjugating enzyme to the COI1-interacting JAZ proteins. Upon polyubiquitination, JAZ proteins are degraded by the 26S proteasome thereby releasing MYC2 and activating expression of JA responsive genes, including those encoding JAZ proteins, while NINJA and TPL could engage in other interactions. Synthesis of new JAZ repressors results in termination of JA gene expression. Modified after [83]. For color detail, please see color plate section.

the degradation of the JAZ repressor(s) *via* the 26S proteasome, which leads to de-repression of MYC2 and thereby allowing expression of JA-responsive genes, and release of NINJA and TPL, which may get engaged in interactions with other proteins [91]. This model is supported by a wealth of experimental evidence that was generated by yeast-two-hybrid protein interaction studies, expression analyses, binding assays, pull-down experiments, analyses employing mutants, and over-expression lines for various signaling components presented [4, 24–26, 73, 76, 86]. The mechanism of JA signaling *via* SCF complex-dependent proteasomal degradation exhibits striking similarities to the signaling mechanisms by auxin, ethylene, and gibberellin, where similar components participate [26, 86, 87].

4.4.6 RECEPTOR OF THE NATIVE BIOACTIVE JASMONATE, JA-ILE

The results obtained so far demonstrate that the COI1–JAZ–JA-Ile complex is the essential module for jasmonate perception and signaling, but the participation of additional components could not be excluded because bioassays were performed with crude extracts or partially purified COI1 and JAZ proteins. It also remained unknown in which order the COI1–JAZ–JA-Ile complex assembled. Unraveling these molecular details required purified components and structural information of the complex, which was established by *in silico* homology modeling of COI1 and crystal structure analysis of the COI1–JAZ coreceptor [60, 92].

From the derived homology model and docking simulations of COI1–jasmonate interactions, the highest binding affinity for (+)-7-*iso*-JA-L-Ile (**10**) and its molecular mimic COR (**22**) to a surface-localized binding pocket of COI1 was inferred [60]. Experimental evidence for direct binding of JA-Ile or COR by COI1 was provided by three independent approaches [60]. (i) Retention of COI1 from crude plant extracts was achieved by affinity chromatography using a matrix with JA coupled to the amino group of a 1,6-diaminohexane spacer on sepharose beads (**24**), generating a jasmonoylamide (Fig. 4.4.5). Although COI1 was selectively eluted from the column by JA-Ile (and not JA), this approach does not exclude the possibility that other proteins contribute to this interaction. (ii) Surface plasmon resonance (SPR) measurements revealed that the three purified components, COI1, JA-Ile, and JAZ1, are sufficient to form a complex, thereby ruling out the possibility that other COI1 copurified proteins are involved in JA-Ile perception. The SPR technology records changes in light refraction on sensor chip surfaces that occur upon interaction between two (or more) binding partners, one of which is covalently linked to the sensor chip surface. This approach also verified that JAZ1 alone (immobilized on the sensor chip) could neither bind JA-Ile/COR nor COI1; but unfortunately, the inverse experimental set-up, COI1 immobilized on chip surface, could not be realized. (iii) To gain direct evidence for COI1 binding JA-Ile/COR, a photoaffinity probe, PACOR (**25**), was designed and synthesized that contained three functional groups, COR, biotin, and the photoreactive phenyldiazirine group (Fig. 4.4.5). Upon incubation with affinity-purified His-tagged COI1 and exposure to UV irradiation, the PACOR probe was covalently linked to the protein through

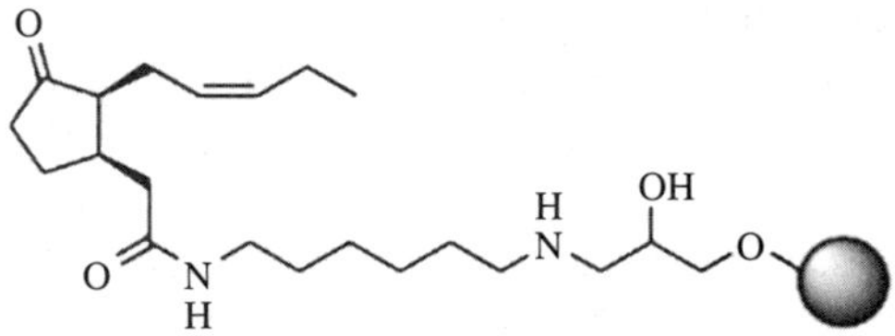

JA linked to sepharose (**24**)

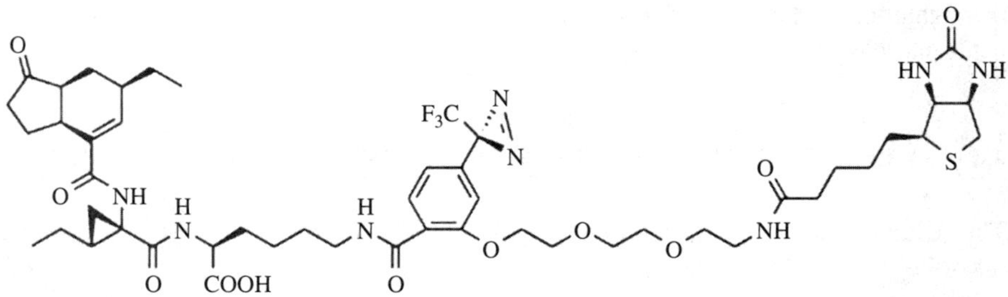

Coronatine linked to photoaffinity tag and biotin (**25**)

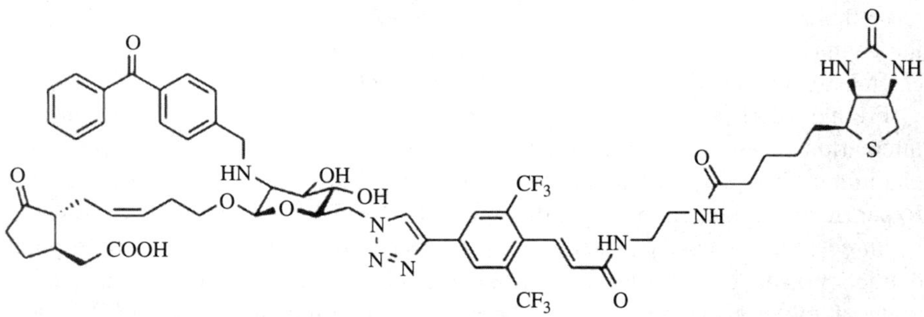

12–*O*–Glc–(-)–JA linked to photoaffinity tag and biotin (**26**)

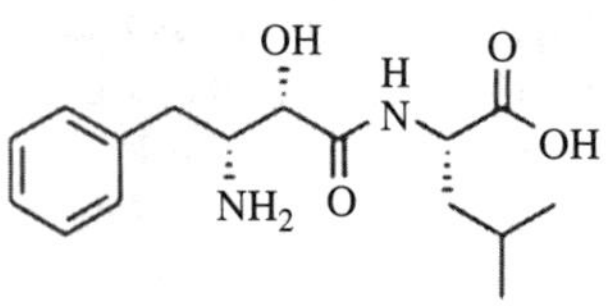

Bestatin (**27**)

FIGURE 4.4.5 Structures of synthetic probes used for jasmonate target identification. Jasmonic acid immobilized on sepharose beads (**24**) served as affinity matrix for purification of JA interacting proteins. The trifluoromethyldiazirine-photoaffinity- and biotin-tagged coronatine, PACOR (**25**), a bioactive jasmonate mimic, was used to demonstrate that COI1 has the capacity to bind COR and functions as jasmonate receptor. Similarly, benzophenone-photoaffinity- and biotin-tagged 12-*O*-β-D-glucopyranosyl-(−)-jasmonic acid [12-*O*-Glc-JA] (**26**) labeled the (−)-LCF-binding membrane protein. Bestatin [*N*-((2*S*,3*R*)-3-amino-2-hydroxy-4-phenylbutyryl)-L-leucine] (**27**) served as tool in chemical genetic screens for novel jasmonate regulators.

photolysis, which allowed simultaneous detection of the probe-labeled receptor/COI1 after SDS–PAGE separation by immunoblotting using anti-His and anti-biotin antibodies. Importantly, PACOR retained reasonable biological activity and binding to COI1 was competed out by COR, suggesting that COI1 directly binds JA-Ile/COR and functions as a receptor for JA-Ile/COR [60]. Taken together, these results demonstrate that COI1 has the capacity to directly bind (+)-7-*iso*-JA-L-Ile (**10**) and COR (**22**) and thus serves as a receptor for jasmonate. The derived model for jasmonate signaling also implies a defined order for assembly of the COI1–JAZ–JA-Ile complex [60]. Once JA-Ile is anchored to COI1 *via* three important functional groups, the Ile moiety of JA-Ile may form new interfaces for interactions with JAZ proteins.

Further insight into the mechanism of jasmonate perception was eventually provided by the crystal structure of COI1 in complex with JA-Ile and a JAZ peptide [92]. First, binding assays using ^{3}H-labeled COR and purified proteins confirmed that the jasmonate receptor is a coreceptor consisting of COI1 and JAZ proteins [92]. High-affinity, saturable binding of ^{3}H-COR to COI1 occurred only in the presence of full-length JAZ protein and (+)-7-*iso*-JA-L-Ile effectively competed for binding (K_i value of 1.8 μM), whereas the inactive stereoisomer (−)-JA-L-Ile was less effective. By contrast, ^{3}H-COR showed no binding to JAZ proteins and only marginal binding to COI1, accounting for less than 2% in comparison to the COI1–JAZ complex. Mapping of the COI1-binding region in the JAZ proteins revealed that a minimal sequence of 21 amino acids (Glu200-Val220 defining the JAZ degron peptide) is sufficient for establishing the COI1–JAZ–JA-Ile complex [92]. The crystal structure of COI1 in complex with the JAZ1 degron and JA-Ile/COR revealed that the ligand is buried deeply in the binding site and only the keto and the carboxyl groups of COR/JA-Ile remain exposed. This provides the surface for interaction with the JAZ degron peptide and upon its binding the COR/JA-Ile ligand is completely covered by the two components, explaining high-affinity binding of the hormone to the COI1–JAZ coreceptor [92]. Unexpectedly, inositol pentakisphosphate ($InsP_5$) was identified as a third critical component of the jasmonate coreceptor complex COI1–JAZ, extending the similarity to the auxin receptor complex TIR–AUX/IAA, which contains a inositol hexakisphosphate ($InsP_6$) molecule bound underneath the auxin-binding site [93]. For the COI1–JAZ1 coreceptor, it was demonstrated that $InsP_5$ is crucial for high-affinity ligand binding, suggesting that interaction of its phosphate residues with basic amino acids of COI1 is involved in shaping the ligand-binding pocket properly [92].

Although the observed binding mode of COR/JA-Ile by the COI1–JAZ1 coreceptor is markedly differed from that predicted by computational modeling, it does not preclude the previously proposed sequential order for assembling the receptor complex [60]. But eventually the bioactive jasmonate is in direct contact with both COI1 and JAZ1 supporting the molecular glue mechanism previously proposed for the auxin system [93]. Despite differences in structural and mechanistic details, jasmonate perception follows the same common theme as previously established for other plant hormones, including auxin, gibberellin, and abscisic acid [87].

4.4.7 SEARCHING FOR ADDITIONAL JA SIGNALING PATHWAYS AND COMPONENTS

Despite the enormous progress in understanding JA perception and signaling, many important questions remain to be answered. Perhaps the most intriguing and captivating is to mechanistically explain the multitude of biological jasmonate responses and the pronounced differential bioactivity of jasmonate derivatives, which are difficult to reconcile with only one common perception and signaling mechanism as outlined earlier for the COI1–JAZ–JA-Ile module. Thus, the search for missing signaling components is not yet over.

For example, OPDA (**4**), the bioactive precursor of JA biosynthesis, has signaling properties that are markedly different from JA, JAMe, or JA-Ile [1, 10, 11, 94, 95]. In *Eschscholtzia californica*, OPDA induced alkaloid biosynthesis, and in *Bryonia dioica* and *Phaseolus vulgaris* it was identified as endogenous signaling molecule mediating tendril coiling [54, 95, 96]. In cultured tobacco cells, OPDA, as well as JA, induced transient increase in cytoplasmic Ca^{2+} concentrations, whereas JA-Ile did not [97], and in the conditional *flu* mutant of *Arabidopsis* OPDA was suggested to antagonize JA-promoted cell death [27, 98]. In the *Arabidopsis opr3* mutant, which is defective in JA biosynthesis, the activation of the complete set of wound-inducible gene required the application of both OPDA and JA, suggesting that two independent signaling pathways exist, which is further supported by the observation that *opr3*, in comparison to *coi1*, showed increased resistance to fungal and insect attack and concomitant activation of COI1-dependent genes that are also regulated by JA, and COI1-independent genes that are not induced by JA [10]. Similarly, global expression profiling experiments using *Arabidopsis* and different jasmonates revealed that OPDA triggers expression of distinct set of genes that are related to the wound response [11]. Many plant species contain high levels of OPDA, both as free acid and bound in chloroplast membranes as galactolipids, that further increase upon wounding or infection, but the mechanisms that initiate distinct signaling cascades from the SCF^{COI1} complex depending on either OPDA or JA/JA-Ile are currently unknown [8, 99]. Similarly, COI1-independent defense responses (to wounding or infection) are only poorly characterized and distinct and informative mutants have not yet emerged from genetic analyses [99–103].

Other examples of JA derivatives with distinct bioactivities are *cis*-jasmone (**18**), which induced expression of a set of genes in *Arabidopsis* that is different from JAMe treatment [104], and 12-*O*-β-D-glucopyranosyljasmonic acid (12-*O*-Glc-JA) (**14**), which has been identified as specific regulator of nycnastic leaf movement, leaf-closing factor (LCF), in the Leguminosae *Samanea saman* and *Albizzia julibrissin* [105–107]. For the latter, chemical synthesis and studies of structure–function relationships revealed that only the (3*R*,7*R*) epimer of 12-*O*-Glc-JA, (–)-LCF, displayed high leaf-closing activity, the corresponding aglycon, (–)-12-OH-JA [also known as tuberonic acid] (**12**), showed low activity, whereas all other compounds tested were inactive, including (+)-*ent*-LCF, JA, JA-Ile, and COR, suggesting that the mechanism of leaf closing operates independent of the COI1–JAZ module [105, 108]. To unravel this signaling pathway, the direct target of (–)-LCF (**14**) was identified as plasma membrane resident protein using a biotin-tagged affinity probe containing a photoreactive benzophenone residue (**26**) [109, 110]. Although the identity of the protein target has not yet been uncovered, its successful and efficient labeling has been made possible by first establishing structure–function relationships and subsequent design and optimization of the affinity probe for photoaffinity cross-linking. This highlights the importance of alternative experimental strategies for dissecting signaling pathways in systems that are not tractable by genetic approaches.

Jasmonate action is integrated in a network with other signaling processes, including biotic and abiotic responses as well as the action of other hormones [87, 111–113]. Although a synergistic and antagonistic interaction between jasmonate, salicylic acid, abscisic acid, and auxin is well documented, the molecular mechanisms and components participating in these processes are not yet identified, although the recent identification of the general repressor TPL, which also acts in auxin signaling, may provide a possible link [83, 84, 91]. Likewise, the participation of mitogen-activated protein (MAP) kinases, as well as other protein kinases and phosphatases in jasmonate signaling, has been established, some of which operate in a COI1-independent manner [114–117]. However, how these regulatory steps are integrated into the current model of JA signaling remains unclear.

The identification of signaling components or their molecular characterization often involves the application of biochemical or other combined chemical and biological methods, when genetic approaches fail and mutants cannot be isolated because of redundant or indispensable gene functions. Identification of selective chemicals, acting as agonist or antagonists of a response, and their application in genetic screens have recently found broader application in plant biology [118–121]. However, in contrast to other plant hormone responses [122–126], jasmonate signaling has not been subject to such alternative chemical genetic interrogation in search for additional signaling components. The only exception documented so far is bestatin (**27**), an inhibitor of aminopeptidases and leukotriene A4 hydrolase in plants and animals [127, 128]. Bestatin was shown to specifically activate expression of JA-responsive genes in tomato and *Arabidopsis* in a COI1-dependent manner, but without strict requirement for JA biosynthesis, which lead to the hypothesis that bestatin exerts its function through modulation of key regulators in JA signaling [128]. To identify these regulators, bestatin was applied in a genetic screen for mutants resistant to root growth inhibition, which bestatin exerted as efficiently as JAMe. Although numerous bestatin resistant (*ber*) mutants were isolated by this approach, novel JA signaling components have not been uncovered [128]. However, the rational for such screen and its feasibility is documented by the finding that one of the *ber* mutants turned out to be a new *jin1/myc2* allele [128].

4.4.8 CONCLUSIONS

Over the past few years the field of jasmonate research has seen exciting developments and many details of JA biosynthesis, metabolism, perception, and signal transduction have been unraveled. And yet, many questions remain open. They include, but are not restricted to, the following: (i) How is selectivity of diverse biological JA responses established? (ii) What is the mode of action and perception mechanism for different bioactive jasmonates/oxylipins? (iii) What are the molecular mechanisms of cross-talk between different hormonal signaling pathways and what are the missing components? Answers to these and other open questions will likely provide an increased mechanistic understanding of plant hormone perception and signaling beyond the specific action of jasmonates. However, important and novel discoveries will require interdisciplinary research activities and combined application of new biological and chemical tools, including chemical genetic approaches aiming at identification of small molecules that specifically impair the functions of or interactions with new signaling components. In contrast to other research areas in plant biology, the potential of this experimental strategy remained unused in jasmonate research so far, providing opportunities for future discoveries.

ACKNOWLEDGMENT

This work was supported by the Max Planck Society, the Deutsche Forschungsgemeinschaft (DFG, KO 1192/9-1), and a PhD fellowship from the International Max Planck Research School program (C.M.).

REFERENCES

1. Wasternack, C. (2007) Jasmonates: An update on biosynthesis, signal transduction and action in plant stress response, growth and development. *Ann. Bot.* **100**, 681–697.
2. Browse, J., and Howe, G. A. (2008) New weapons and a rapid response against insect attack. *Plant Physiol.* **146**, 832–838.
3. Dombrecht, B., et al. (2007) MYC2 differentially modulates diverse jasmonate-dependent functions in *Arabidopsis*. *Plant Cell* **19**, 2225–2245.
4. Fonseca, S., Chico, J. M., and Solano, R. (2009) The jasmonate pathway: The ligand, the receptor and the core signalling module. *Curr. Opin. Plant Biol.* **12**, 539–547.
5. Lorenzo, O., et al. (2004) *JASMONATE-INSENSITIVE1* encodes a MYC transcription factor essential to discriminate between different jasmonate-regulated defense responses in Arabidopsis. *Plant Cell* **16**, 1938–1950.
6. Kramell, R., et al. (1997) Amino acid conjugates of jasmonic acid induce jasmonate-responsive gene expression in barley (*Hordeum vulgare* L.) leaves. *FEBS Lett.* **414**, 197–202.
7. Krumm, T., Bandemer, K., and Boland, W. (1995) Induction of volatile biosynthesis in the Lima bean (*Phaseolus lunatus*) by leucine- and isoleucine conjugates of 1-oxo- and 1-hydroxyindan-4-carboxylic acid: Evidence for amino acid conjugates of jasmonic acid as intermediates in the octadecanoid signalling pathway. *FEBS Lett.* **377**, 523–529.
8. Miersch, O., et al. (2008) Hydroxylated jasmonates are commonly occuring metabolites of jasmonic acid and contribute to a partial switch-off in jasmonate signaling. *New Phytol.* **117**, 114–127.
9. Staswick, P. E., and Tiryaki, I. (2004) The oxylipin signal jasmonic acid is activated by an enzyme that conjugates it to isoleucine in Arabidopsis. *Plant Cell* **16**, 2117–2127.

10. Stintzi, A., et al. (2001) Plant defense in the absence of jasmonic acid: The role of cyclopentenones. *Proc. Natl. Acad. Sci. U.S.A.* **98**, 12837–12842.

11. Taki, N., et al. (2005) 12-Oxo-phytodienoic acid triggers expression of a distinct set of genes and plays a role in wound-induced gene expression in Arabidopsis. *Plant Physiol.* **139**, 1268–1283.

12. Chini, A., et al. (2007) The JAZ family of repressors is the missing link in jasmonate signalling. *Nature* **448**, 666–671.

13. Fonseca, S., et al. (2009) (+)-7-iso-Jasmonoyl-L-isoleucine is the endogenous bioactive jasmonate. *Nat. Chem. Biol.* **5**, 344–350.

14. Thines, B., et al. (2007) JAZ repressor proteins are targets of the SCFCOI1 complex during jasmonate signalling. *Nature* **448**, 661–665.

15. Berger, S. (2002) Jasmonate-related mutants of *Arabidopsis* as tools for studying stress signaling. *Planta* **214**, 497–504.

16. Browse, J. (2009) The power of mutants for investigating jasmonate biosynthesis and signaling. *Phytochemistry* **70**, 1539–1546.

17. Wasternack, C. (2006) Oxylipins: Biosynthesis, signal transduction and action. In *Plant hormone signaling* (Hedden, P., and Thomas, S., eds), Blackwell Publishing Ltd., Oxford, pp. 185–228.

18. Dathe, W., et al. (1981) Endogenous plant hormones of the broad bean, *Vicia faba* L. (−)-jasmonic acid, a plant growth inhibitor in pericarp. *Planta* **153**, 530–535.

19. Ueda, J., and Kato, J. (1980) Isolation and identification of a senescence-promoting substance from wormwood (*Artemisia absinthium* L.). *Plant Physiol.* **66**, 246–249.

20. Wasternack, C., and Kombrink, E. (2010) Jasmonates: Structural requirements for lipid-derived signals active in plant stress responses and development. *ACS Chem. Biol.* **5**, 63–77.

21. Mithöfer, A., Maitrejean, M., and Boland, W. (2004) Structural and biological diversity of cyclic octadecanoids, jasmonates, and mimetics. *J. Plant Growth Regul.* **23**, 170–178.

22. Schüler, G., et al. (2004) Coronalon: A powerful tool in plant stress physiology. *FEBS Lett.* **563**, 17–22.

23. Feys, B. J. F., et al. (1994) Arabidopsis mutants selected for resistance to the phytotoxin coronatine are male sterile, insensitive to methyl jasmonate, and resistant to a bacterial pathogen. *Plant Cell* **6**, 751–759.

24. Browse, J. (2009) Jasmonate passes muster: A receptor and targets for the defense hormone. *Annu. Rev. Plant Biol.* **60**, 183–205.

25. Kazan, K., and Manners, J. M. (2008) Jasmonate signaling: Toward an integrated view. *Plant Physiol.* **146**, 1459–1468.

26. Pauwels, L., and Goossens, A. (2011) The JAZ proteins: A crucial interface in the jasmonate signaling cascade. *Plant Cell* **23**, 3089–3100.

27. Reinbothe, C., et al. (2009) Plant oxylipins: Role of jasmonic acid during programmed cell death, defence and leaf senescence. *FEBS J.* **276**, 4666–4681.

28. Schaller, A., and Stintzi, A. (2009) Enzymes in jasmonate biosynthesis – Structure, function, regulation. *Phytochemistry* **70**, 1532–1538.

29. Hyun, Y., et al. (2008) Cooperation and functional diversification of two closely related galactolipase genes for jasmonate biosynthesis. *Dev. Cell* **14**, 183–192.

30. Ishiguro, S., et al. (2001) The *DEFECTIVE IN ANTHER DEHISCENCE1* gene encodes a novel phospholipase A1 catalyzing the initial step of jasmonic acid biosynthesis, which synchronizes pollen maturation, anther dehiscence, and flower opening in Arabidopsis. *Plant Cell* **13**, 2191–2209.
31. Feussner, I., and Wasternack, C. (2002) The lipoxygenase pathway. *Annu. Rev. Plant Biol.* **53**, 275–297.
32. Mosblech, A., Feussner, I., and Heilmann, I. (2009) Oxylipins: Structurally diverse metabolites from fatty acid oxidation. *Plant Physiol. Biochem.* **47**, 511–517.
33. Park, J.-H., et al. (2002) A knock-out mutation in allene oxide synthase results in male sterility and defective wound signal transduction in *Arabidopsis* due to a block in jasmonic acid biosynthesis. *Plant J.* **31**, 1–12.
34. von Malek, B., et al. (2002) The *Arabidopsis* male-sterile mutant *dde2-2* is defective in the *ALLENE OXIDE SYNTHASE* gene encoding one of the key enzymes of the jasmonic acid biosynthesis pathway. *Planta* **216**, 187–192.
35. Stenzel, I., et al. (2003) Jasmonate biosynthesis and the allene oxide cyclase family of *Arabidopsis thaliana*. *Plant Mol. Biol.* **51**, 895–911.
36. Hofmann, E., Zerbe, P., and Schaller, F. (2006) The crystal structure of *Arabidopsis thaliana* allene oxide cyclase: Insights into the oxylipin cyclization reaction. *Plant Cell* **18**, 3201–3217.
37. Weber, H., Vick, B. A., and Farmer, E. E. (1997) Dinor-oxo-phytodienoic acid: A new hexadecanoid signal in the jasmonate family. *Proc. Natl. Acad. Sci. U.S.A.* **94**, 10473–10478.
38. Andersson, M. X., et al. (2006) Oxylipin profiling of the hypersensitive response in *Arabidopsis thaliana*. Formation of a novel oxo-phytodienoic acid-containing galactolipid, arabiodopside E. *J. Biol. Chem.* **281**, 31528–31537.
39. Stelmach, B. A., et al. (2001) A novel class of oxylipins, *sn*1-*O*-(12-oxophytodienoyl)-*sn*2-*O*-(hexadecatrienoyl)-monogalactosyl diglyceride, from *Arabidopsis thaliana*. *J. Biol. Chem.* **276**, 12832–12838.
40. Theodoulou, F. L., et al. (2005) Jasmonic acid levels are reduced in COMATOSE ATP-binding cassette transporter mutants. Implications for transport of jasmonate precursors into peroxisomes. *Plant Physiol.* **137**, 835–840.
41. Sanders, P. M., et al. (2000) The Arabidopsis *DELAYED DEHISCENCE1* gene encodes an enzyme in the jasmonic acid synthesis pathway. *Plant Cell* **12**, 1041–1062.
42. Stintzi, A., and Browse, J. (2000) The *Arabidopsis* male-sterile mutant, opr3, lacks the 12-oxophytodienoic acid reductase required for jasmonate synthesis. *Proc. Natl. Acad. Sci. U.S.A.* **97**, 10625–10630.
43. Kienow, L., et al. (2008) Jasmonates meet fatty acids: Functional analysis of a new acyl-coenzyme A synthetase protein family from *Arabidopsis thaliana*. *J. Exp. Bot.* **59**, 403–419.
44. Koo, A. J. K., et al. (2006) Identification of a peroxisomal acyl-activating enzyme involved in the biosynthesis of jasmonic acid in Arabidopsis. *J. Biol. Chem.* **281**, 33511–33520.
45. Schneider, K., et al. (2005) A new type of peroxisomal acyl-coenzyme A synthetase from *Arabidopsis thaliana* has the catalytic capacity to activate biosynthetic precursors of jasmonic acid. *J. Biol. Chem.* **280**, 13962–13972.
46. Afitlhile, M. M., et al. (2005) A defect in glyoxysomal fatty acid β-oxidation reduces jasmonic acid accumulation in *Arabidopsis*. *Plant Physiol. Biochem.* **43**, 603–609.

47. Castillo, M. C., et al. (2004) Gene-specific involvement of βboxidation in wound-activated responses in Arabidopsis. *Plant Physiol.* **135**, 85–94.
48. Li, C., et al. (2005) Role of β-oxidation in jasmonate biosynthesis and systemic wound signaling in tomato. *Plant Cell* **17**, 971–986.
49. Schilmiller, A. L., Koo, A. J. K., and Howe, G. A. (2007) Functional diversification of acyl-coenzyme A oxidases in jasmonic acid biosynthesis and action. *Plant Physiol.* **143**, 812–824.
50. Göbel, C., and Feussner, I. (2009) Methods for the analysis of oxylipins in plants. *Phytochemistry* **70**, 1485–1503.
51. Glauser, G., et al. (2010) Mass spectrometry-based metabolomics oriented by correlation analysis for wound-induced molecule discovery: Identification of a novel jasmonate glucoside. *Phytochem. Anal.* **21**, 95–101.
52. Glauser, G., et al. (2008) Spatial and temporal dynamics of jasmonate synthesis and accumulation in *Arabidopsis* in response to wounding. *J. Biol. Chem.* **283**, 16400–16407.
53. Blechert, S., et al. (1999) Structure–activity analyses reveal the existence of two separate groups of active octadecanoids in elicitation of the tendril-coiling response of *Bryonia dioica* Jacq. *Planta* **207**, 470–479.
54. Blechert, S., et al. (1995) The octadecanoic pathway: Signal molecules for the regulation of secondary pathways. *Proc. Natl. Acad. Sci. U.S.A.* **92**, 4099–4105.
55. Koda, Y., et al. (1992) Comparisons of various biological activities of stereoisomers of methyl jasmonate. *Phytochemistry* **31**, 1111–1114.
56. Miersch, O., et al. (1999) Structure–activity relations of substituted, deleted or stereospecifically altered jasmonic acid in gene expression of barley leaves. *Phytochemistry* **50**, 353–361.
57. Staswick, P., Su, W., and Howell, S. (1992) Methyl jasmonate inhibition of root growth and induction of a leaf protein are decreased in an *Arabidopsis thaliana* mutant. *Proc. Natl. Acad. Sci. U.S.A.* **89**, 6837–6840.
58. Gidda, S. K., et al. (2003) Biochemical and molecular characterization of a hydroxyjasmonate sulfotransferase from *Arabidopsis thaliana*. *J. Biol. Chem.* **278**, 17895–17900.
59. Katsir, L., et al. (2008) COI1 is a critical component of a receptor for jasmonate and the bacterial virulence factor coronatine. *Proc. Natl. Acad. Sci. U.S.A.* **105**, 7100–7105.
60. Yan, J., et al. (2009) The *Arabidopsis* CORONATINE INSENSITIVE1 protein is a jasmonate receptor. *Plant Cell* **21**, 2220–2236.
61. Holbrook, L., et al. (1997) Importance of the chiral centers of jasmonic acid in the responses of plants – Activities and antagonism between natural and synthetic analogs. *Plant Physiol.* **114**, 419–428.
62. Kitaoka, N., et al. (2009) Potato tuber cell expansion-inducing activity of stereochemically restricted JA analogs *Biosci. Biotechnol. Biochem.* **73**, 1872–1876.
63. Schulze, B., et al. (2006) Profiling of structurally labile oxylipins in plants by in situ derivatization with pentafluorobenzyl hydroxylamine. *Anal. Biochem.* **348**, 269–283.
64. Suza, W., et al. (2010) A tomato enzyme synthesizes (+)-7-iso-jasmonoyl-L-isoleucine in wounded leaves. *Planta* **231**, 717–728.
65. Suza, W., and Staswick, P. (2008) The role of JAR1 in jasmonoyl-L-isoleucine production during *Arabidopsis* wound response. *Planta* **227**, 1221–1232.
66. Westfall, C. S., et al. (2012) Structural basis for prereceptor modulation of plant hormones by GH3 proteins. *Science* **336**, 1708–1711.

67. Heitz, T., et al. (2012) Cytochromes P450 CYP94C1 and CYP94B3 catalyze two successive oxidation steps of plant hormone jasmonoyl-isoleucine for catabolic turnover. *J. Biol. Chem.* **287**, 6296–6306.

68. Koo, A. J. K., Cooke, T. F., and Howe, G. A. (2011) Cytochrome P450 CYP94B3 mediates catabolism and inactivation of the plant hormone jasmonoyl-L-isoleucine. *Proc. Natl. Acad. Sci. U.S.A.* **108**, 9298–9303.

69. Hause, B., et al. (2000) Tissue-specific oxylipin signature of tomato flowers: Allene oxide cyclase is highly expressed in distinct flower organs and vascular bundles. *Plant J.* **24**, 113–126.

70. McConn, M., and Browse, J. (1996) The critical requirement for linolenic acid is pollen development, not photosynthesis, in an Arabidopsis mutant. *Plant Cell* **8**, 403–416.

71. McConn, M., et al. (1997) Jasmonate is essential for insect defense in *Arabidopsis*. *Proc. Natl. Acad. Sci. U.S.A.* **94**, 5473–5477.

72. Staswick, P. E., Tiryaki, I., and Rowe, M. L. (2002) Jasmonate response locus *JAR1* and several related Arabidopsis genes encode enzymes of the firefly luciferase superfamily that show activity on jasmonic, salicylic, and indole-3-acetic acids in an assay for adenylation. *Plant Cell* **14**, 1405–1415.

73. Staswick, P. E. (2008) JAZing up jasmonate signaling. *Trends Plant Sci.* **13**, 66–71.

74. Xie, D.-X., et al. (1998) *COI1*: An *Arabidopsis* gene required for jasmonate-regulated defense and fertility. *Science* **280**, 1091–1094.

75. Yan, Y., et al. (2007) A downstream mediator in the growth repression limb of the jasmonate pathway. *Plant Cell* **19**, 2470–2483.

76. Chung, H. S., et al. (2009) Top hits in contemporary JAZ: An update on jasmonate signaling. *Phytochemistry* **70**, 1547–1559.

77. Grunewald, W., et al. (2009) Expression of the *Arabidopsis* jasmonate signalling repressor *JAZ1/TIFY10A* is stimulated by auxin. *EMBO Rep.* **10**, 923–928.

78. Chung, H. S., and Howe, G. A. (2009) A critical role for the TIFY motif in repression of jasmonate signaling by a stabilized splice variant of the JASMONATE ZIM-domain protein JAZ10 in *Arabidopsis*. *Plant Cell* **21**, 131–145.

79. Chini, A., et al. (2009) The ZIM domain mediates homo- and heteromeric interactions between Arabidopsis JAZ proteins. *Plant J.* **59**, 77–87.

80. Fernández-Calvo, P., et al. (2011) The *Arabidopsis* bHLH transcription factors MYC3 and MYC4 are targets of JAZ repressors and act additively with MYC2 in the activation of jasmonate responses. *Plant Cell* **23**, 701–715.

81. Qi, T., et al. (2011) The jasmonate-ZIM-domain proteins interact with the WD-repeat/bHLH/MYB complexes to regulate jasmonate-mediated anthocyanin accumulation and trichome initiation in *Arabidopsis thaliana*. *Plant Cell* **23**, 1795–1814.

82. Song, S., et al. (2011) The jasmonate-ZIM domain proteins interact with the R2R3-MYB transcription factors MYB21 and MYB24 to affect jasmonate-regulated stamen development in *Arabidopsis*. *Plant Cell* **23**, 1000–1013.

83. Pauwels, L., et al. (2010) NINJA connects the co-repressor TOPLESS to jasmonate signalling. *Nature* **464**, 788–791.

84. Szemenyei, H., Hannon, M., and Long, J. A. (2008) TOPLESS mediates auxin-dependent transcriptional repression during *Arabidopsis* embryogenesis. *Science* **319**, 1384–1386.

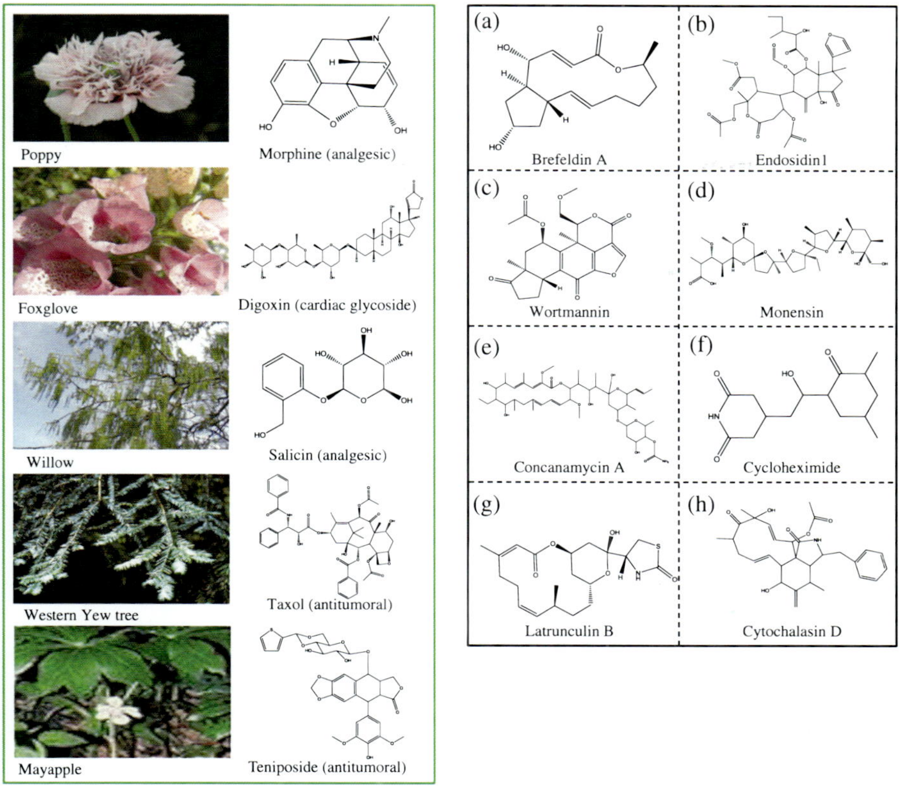

FIGURE 1.1.1 Many plant metabolites have been utilized as pharmaceuticals, such as morphine, digitalis, and taxol. Several compounds have been utilized in cell biology research. These include compounds that inhibit protein trafficking (**a**)–(**e**), protein synthesis inhibitor (**f**), and cytoskeleton-inhibiting drugs (**g**)–(**h**).

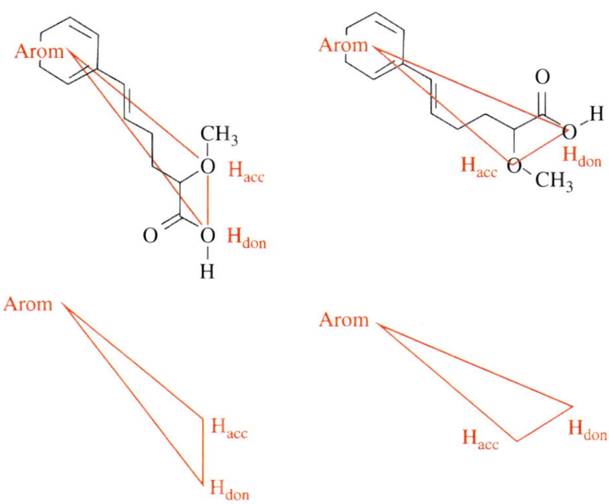

FIGURE 2.1.5 The multiple pharmacophore concept from a single molecule.

Plant Chemical Biology, First Edition. Edited by Dominique Audenaert and Paul Overvoorde.

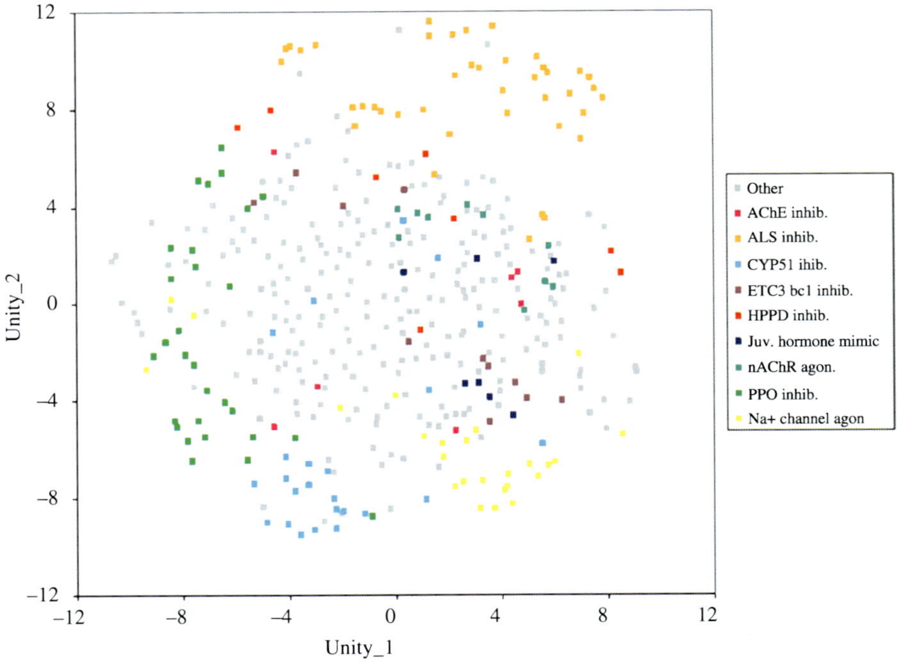

FIGURE 2.2.5 A set of 500 agrochemicals with known modes of action mapped into a 2D chemical space obtained using a PCA-NLM transformation of the Unity fingerprints.

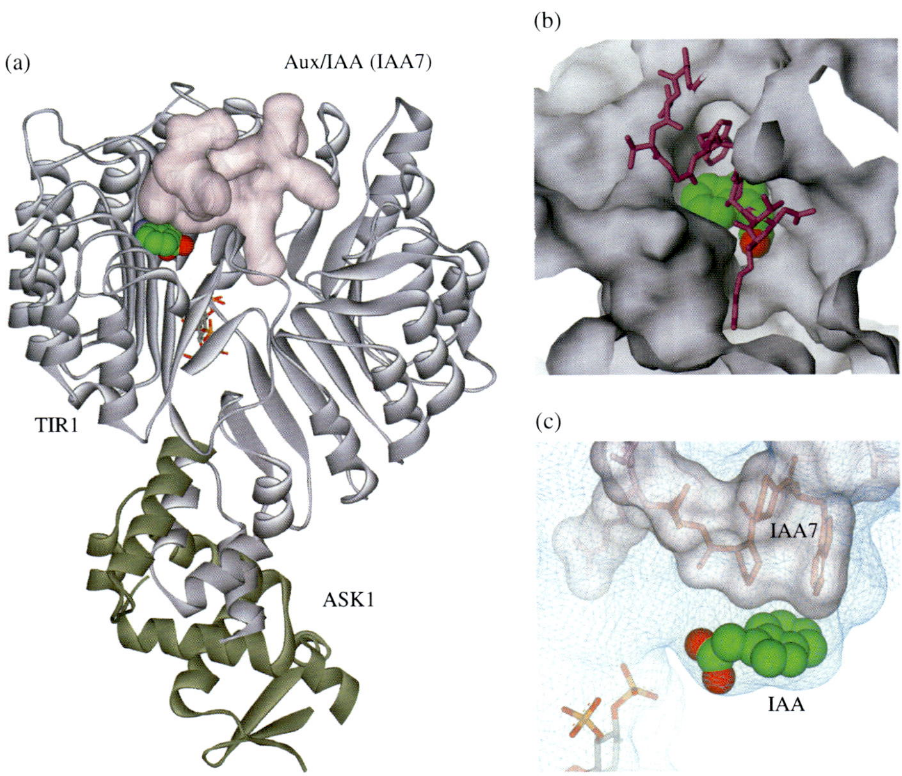

FIGURE 4.1.6 Structural highlights of the TIR1-IAA-Aux/IAA complex. For full caption, please refer to text.

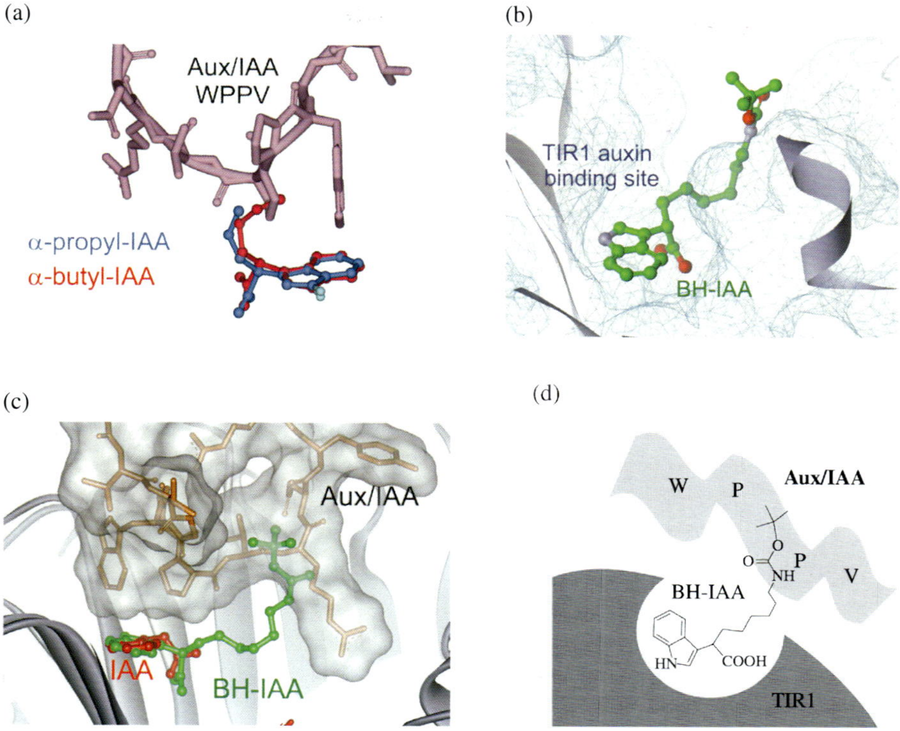

FIGURE 4.1.7 Structural constraints imposed by α-alkyl-IAA derivatives.
(**a**) The coordinates of α-propyl-IAA (blue) and α-butyl-IAA (red) were superimposed on the TIR1-IAA7 crystal structure (PDB ID 2P1Q) to show their hypothesized interaction with the WPPV motif of IAA7 (pink). The terminal region of α-butyl-IAA clashes with the second proline in the WPPV motif and thus would block access of Aux/IAA to the TIR1 docking site. In contrast, the shorter chain of α-propyl-IAA does not prevent docking of the Aux/IAA. These observations account for why the α-butyl-IAA serves as an auxin antagonist and the α-propyl-IAA functions as an auxin agonist. (**b**)–(**d**) Images based on the crystal structure of BH-IAA interacting with TIR1 to prevent Aux/IAA protein docking. (**b**) BH-IAA (green and red) is shown interacting with the auxin-binding site of TIR1. (**c**) IAA (red) and BH-IAA (green) shown interacting with the TIR1 (shown as ribbon structure). The Aux/IAA protein coordinates (tan) were superimposed to highlight how BH-IAA would exclude the docking of domain II. (**d**) A schematic summary of how the long alkyl chain and bulky terminal groups effectively block the access of Aux/IAA proteins to the TIR1 docking site.

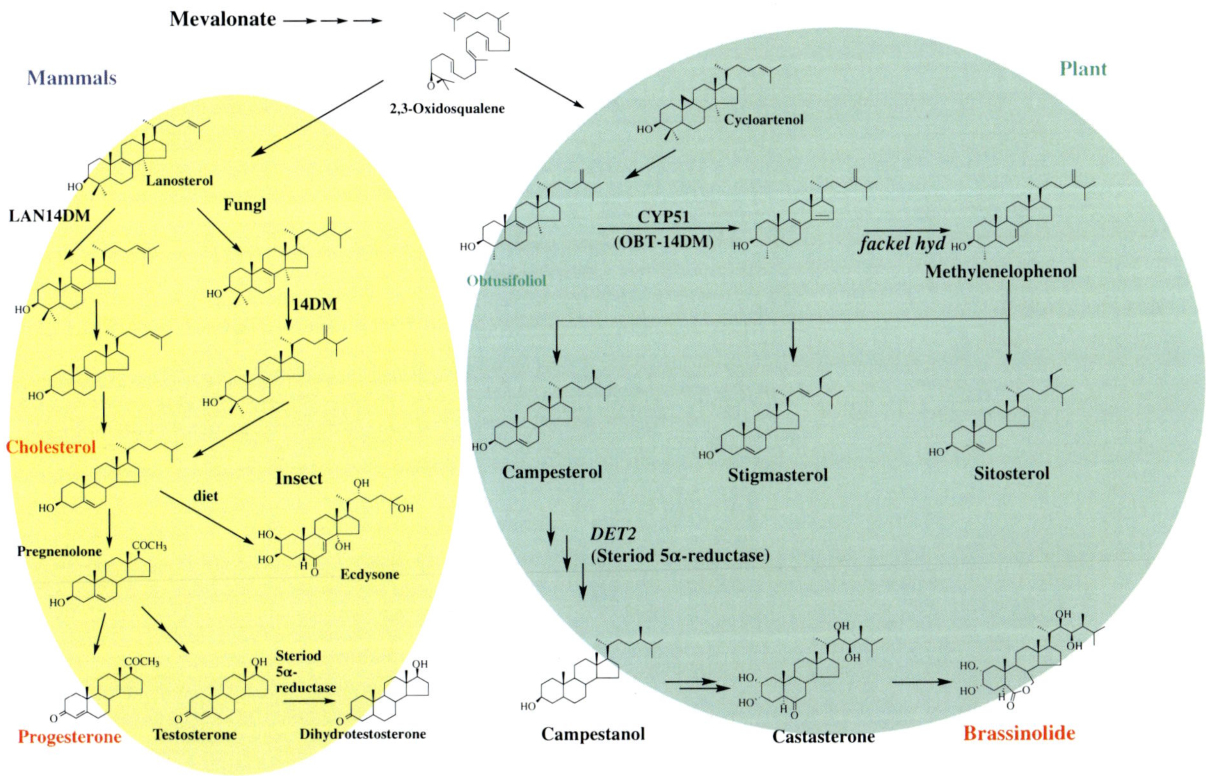

FIGURE 4.2.1 Comparison of biosynthesis pathways between plant brassinosteroid and animal or insect steroid hormone.

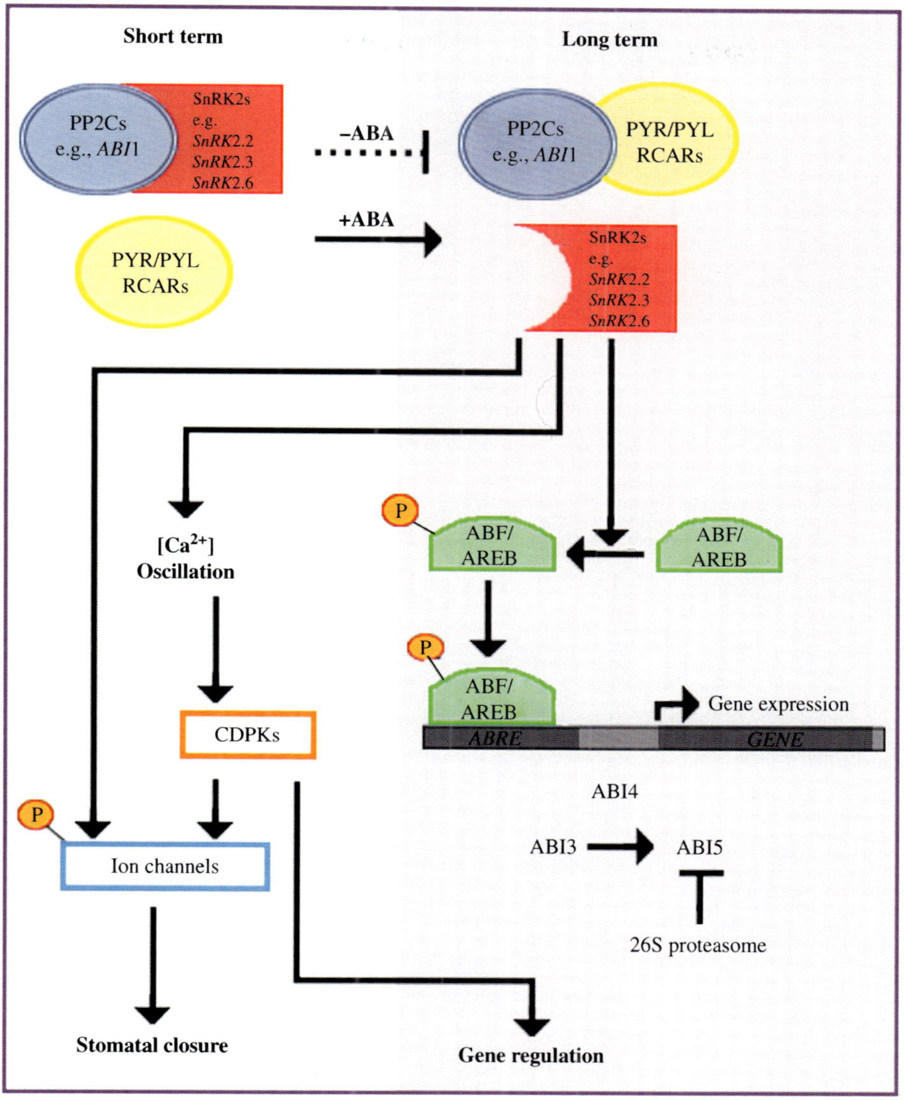

FIGURE 4.3.1 ABA signaling pathway from ABA perception to downstream responses. Short term stomatal closure responses and long term gene expression responses are presented here as ABA responses. PP2Cs negatively regulate SnRK2s in normal condition. Upon ABA perception under abiotic stress condition, such as drought, ABA activates the signaling pathway by promoting PP2Cs and PYR/PYL/RCARs interaction. As a result, unbound SnRK2s stimulate phosphorylation of ABR/AREBs and downstream genes. At the same time, activated SnRK2s regulate stomatal closure through targeting Ca^{2+} oscillation and ion channels.

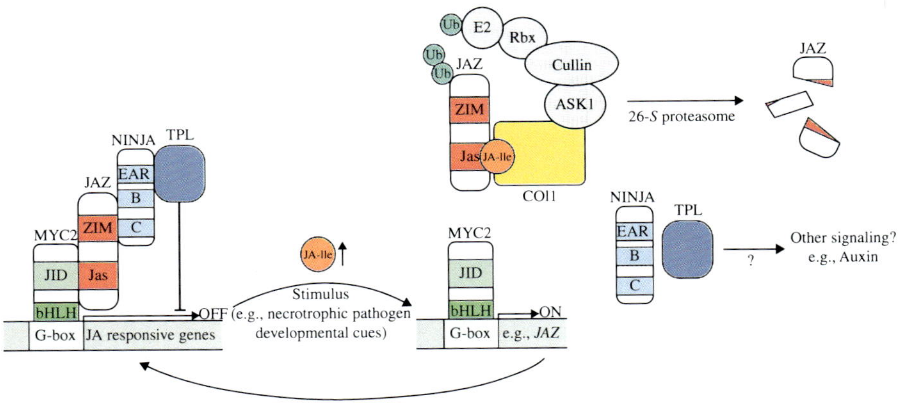

FIGURE 4.4.4 Model of jasmonate action in regulation of gene expression. For full caption, please refer to text.

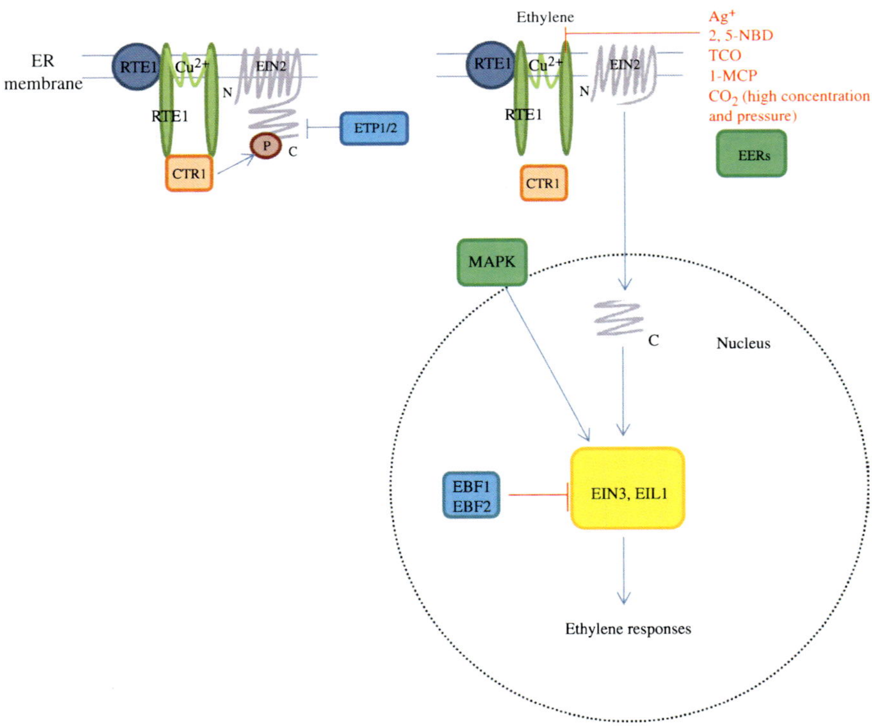

FIGURE 4.5.2 Central points of ethylene signaling including antagonists. For full caption, please refer to text.

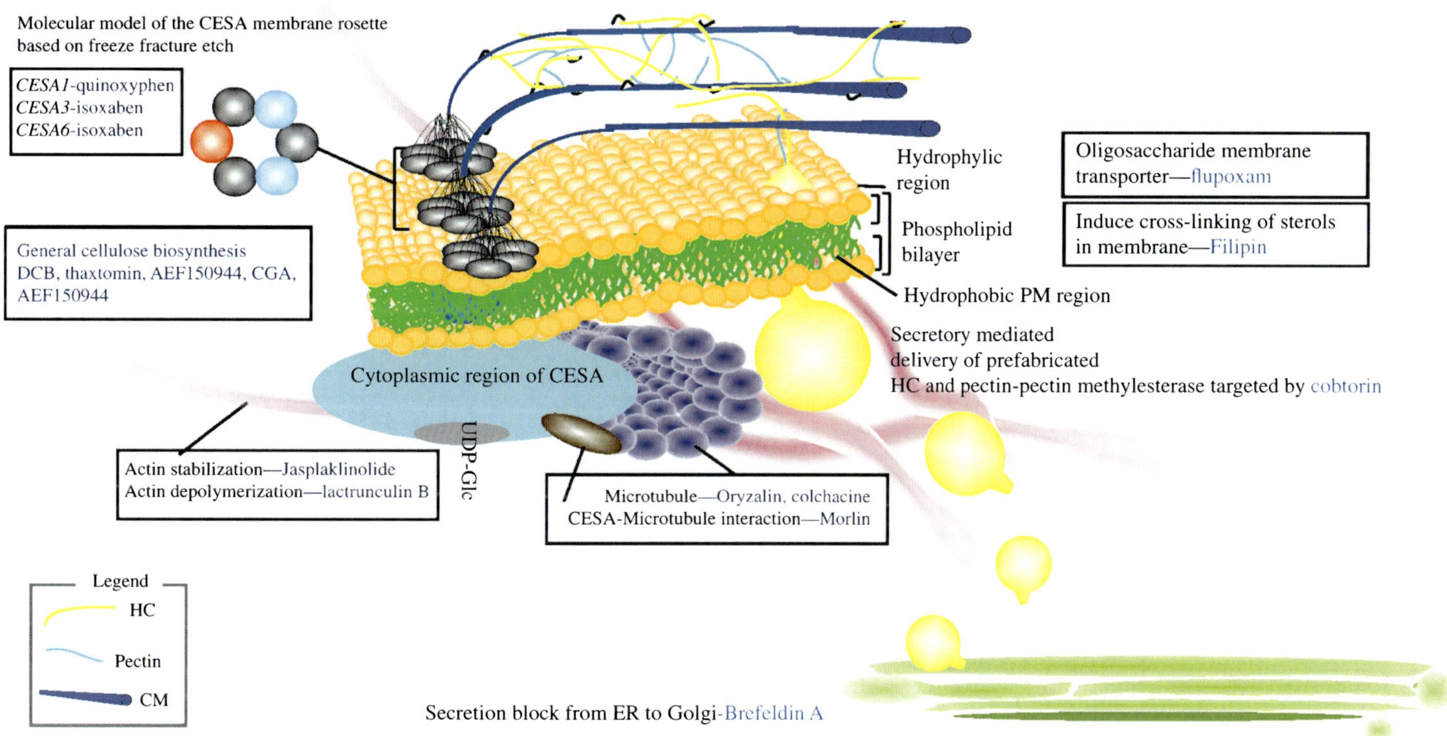

FIGURE 5.1.1 Chemical toolbox for dissecting cellulose biosynthesis. Numerous cellulose synthesis inhibitors are classified as being mapped to a target gene for forward chemical genetics such as isoxaben (*CESA3* and *CESA6*) or quinoxyphen (*CESA1*). Furthermore, DCB, CGA, thaxtomin, and AE F150944 have not been mapped to a target gene, although DCB is proposed to interact with a MAP [47]. Morlin putatively targets the interaction between MTs and cellulose synthase. Oryzalin and colchicine are MT-targeting compounds and have been used to illustrate the molecular rail hypothesis for guidance of cellulose synthase [55]. Actin depolymerization by latrunculin B has been used to show the requirement for actin-mediated trafficking in placing cellulose synthase uniformly at the plasma membrane [57]. Stabilization of actin by jasplakinolide reveals functional association between MTs and actin, which has yet to be examined in context of cellulose biosynthesis [59]. Cobtorin, a cellulose inhibitor [61] was examined genetically using FOX lines, and resistance was gained via mutations in a lectin family protein, a pectin methylesterase (AtPME1), and a putative polygalacturonase [62]. Furthermore, examining the plasma membrane cell wall continuum in relation to cellulose synthase may utilize tools such as filipin or the new drug flupoxam, which appears to target an oligosaccharide membrane transporter [24].

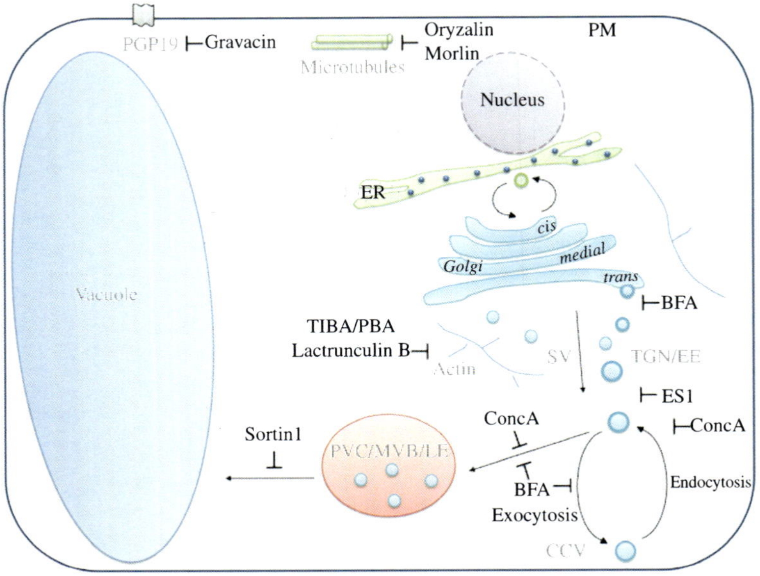

FIGURE 5.2.1 The plant endomembrane system. The trafficking machinery consists of several major compartments: the endoplasmic reticulum (ER), the Golgi complex, the *trans*-Golgi network/early endosome (TGN/EE), the pre-vacuolar compartment (PVC), the vacuole, and the plasma membrane (PM). Proteins are transported from the ER via Golgi/TGN to the PM or vacuole. Newly synthesized proteins are transported from the ER to the Golgi where they pass sequentially through the *cis*-, *medial*-, and *trans*-cisterna as they mature and finally onto the TGN. In the TGN, proteins are sorted into pathways leading either to the vacuole via the PVC (also known as a late endosome (LE) or the multivesicular body (MVB)) or to the PM. Proteins targeted to the PM are alternatively sorted in secretory vesicles (SV), uncoated vesicles or clathrin-coated vesicles (CCV). Proteins destined for the vacuole are sorted into CCV and transported to the PVC. Also illustrated are the points of action for several chemical agents used in subcellular trafficking research. Brefeldin A (BFA) inhibits exocytosis by preventing vesicle formation, resulting in the formation of large compartments. Concanamycin A (ConcA) inhibits vesicular trafficking between Golgi and TGN. Sortin1 affects the biogenesis of vacuoles and vacuolar transport. Gravacin interacts with PGP19, a membrane-localized auxin transporter. Oryzalin causes microtubule depolymerization, affecting vesicle trafficking. Similarly, latrunculin B and the auxin transport inhibitors *N*-1-naphthylphthalamic acid (NPA), PBA, and TIBA disrupt polymerization of actin and vesicle trafficking. Endosidin 1 (ES1) disrupts endocytosis via interaction with an unknown target.

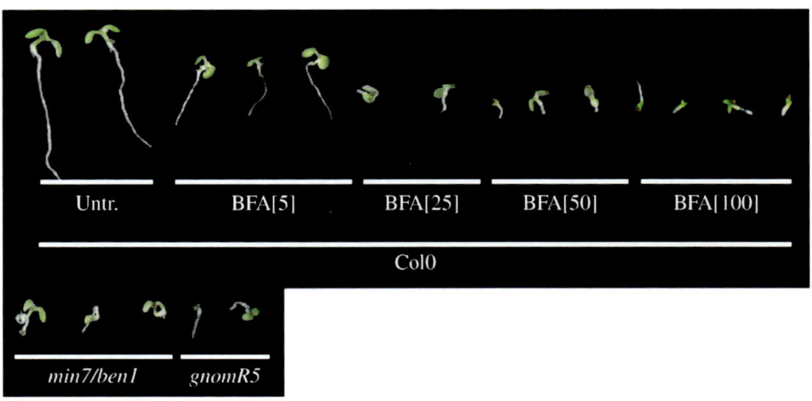

FIGURE 5.2.3 Brefeldin A activity is tunable. (**a**) Five-day-old Col-0 seedlings grown in presence of 5, 25, 50, or 100 μM BFA. (**b**) Phenotype of 5-day-old mutants (*gnomR5*, *min7/ben1*) impaired in the function of ARF-GEF protein with an SEC7 domain.

85. Shyu, C., et al. (2012) JAZ8 lacks a canonical degron and has an EAR motif that mediates transcriptional repression of jasmonate responses in *Arabidopsis*. *Plant Cell* **24**, 536–550.

86. Chini, A., Boter, M., and Solano, R. (2009) Plant oxylipins: COI1/JAZs/MYC2 as the core jasmonic acid-signalling module. *FEBS J.* **276**, 4682–4692.

87. Santner, A., and Estelle, M. (2009) Recent advances and emerging trends in plant hormone signalling. *Nature* **459**, 1071–1078.

88. Chung, H. S., et al. (2008) Regulation and function of Arabidopsis *JASMONATE ZIM*-domain genes in response to wounding and herbivory. *Plant Physiol.* **146**, 952–964.

89. Koo, A. J. K., et al. (2009) A rapid wound signal activates the systemic synthesis of bioactive jasmonates in Arabidopsis. *Plant J.* **59**, 974–986.

90. Koo, A. J. K., and Howe, G. A. (2009) The wound hormone jasmonate. *Phytochemistry* **70**, 1571–1580.

91. Causier, B., et al. (2012) The TOPLESS interactome: A framework for gene repression in *Arabidopsis*. *Plant Physiol.* **158**, 423–438.

92. Sheard, L. B., et al. (2010) Jasmonate perception by inositol-phosphate-potentiated COI1-JAZ co-receptor. *Nature* **468**, 400–405.

93. Tan, X., et al. (2007) Mechanism of auxin perception by the TIR1 ubiquitin ligase. *Nature* **446**, 640–645.

94. Böttcher, C., and Pollmann, S. (2009) Plant oxylipins: Plant responses to 12-oxo-phytodienoic acid are governed by its specific structural and functional properties. *FEBS J.* **276**, 4693–4704.

95. Weiler, E. W., et al. (1994) The *Pseudomonas* phytotoxin coronatine mimics octadecanoid signalling molecules of higher plants. *FEBS Lett.* **345**, 9–13.

96. Stelmach, B. A., et al. (1998) Quantitation of the octadecanoid 12-oxo-phytodienoic acid, a signalling compound in plant mechanotransduction. *Phytochemistry* **47**, 539–546.

97. Walter, A., et al. (2007) Structural requirements of jasmonates and synthetic analogues as inducers of Ca^{2+} signals in the nucleus and the cytosol of plant cells. *Angew. Chem Int. Ed.* **46**, 4783–4785.

98. Danon, A., et al. (2005) Concurrent activation of cell death-regulating signaling pathways by singlet oxygen in *Arabidopsis thaliana*. *Plant J.* **41**, 68–80.

99. Ribot, C., et al. (2008) Induction of the Arabidopsis *PHO1;H10* gene by 12-oxo-phytodienoic acid but not jasmonic acid via a CORONATINE INSENSITIVE1-dependent pathway. *Plant Physiol.* **147**, 696–706.

100. Devoto, A., et al. (2005) Expression profiling reveals *COI1* to be a key regulator of genes involved in wound- and methyl jasmonate-induced secondary metabolism, defence, and hormone interactions. *Plant Mol. Biol.* **58**, 497–513.

101. Reymond, P., et al. (2004) A conserved transcript pattern in response to a specialist and a generalist herbivore. *Plant Cell* **16**, 3132–3147.

102. Reymond, P., et al. (2000) Differential gene expression in response to mechanical wounding and insect feeding in Arabidopsis. *Plant Cell* **12**, 707–720.

103. Stotz, H. U., et al. (2011) Jasmonate-dependent and COI1-independent defense responses against *Sclerotinia sclerotiorum* in *Arabidopsis thaliana*: Auxin is part of COI1-independent defense signaling. *Plant Cell Physiol.* **52**, 1941–1956.

104. Bruce, T. J. A., et al. (2008) *cis*-Jasmone induces *Arabidopsis* genes that affect the chemical ecology of multitrophic interactions with aphids and their parasitoids. *Proc. Natl. Acad. Sci. U.S.A.* **105**, 4553–4558.

105. Nakamura, Y., et al. (2008) Synthesis and bioactivity of potassium β-D-glucopyranosyl 12-hydroxy jasmonate and related compounds. *Biosci. Biotechnol. Biochem.* **72**, 2867–2876.

106. Ueda, M., and Nakamura, Y. (2007) Chemical basis of plant leaf movement. *Plant Cell Physiol.* **48**, 900–907.

107. Ueda, M., et al. (2000) A leaf-closing substance of *Albizzia julibrissin* Durazz. *Tetrahedron* **56**, 8101–8105.

108. Nakamura, Y., et al. (2011) 12-Hydroxyjasmonic acid glucoside is a COI1-JAZs independent activator of leaf closing movement in *Samanea saman*. *Plant Physiol.* **115**, 1226–1236.

109. Nakamura, Y., et al. (2011) "Click-made" biaryl-linker improving efficiency in protein labelling for the membrane target protein of a bioactive compound. *Org. Biomol. Chem.* **9**, 83–85.

110. Nakamura, Y., Miyatake, R., and Ueda, M. (2008) Enantiodifferential approach for the detection of the target membrane protein of the jasmonate glycoside that controls the leaf movement of *Albizzia saman*. *Angew. Chem. Int. Ed.* **47**, 7289–7292.

111. Glazebrook, J. (2005) Contrasting mechanisms of defense against biotrophic and necrotrophic pathogens. *Annu. Rev. Phytopathol.* **43**, 205–227.

112. Lorenzo, O., and Solano, R. (2005) Molecular players regulating the jasmonate signalling network. *Curr. Opin. Plant Biol.* **8**, 532–540.

113. Pieterse, C. M. J., et al. (2009) Networking by small-molecule hormones in plant immunity. *Nat. Chem. Biol.* **5**, 308–316.

114. Brodersen, P., et al. (2006) Arabidopsis MAP kinase 4 regulates salicylic acid- and jasmonic acid/ethylene-dependent responses via EDS1 and PAD4. *Plant J.* **47**, 532–546.

115. Jensen, A. B., Raventos, D., and Mundy, J. (2002) Fusion genetic analysis of jasmonate-signalling mutants in *Arabidopsis*. *Plant J.* **29**, 595–606.

116. Ortiz-Masia, D., et al. (2007) Diverse stress signals activate the C1 subgroup MAP kinases of *Arabidopsis*. *FEBS Lett.* **581**, 1834–1840.

117. Takahashi, F., et al. (2007) The mitogen-activated protein kinase cascade MKK3-MPK6 is an important part of the jasmonate signal transduction pathway in Arabidopsis. *Plant Cell* **19**, 805–818.

118. Blackwell, H. E., and Zhao, Y. (2003) Chemical genetic approaches to plant biology. *Plant Physiol.* **133**, 448–455.

119. Hicks, G. R., and Raikhel, N. V. (2009) Opportunities and challenges in plant chemical biology. *Nat. Chem. Biol.* **5**, 268–272.

120. McCourt, P., and Desveaux, D. (2010) Plant chemical genetics. *New Phytol.* **185**, 15–26.

121. Walsh, T. A. (2007) The emerging field of chemical genetics: Potential applications for pesticide discovery. *Pest Manag. Sci.* **63**, 1165–1171.

122. Armstrong, J. I., et al. (2004) Identification of inhibitors of auxin transcriptional activation by means of chemical genetics in *Arabidopsis*. *Proc. Natl. Acad. Sci. U.S.A.* **101**, 14978–14983.

123. De Rybel, B., et al. (2009) Chemical inhibition of a subset of *Arabidopsis thaliana* GSK3-like kinases activates brassinosteroid signaling. *Chem. Biol.* **16**, 594–604.

124. Gendron, J. M., et al. (2008) Chemical genetic dissection of brassinosteroid-ethylene interaction. *Mol. Plant* **1**, 368–379.

125. Lin, L.-C., Hsu, J.-H., and Wang, L.-C. (2010) Identification of novel inhibitors of 1-aminocyclopropane-1-carboxylic acid synthase by chemical screening in *Arabidopsis thaliana*. *J. Biol. Chem.* **285**, 33445–33456.

126. Park, S.-Y., et al. (2009) Abscisic acid inhibits type 2C protein phosphatases via the PYR/PYL family of START proteins. *Science* **324**, 1068–1071.

127. Harbut, M. B., et al. (2011) Bestatin-based chemical biology strategy reveals distinct roles for malaria M1- and M17-family aminopeptidases. *Proc. Natl. Acad. Sci. U.S.A.* **108**, E526–E534.

128. Zheng, W., et al. (2006) Bestatin, an inhibitor of aminopeptidases, provides a chemical genetics approach to dissect jasmonate signaling in *Arabidopsis*. *Plant Physiol.* **141**, 1400–1413.

4.5

CHEMICAL GENETICS AS A TOOL TO STUDY ETHYLENE BIOLOGY IN PLANTS

YUMING HU, FILIP VANDENBUSSCHE, AND DOMINIQUE VAN DER STRAETEN

4.5.1 INTRODUCTION

Ethylene (C_2H_4) is the only gaseous plant hormone with profound effects throughout plant growth and development. It affects seed germination, vegetative development, leaf and flower senescence, fruit ripening, and stress and pathogen responses [1, 2]. Morphological changes in dark-grown seedlings in the presence of ethylene or its metabolic precursor 1-aminocyclopropane-1-carboxylic acid (ACC), the so-called triple response, has been used to identify mutants defective in ethylene biosynthesis or response. In *Arabidopsis thaliana*, the triple response phenotype consists of a short hypocotyl and root, a radially swollen hypocotyl, and an exaggerated apical hook [3]. Studies of ethylene mutants by conventional genetic and molecular approaches resulted in the cloning and characterization of genes in the ethylene pathway and have led to the elucidation of many aspects related to the regulation of ethylene synthesis and signal transduction. To further understand the role of ethylene in plant function, chemical genetics is emerging to support the discovery of novel bioactive molecules interfering with the ethylene response [4, 5]. The high-throughput screening of small molecules as agonists or antagonists of hormones represents a new approach at the interface of chemistry and biology [6]. In contrast to classical genetics, where mutations are introduced at the DNA or RNA level to disturb a biological process, chemical genetics is a ligand-based methodology to study gene or

Plant Chemical Biology, First Edition. Edited by Dominique Audenaert and Paul Overvoorde.

protein function, which uses low-molecular-mass organic compounds to modify or disrupt the function of specific proteins [7–9]. One of the major advantages over traditional genetic approaches is the possibility to overcome loss-of-function lethality (since the ligands only have an effect upon chemical treatment) and gene redundancy (since ligands may act as general antagonists inhibiting all isoforms of a protein, or as specific agonists activating a particular isoform). A second important advantage is that the chemical genetics approach allows reversible and conditional control of a phenotype by addition and subsequent removal of a compound, which enables a kinetic analysis of the *in vivo* consequences of protein function changes in a dose-dependent manner. Last but not least, well-characterized bioactive chemicals and their targets identified in Arabidopsis can be used in non-model species to improve agronomic traits and increase crop value [10]. Hence, chemical genetics allows the dissection of biological mechanisms and gene networks, in a way complementary to a classical, mutation-based approach.

Here, we describe approaches to screen libraries for bioactive chemicals that directly or indirectly affect ethylene-related processes in Arabidopsis. Considerations related to the characterization of these compounds and identification of their targets will be discussed. We also provide an overview of agonists and antagonists of ethylene biosynthesis, perception, and signaling, which have been identified previously, and are used to control ethylene effects in plants. Finally, we highlight the current issues and future perspectives of a chemical genetics approach to further elucidate the ethylene pathway and its cross talk with other hormones.

4.5.2 SMALL MOLECULES IN ETHYLENE BIOSYNTHESIS AND SIGNALING

For a recent review of the ethylene pathway, we refer to References [2, 11].

Ethylene biosynthesis in plant tissues is initiated from methionine, which is converted to *S*-adenosylmethionine (AdoMet) by AdoMet synthetase (Fig. 4.5.1). Subsequently, the methionyl side chain of AdoMet undergoes cyclization to form ACC by ACC synthase (ACS), which is the major rate-limiting step in the pathway. The by-product 5′-methylthioadenosine is recycled to methionine in the Yang cycle, feeding another round of ethylene synthesis [12]. As the direct precursor of ethylene, ACC is further oxidized to ethylene by ACC oxidase (ACO), a reaction that does not proceed under anaerobic conditions, resulting in ACC accumulation. Recently, it was hypothesized that ACC might act as a signal independently of ethylene receptors or the canonical pathway downstream thereof [13, 14], a hypothesis that needs further investigation. ACS and ACO are encoded by multigene families that are differentially regulated [1]. ACS belongs to pyridoxal 5′-phosphate (PLP)-dependent aminotransferases [15], encoding eight functional genes (*ACS2*, *ACS4–9*, *ACS11*) and one nonfunctional gene (*ACS1*) in Arabidopsis [16, 17] (*ACS1/2* in Reference [16] are called *ACS2/4* in Reference [17]). Ethylene overproducer mutants, *eto1*, *eto2*, *eto3*, defective in the regulation of ethylene biosynthesis, emanate up to 50-fold more ethylene than the wild type by affecting ACS stability [18–20]. ETO2 and ETO3

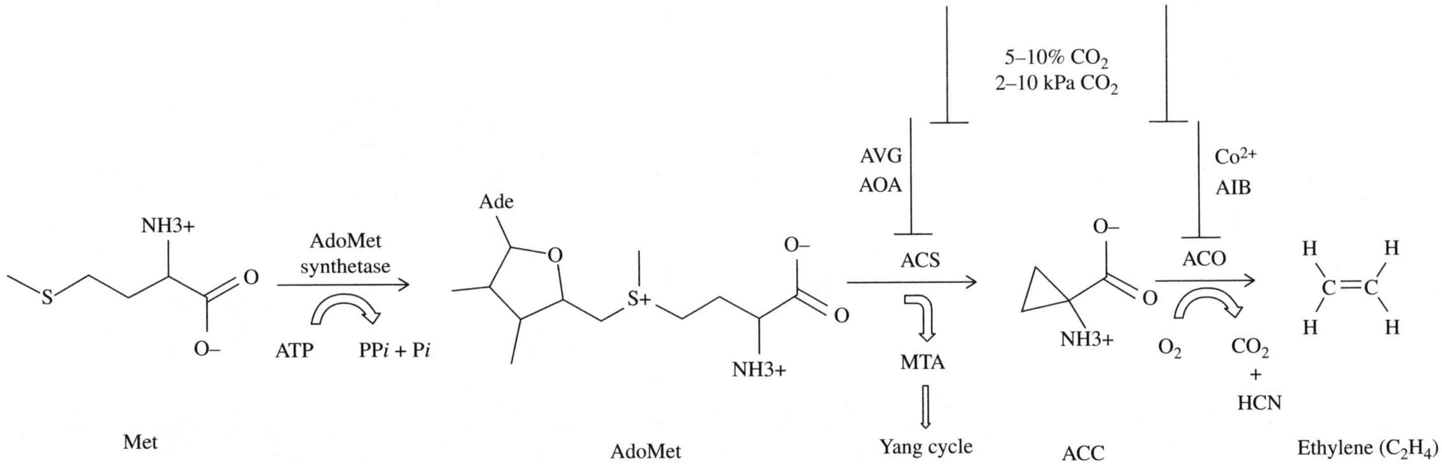

FIGURE 4.5.1 Central points of ethylene biosynthesis including antagonists. The enzymes catalyzing steps are shown with the arrows. Met: methionine; AdoMet: *S*-adenyl-methionine; MTA: methylthioadenine; ACC: 1-amniocyclopropane-1-carboxylic acid. Antagonists: Aminoethoxyvinylglycine (AVG); aminooxyacetic acid (AOA); Co^{2+}, α-aminoisobutyric acid (AIB).

encode ACS5 and ACS9, respectively. Being part of an E3 ligase, ETO1 and ETO1-like proteins (EOL) regulate protein stability of ETO2 and ETO3 by the ubiquitin proteasome pathway [21, 22]. Use of these mutants in chemical genetics studies, either as a control or to aim at phenotypic reversion, can help to relate the compound function to ethylene biosynthesis. Amongst the four genes that bear homology to *ACO*, *ACO1*, *ACO2*, and *ACO4* (ethylene-forming enzyme) are ethylene-related [23–25].

Several small chemicals have been characterized as inhibitors of ethylene biosynthesis and have been tested for agricultural or horticultural applications. Aminoethoxyvinylglycine (AVG) is a widely used irreversible inhibitor of ACS that decreases ethylene production by forming a conjugated enol ether with its cofactor, pyridoxal phosphate (PLP), blocking the active site of the enzyme [26]. It is commercially named Retain (containing 15% w/w AVG, Valent Biosciences Crop, USA), and is applied preharvest to slow down maturation and ripening of fruits and vegetables. The effectiveness is dose-dependent and differs between species and cultivars [27, 28] as is the case for application methods [29]. On lab scale, the constitutive triple-response phenotype of the Arabidopsis *eto* mutants can be suppressed by 0.5–10 μM AVG [4]. However, AVG is not specific to ACS, and likely inhibits most PLP-dependent enzymes. For example, AVG has recently been reported to be an inhibitor of auxin biosynthesis by blocking Tryptophan (Trp) aminotransferase activity [30]. This makes the need for a specific ethylene biosynthesis blocker more eminent. Another type of ACS inhibitors are hydroxylamine analogs that react with PLP to form stable oximes. Aminooxyacetic acid (AOA) belongs to this group of inhibitors, which are applied in a concentration of 100–500 μM [13, 31]. Cobalt ion (Co^{2+}) [32] and α-aminoisobutyric acid (AIB) [33] are also inhibitors of the ethylene biosynthesis; however, these compounds disrupt ACO activity. AIB is the only known ACC analog that significantly and competitively inhibits ACO. This inhibition is by far not as strong as the ACS inhibitors mentioned earlier. Carbon dioxide (CO_2) at higher pressure (2–10 kPa) and concentration (5–10%) inhibits ethylene effects in climacteric fruit [34, 35]. It was suggested that inhibition by CO_2 is due to competition with ethylene at the ethylene receptor site [36, 37]. However, CO_2 suppresses the expression of ethylene-independent and ethylene-dependent ripening genes [38], while it does not inhibit wound-induced ethylene or autocatalytic ethylene via antagonizing ethylene perception [39–41], but rather before the conversion of ACC to ethylene.

In contrast to the use of antagonists of ethylene biosynthesis, ethylene itself can also be applied to accelerate (postharvest) ripening, for instance, in tomato and banana, or to prevent lodging in grain crops [42]. However, as a gas it is difficult to apply in the field. Ethephon (2-chloroethylphosphonic acid) was discovered as an ethylene-releasing compound that can be absorbed and transported within the plant and has been marketed by Bayer CropScience as Ethrel [43]. In some cases, the effect of ethephon has been shown to be independent of ethylene generation [44].

Current horticultural applications of inhibitors of ethylene action, such as inhibition of wilting or ripening, largely rely on the interference with ethylene perception at the receptor level [45, 46] (Fig. 4.5.2). In Arabidopsis, ethylene is perceived by a family of transmembrane receptors Ethylene-Resistant 1 and 2 (ETR1, ETR2),

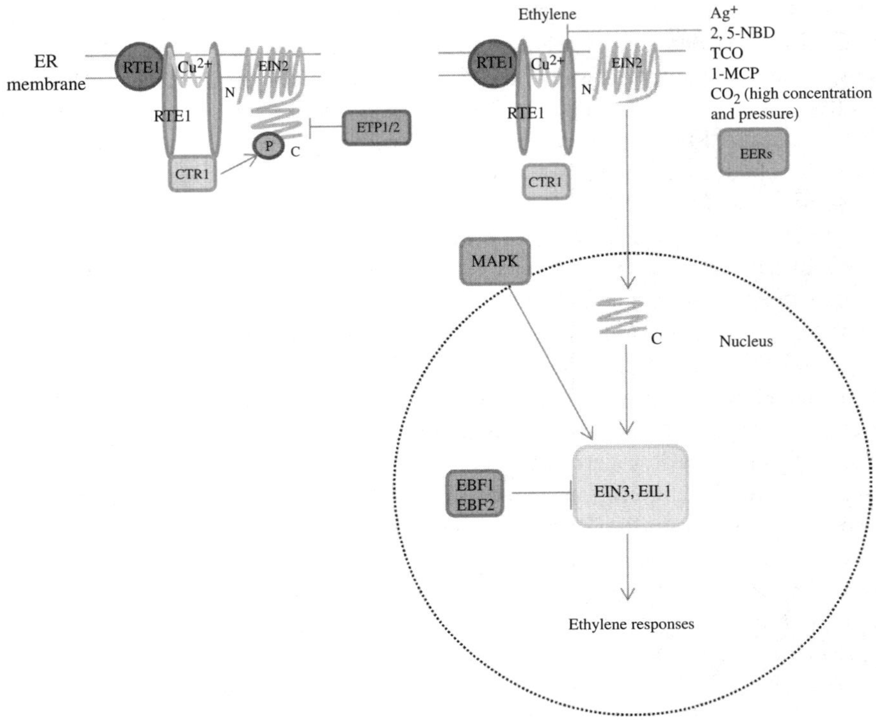

FIGURE 4.5.2 Central points of ethylene signaling including antagonists. In the absence of ethylene, CONSTITUTIVE TRIPLE RESPONSE 1 (CTR1) is tightly associated with the receptor complex, and phosphorylates the C-terminal domain of ETHYLENE-INSENSITIVE 2 (EIN2), preventing signal transduction to the nucleus. The abundance of EIN2 is regulated by EIN2 targeting proteins 1 and 2 (ETP1/2). Ethylene-Resistant 1 and 2 (ETR1, ETR2), Ethylene Response Sensor 1 and 2 (ERS1, ERS2), and Ethylene-Insensitive 4 (EIN4) are the receptors residing at the endoplasmic reticulum membrane, which are regulated by Reversion to Ethylene Sensitivity (RTE1). In the presence of ethylene, the receptors and CTR1 are inactivated, leading to EIN2 dephosphorylation, which results in C-terminal cleavage and relocalization of the C-end to the nucleus. Subsequently, the C-terminal end of EIN2 transfers the signal to the EIN3 and EIN3-LIKE1 (EIL1) transcription factors. EIN3 and EIL1 protein stability is regulated by EIN3 BINDING F-BOX PROTEINS (EBF1/2). The aforementioned route is the linear ethylene-signaling pathway that is highly conserved in different plant species. Furthermore, MAPK (mitogen-activated protein kinase) cascades are involved in autocatalytic ethylene production under stress by enhancing ACS stability. Finally, some loci identified as *enhanced ethylene response (eer)* mutants showing an enhanced *ctr1* phenotype are assumed to play a negative role in ethylene signaling. Antagonists: Ag+; 2,5-norbornadiene (2,5-NBD); trans-cyclooctene (TCO); 1-methylcyclopropene (1-MCP); high concentration of CO_2 at high pressure. For color detail, please see color plate section.

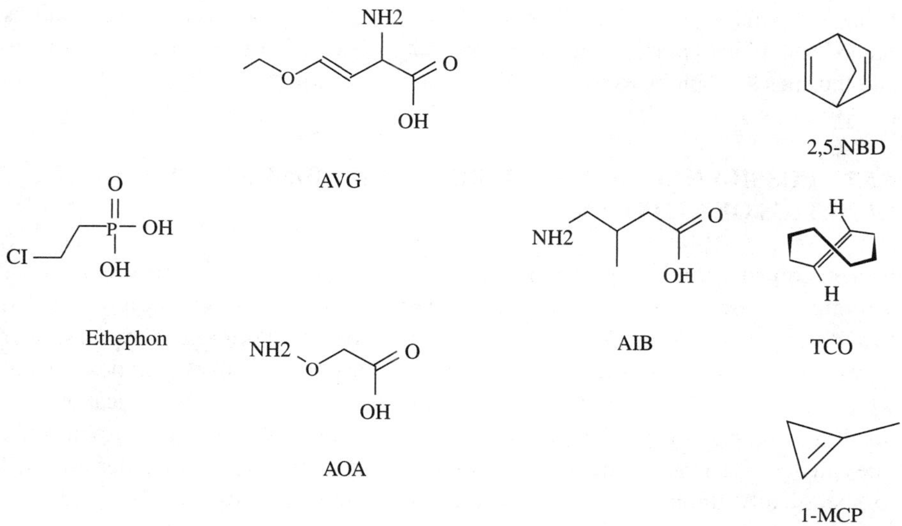

FIGURE 4.5.3 Structures of ethylene agonist and antagonists. Agonist: Ethephon (2-chloroethylphosphonic acid). Antagonists: ACC synthase inhibitors: Aminoethoxyvinylglycine (AVG) and Aminooxyacetic acid (AOA); ACC oxidase inhibitor: α-aminoisobutyric acid (AIB); ethylene perception inhibitors: 2,5-norbornadiene (2,5-NBD); trans-cyclooctene (TCO); 1-methylcyclopropene (1-MCP).

Ethylene Response Sensor 1 and 2 (ERS1, ERS2), and Ethylene-Insensitive 4 (EIN4), residing at the ER membrane. The receptors are regulated by Reversion to Ethylene Sensitivity (RTE1) [47]. The ethylene receptors require copper ions as a cofactor to bind ethylene [48]. Some amino acid residues (D25, Y32, I35, and P36 in Helix I and I62, C65, and H69 in Helix II) at the ligand-binding domain in the N-terminus of the receptors are important for ethylene binding [49]. Many ethylene antagonists interfere with the binding of ethylene to its copper-containing receptors. Even before the receptors were identified, silver ions, applied as silver nitrate ($AgNO_3$) or as silver thiosulfate ($Ag(S_2O_3)_2^{3-}$ (STS)) have been known as ethylene inhibitors [50]. Beyer proposed that silver ions replaced another metal, at that time suggested to be copper or zinc, in the receptors, thus blocking ethylene perception by occupying the copper-binding site of the receptors and interacting with ethylene, obviously inhibiting response [48, 51]. Gold ions can substitute silver ions for ethylene binding, but do not block its action in plants, albeit that it can affect seedling growth independently of ethylene signaling [52]. Recently, silver ions have been demonstrated to promote IAA efflux independently of ethylene perception, in addition to block ethylene signaling, so that using silver ions to block ethylene signaling needs caution [53]. Other antagonists of ethylene receptors are strained alkenes that have greater affinity to metal ion π-complexation than ethylene, including 2,5-norbornadiene (2,5-NBD), *trans*-cyclooctene (TCO), and 1-methylcyclopropene (1-MCP) [54, 55] (Fig. 4.5.3). Some of these compounds have been used for basic research. The most ideal one is 1-MCP because of its effectiveness, stability, and its lack of odor and toxicity [56].

It has been marketed as EthylBloc to increase the shelf life of cut flowers and as SmartFresh to help preserve the freshness and quality of fruits; however, it is volatile and thus has limitations for some agricultural applications.

4.5.3 POSSIBLE SCREENS AND RECENT FINDINGS IN CHEMICAL GENETICS OF ETHYLENE

In a chemical genetics approach, the first essential step is the development of a robust bioassay that reports on the process of interest and can be downscaled to a 96-well plate format. A screening method based on reporter gene expression or a phenotype-based assay can be defined as a forward chemical genetics screen, where thousands of compounds are tested for their ability to alter a biological process resulting in a phenotype change and an altered reporter expression. Alternatively, one can design a reverse chemical genetics screen, where small molecules are used to specifically inhibit or activate known selected targets in order to study the functional consequences [6].

Lin et al. [4] used a phenotype-based strategy based on suppression of the triple response phenotype in etiolated seedlings of the ethylene overproducer mutant *eto1–4*, on 10,000 structurally diverse compounds in the DIVERSet library (ChemBridge Inc.). This library represents a maximal chemical space with a minimal number of compounds chosen by a range of filtering methods. Moreover, the compounds are stable and nontoxic. The latter is important if compounds are to be used for agricultural purposes in the future. The primary screening was performed in 96-well micro-titer plates at a concentration of 50 μM of chemicals (dissolved in DMSO) in MS/2 agar medium to score the long hypocotyl phenotype. A picture database of the phenotypes of seedlings exposed to these 10,000 small molecules might be useful for future purposes. In fact, once publically available, such database could help researchers worldwide to determine characteristics and specificities of compounds of their interest.

For the secondary and tertiary screen, the number of selected compounds can be narrowed down from around 100 to about 10, by selecting highly active and specific compounds. Lin et al. [4] finally selected two compounds, the effectiveness of which is comparable with silver nitrate, therefore being potential suppressors of ethylene response. To determine whether they inhibit ethylene biosynthesis or a step upstream of ethylene signaling, they analyzed the effect of hit compounds in the constitutive ethylene mutant *ctr1* and mutants overexpressing EIN3 (ETHYLENE INSENSITIVE 3) under control of the cauliflower mosaic virus 35S promoter [57]. However, any seedling with constitutive or enhanced ethylene response could serve the purpose, including 35S::EIN2 and mutants *ebf1–1 ebf2–1* (*EIN3 BINDING F-BOX PROTEINS*) [58], *eer(s)* (*enhanced ethylene response*), and the recently identified *sar1–7* (*suppressor of auxin resistance 1*) [59]. Conversely, when looking for compounds that induce or enhance ethylene responses, ethylene-resistant *etr1–1* [60], ethylene-insensitive *ein2–1* [61], and *ein3* or *ein3–1 eil1–1*(*EIL: EIN3-LIKE*) [62] can be used. These compounds can be further tested for phenotypic reversal of the action of ethylene synthesis inhibitors or antagonists. In addition, Lin et al. used a reporter line of *eto1–4* harboring 5xEBS::LUC to confirm the suppression of the ethylene response. Both LUC (luciferase gene) and GUS (beta-glucuronidase) reporter systems have been fused to the EIN3-binding-sequence, and report EIN3 transcription factor activity, which acts as an indicator for the presence of an ethylene signal [63]. The LUC activity or GUS-expression in the apical hook, hypocotyl, and root can be scored in comparison with the ACC/non-ACC-treated or ethylene-inhibitor-treated controls. Using an EBS::reporter line in the primary screen can be a rewarding approach for rapidly selecting ethylene-specific effects brought about by the compounds.

As an alternative to the use of ethylene overproducer mutants, the triple response phenotype can be induced by adding exogenous ACC to wild-type plants, or applying ethylene in an open flow gassing system to prove that the observed phenotypes are truly ethylene-related.

Comparison of structural analogs is a valuable tool to discover novel active compounds. Lin et al. [4] discovered a compound more effective than that identified in their initial screen. Structural comparison of compounds with similar biological effects or with known function can give further hints on their function [64–67]. Ultimately, the selected compounds should be stable and effective at low concentrations to reduce off-target effects [68, 69]. It is noteworthy that decreased signaling causes positive feedback on the biosynthesis, as in *etr1* or *ein2*, where the level of ethylene production is high, while in *ctr1–1* [70] it is low compared to dark-grown, wild-type plants [2]. This overproduction can be referred to an increased ACS activity (octuple ACS mutant have ethylene levels 10 times lower than the wild type) [14]. Thus, ethylene production can be measured to dissect the pathway and the function of the compound in a more detailed manner.

The aforementioned screen [4] identified a group of quinazolinones (Fig. 4.5.4**a**) that function as novel ACS inhibitors. These novel inhibitors are noncompetitive inhibitors of ACS, albeit structurally and mechanistically unrelated to AVG according to *in vitro* activity assay and enzyme kinetics [71]. However, a microarray analysis revealed that more than 40% of the genes in Arabidopsis are commonly regulated by the hit compounds and AVG, including possible factors that are required to establish the triple response during etiolated growth in elevated ethylene. Furthermore, it provides an alternative to investigate the role of ACC in ethylene-independent processes [72]. It is worthwhile to point out that these compounds are different from naturally occurring quinazolinones [73].

4.5.4 CHEMICAL GENETICS IN ETHYLENE—HORMONE INTERACTION STUDIES

Ethylene is known to functionally interact with various hormones [74]. Reporter lines specific for these other hormonal pathways can assess the effect of chemicals identified in an ethylene-related screen, revealing hormonal cross talk. The following reporters

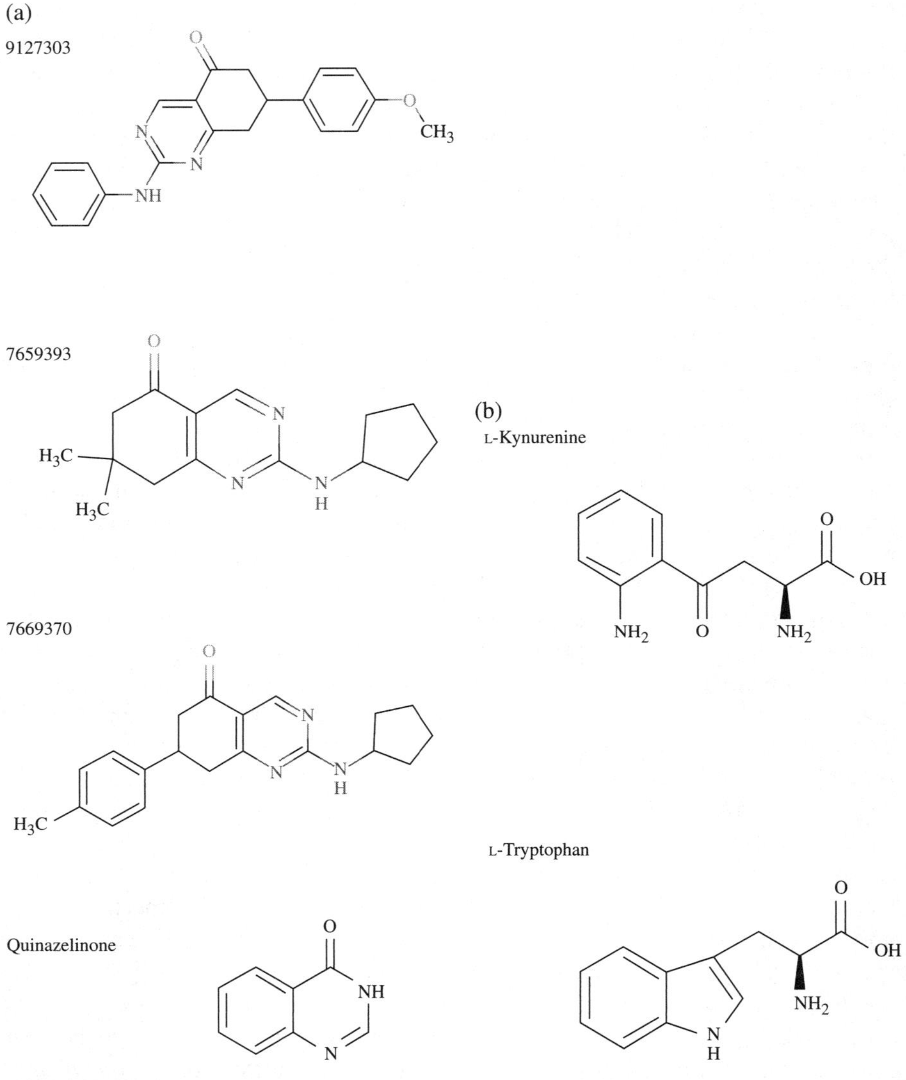

FIGURE 4.5.4 Recent chemical genetics findings in ethylene research. (**a**) Group of quinazolinones that function as novel ACC synthase inhibitors: 9127303 (Chembridge ID, 2-anilino-7-(4-methoxyphenyl)-7,8-dihydro-5(6H)-quinazolinone); 7659393(2-(cyclopentylamino)-7,7-dimethyl-7,8-dihydro-5(6H)-quinazolinone); 7669370 (2-(cyclopentylamino)-7-(4-methylphenyl)-7,8-dihydro-5(6H)-quinazolinone). (**b**) L-Kynurenine is an analog of L-Tryptophon that functions as competitive inhibitor of TAA1/TARs-mediated auxin biosynthesis.

can be used: *DR5::GUS* (auxin responsive-element, AuxRE) for auxins [75]), *TCS::GUS* (two-component system) for cytokinins [76], *CPD::GUS* (cytochrome P450) for brassinolide [77], *RD29A::GUS* (responsive to dessication 29A) for abscisic acid [78], and *Thi2.1::GUS* (thionein 2.1) for jasmonic acid [79].

Interactions between ethylene and auxin are complex. Both agonist and antagonist effects have been reported. As agonists, they synergistically affect root elongation and root hair formation [80, 81]. As antagonists, they affect abscission of flower and fruits [82]. In addition, *WEI2*, *WEI7*, and *WEI8* genes that mediate ethylene responses during plant development are also key elements in the regulation of auxin biosynthesis [83, 84]; as well as *WEI8*, which encodes Trp aminotransferase of Arabidopsis 1 (TAA1), the key enzyme catalyzing the indole-3-pyruvic acid pathway for auxin biosynthesis (IPyA pathway). Its close relative TAR2 and the flavin monooxygenase YUC1 reporter lines can also link to ethylene through newly found L-Kynurenine [5, 85]. Defects in auxin biosynthesis or transport also result in ethylene insensitivity. For instance, mutation of the auxin transporters auxin-resistant AUX1 [86] and ethylene-insensitive root EIR1/PIN2 [87, 88] both confer insensitivity to ethylene. Linking the aforementioned elements, a first step to monitor whether ethylene inhibits root growth by modulating auxin biosynthesis and transport, auxin reporter *DR5::GUS* in wild type and ethylene mutant background can be used. The next step will be to determine which pathway of the auxin action is disturbed, biosynthesis or transport. Several reporter lines such as WEI2, WEI7, or TAR2 promoters driving GUS or GFP can be used to check whether the chemicals affect auxin biosynthesis; while reporter lines for auxin influx and efflux transporters AUX1 and PIN2 can be used to assess the effect on auxin transport. Moreover, the phenotypes of mutants treated with chemicals can be examined, and exogenous auxins can be applied in addition to the chemicals to further demonstrate the relations between auxin and ethylene.

The work by He et al. [5] is a good example of how the search for ethylene-related compounds can lead to an auxin–ethylene link; in this case, that auxin positively regulates the ethylene-signaling pathway to accelerate its own biosynthesis. The screen was initially done based on reversions of the ethylene-related short-root phenotype of *eto1–2* and *ctr1–1* etiolated seedlings by compounds from a combinatorial SP2000 chemical library (www.msdiscovery.com). L-Kynurenine (Kyn) (Fig. 4.5.4**b**) was identified as a new auxin biosynthetic inhibitor based on the suppression of the expression of a series of auxin biosynthesis reporters upon ACC treatment. Kyn selectively targets TAA1-like Trp aminotransferases, providing a new tool to explore TAA1/TAR-mediated auxin biosynthesis [85]. These findings illustrate how strongly interwoven ethylene and auxin responses are. Secondary screens for auxin characteristics are therefore highly advisable when primary screens are based on ethylene-related phenotypes. Kyn is one of the few small molecules inhibitors with identified targets [89], the finding of which is based on the integration of the analysis of representative mutant lines and hormone marker gene, enzyme assays (if the potential target is an enzyme) and computational modeling. Structural analysis of the chemical in conjunction with the analysis of potential metabolites is also important in order to discover the direct target. Kyn is a natural product that has been reported in a non-plant species, and the metabolic pathway was investigated. Unlike Kyn, its derivatives such as quinolinic acid, NAD^+, nicotinamide, and nicotinic acid [90] did not promote root elongation of *ctr1–1*.

In certain cases, compounds may also have double (or multiple targets). For example, a BR inhibitor screen identified brassinopride (BRP), which besides inhibiting BR, also activates ethylene responses [91]. BRP activates expression of the BR-inhibited reporter gene *CPD::GUS*, inhibits hypocotyl length, and causes an exaggeration of the apical hook. The two latter features are characteristic of the triple response to ethylene in darkness, and were indicative for a double function of BRP. Further analysis using BR and ethylene mutants, treatment with ACC and an ethylene-perception inhibitor, support the view that BRP promotes ethylene action at the step of ethylene perception or upstream. The latter is consistent with the finding that ethylene controls the biosynthesis of BRs and establishes a gradient of BR in the apical hook to the hook formation [92].

4.5.5 TARGET IDENTIFICATION

Many techniques have been developed to facilitate chemical genetic screens, such as synthesis of various chemical libraries [93], automated preparation, and phenotyping tools [94], and the development of cell-based assays [95]. However, in order to understand the mode-of-action of the ligands, the target identification remains the biggest challenge. Target identification can be roughly grouped into genetics-based (genetic), transcriptional profiling-based (molecular), and affinity-based (biochemical) [96]. Here, we will discuss possible target identification for ethylene research in Arabidopsis.

Studies of structure–activity relationships (SAR) are a first step toward identification of possible target proteins. Structure-based clustering can be performed through the ChemMine interface using external or internal compounds or a combination of both [97]. One can run a cluster analysis for all compounds identified, and test structurally related and unrelated compounds including functional cores in different bioassays. In some cases, SAR studies resulted in the discovery of antagonists or in uncoupling different targets of a given compound, as for BRP derivatives [91]. Furthermore, there are several examples where the compounds are enzymatically converted to the actual bioactive molecules, for example, by the action of amidases, esterases, aldehyde oxidases, or glucosyltransferases [89]. This also may give a clue toward identification of the active principle. Thus, possible metabolization reactions of a given active compound should be investigated, using sensitive analytical methods such as liquid chromatography or gas chromatography combined with mass spectrometry (MS), in order to get a clue on the bioactive compounds.

To further identify the target protein, transcriptome analysis with either microarrays or RNA sequencing can be used. Genome-wide transcriptome analysis of a wild type treated with the ligand, can help to diagnose which processes are affected by this compound, and thus help in identification of the targets. For example, transcriptional changes induced upon treatment with bikinin overlapped almost 90% with those induced by BR treatment, placing the bikinin targets in the BR signaling cascade [69]. Likewise, in a study of compounds affecting ethylene physiology, panoramic transcriptome analysis upon compound treatment can be compared with the effect of either ethylene or AVG treatment, thus identifying whether the compound shows an agonist or antagonist action.

In parallel, EMS mutants resistant/hypersensitive to the compounds can be isolated, particularly if the aforementioned approaches have generated an interesting lead. As shown in the past, seedling responses to ethylene are extremely useful to identify mutants [3]. T-DNA insertion mutants will be analyzed if there is an unequivocal indication for a particular gene. Next-generation sequencing approaches will allow rapid and relatively low-cost identification of the mutations, and corresponding genes that are affected. The method mentioned earlier is genetics-based rather than biochemical-based. The genetics-based approach could lead to indirect target identification. A biochemical-based approach using affinity purification based on drug-affinity chromatography and high-resolution MS analysis can be applied to complement the genetic approach to discover the direct target [98]. Moreover, several strategies have been developed to optimize the biochemical-based approach, for instance, yeast-three-hybrid (Y3H), protein microarrays, and NMR-based metabolomics, allowing detection of low-abundant targets [99–101]. Another tool that has arisen in chemical genetics is "click chemistry," an organic synthesis method based on joining small units together with a covalent bond [102]. It offers promising possibilities to allow discovery of perfectly customized enzyme inhibitors. In this context, enzymes participate in the "discovery" of their own inhibitors: by acting as a template that brings click reagents together, the binding pocket of a given target enzyme catalyzes the in situ formation of its own, perfectly customized inhibitor. In plant research, click chemistry was used to identify targets of the cysteine protease inhibitor E-64 in Arabidopsis [103]. Moreover, X-ray crystallography is an important tool to determine the interaction between small molecules and their targets, as for instance, for the auxin receptor TIR1 [104].

4.5.6 FUTURE PERSPECTIVES

The examples demonstrate the feasibility and power of chemical genetics in ethylene research. Inhibitors of ethylene biosynthesis and signaling and ethylene-enhancing compounds are invaluable tools to further dissect the ethylene pathway or to distinguish between different hormones that have similar phenotypic effects. The control of ethylene production and action is also an important component in pre- and postharvest management of crops. For instance, discovery of specific inhibitors of ACS/ACO can reduce ethylene synthesis and thus help to control postharvest losses, perhaps complementary to genetic modification approaches in countries where GMOs are withheld from commercialization. Thus, the discovery of novel chemical compounds will be useful in fundamental ethylene research and can offer potentially useful agrochemicals for quality improvement in both horticulture and agriculture.

REFERENCES

1. Johnson, P. R. & Ecker, J. R. (1998) The ethylene gas signal transduction pathway: a molecular perspective, *Annual Review of Genetics.* **32**, 227–254.
2. Vandenbussche, F., et al. (2012) Ethylene in vegetative development: a tale with a riddle, *The New Phytologist.* **194**, 895–909.
3. Guzman, P. & Ecker, J. R. (1990) Exploiting the triple response of Arabidopsis to identify ethylene-related mutants, *Plant Cell.* **2**, 513–523.
4. Lin, L. C., Hsu, J. H. & Wang, L. C. (2010) Identification of novel inhibitors of 1-aminoc yclopropane-1-carboxylic acid synthase by chemical screening in *Arabidopsis thaliana*, *Journal of Biological Chemistry.* **285**, 33445–33456.
5. He, W., et al. (2011) A small-molecule screen identifies L-kynurenine as a competitive inhibitor of TAA1/TAR activity in ethylene-directed auxin biosynthesis and root growth in Arabidopsis, *The Plant Cell Online.* **23**, 3944–3960.
6. Blackwell, H. E. & Zhao, Y. (2003) Chemical genetic approaches to plant biology, *Plant Physiology.* **133**, 448–455.
7. Dobson, C. M. (2004) Chemical space and biology, *Nature.* **432**, 824–828.
8. Lipinski, C. & Hopkins, A. (2004) Navigating chemical space for biology and medicine, *Nature.* **432**, 855–861.
9. Stockwell, B. R. (2000) Chemical genetics: ligand-based discovery of gene function, *Nature Reviews Genetics.* **1**, 116–125.
10. Raikhel, N. & Pirrung, M. (2005) Adding precision tools to the plant biologists' toolbox with chemical genomics, *Plant Physiology.* **138**, 563–564.
11. Zhao, Q. & Guo, H. W. (2011) Paradigms and paradox in the ethylene signaling pathway and interaction network, *Molecular Plant.* **4**, 626–634.
12. Yang, S. F. & Hoffman, N. E. (1984) Ethylene biosynthesis and its regulation in higher-plants, *Annual Review of Plant Physiology.* **35**, 155–189.
13. Xu, S. L., et al. (2008) Two leucine-rich repeat receptor kinases mediate signaling, linking cell wall biosynthesis and ACC synthase in Arabidopsis, *The Plant Cell Online.* **20**, 3065–3079.
14. Tsuchisaka, A., et al. (2009) A combinatorial interplay among the 1-aminocyclopropane-1-carboxylate isoforms regulates ethylene biosynthesis in *Arabidopsis thaliana*, *Genetics.* **183**, 979–1003.
15. Mehta, P. K., Hale, T. I. & Christen, P. (1993) Aminotransferases – demonstration of homology and division into evolutionary subgroups, *European Journal of Biochemistry.* **214**, 549–561.
16. Rodrigues-Pousada, R. A., et al. (1993) The Arabidopsis 1-aminocyclopropane-1-carboxylate synthase gene-1 is expressed during early development, *Plant Cell.* **5**, 897–911.
17. Tsuchisaka, A. & Theologis, A. (2004) Unique and overlapping expression patterns among the arabidopsis 1-amino-cyclopropane-1-carboxylate synthase gene family members, *Plant Physiology.* **136**, 2982–3000.
18. Vogel, J. P., et al. (1998) Recessive and dominant mutations in the ethylene biosynthetic gene ACS5 of Arabidopsis confer cytokinin insensitivity and ethylene overproduction, respectively, *Proceedings of the National Academy of Sciences of the United States of America.* **95**, 4766–4771.

19. Woeste, K. E., Ye, C. & Kieber, J. J. (1999) Two Arabidopsis mutants that overproduce ethylene are affected in the posttranscriptional regulation of 1-aminocyclopropane-1-carboxylic acid synthase, *Plant Physiology.* **119**, 521–529.

20. Chae, H. S., Faure, F. & Kieber, J. J. (2003) The eto1, eto2, and eto3 mutations and cytokinin treatment increase ethylene biosynthesis in Arabidopsis by increasing the stability of ACS protein, *Plant Cell.* **15**, 545–559.

21. Wang, K. L. C., et al. (2004) Regulation of ethylene gas biosynthesis by the Arabidopsis ETO1 protein, *Nature.* **428**, 945–950.

22. Christians, M. J., et al. (2009) The BTB ubiquitin ligases ETO1, EOL1 and EOL2 act collectively to regulate ethylene biosynthesis in Arabidopsis by controlling type-2 ACC synthase levels, *The Plant Journal.* **57**, 332–345.

23. Gomezlim, M. A., et al. (1993) Isolation and characterization of a gene involved in ethylene biosynthesis from *Arabidopsis Thaliana*, *Gene.* **134**, 217–221.

24. Bovy, A. G., et al. (1999) Heterologous expression of the Arabidopsis etr1–1 allele inhibits the senescence of carnation flowers, *Molecular Breeding.* **5**, 301–308.

25. Raz, V. & Ecker, J. R. (1999) Regulation of differential growth in the apical hook of Arabidopsis, *Development.* **126**, 3661–3668.

26. Rando, R. R. (1974) Irreversible inhibition of aspartate-aminotransferase by 2-amino-3-butenoic-acid acid, *Biochemistry.* **13**, 3859–3863.

27. Baker, J. E., etal. (1982) Biosynthesis of ethylene from methionine in aminoethoxyvinylglycine-resistant avocado tissue, *Plant Physiology.* **69**, 93–97.

28. Byers, R. E. (1997) Peach and nectarine fruit softening following aminoethoxyvinylglycine sprays and dips, *Hortscience.* **32**, 86–88.

29. Saltveit, M. E. (2004) Effect of 1-methylcyclopropene on phenylpropanoid metabolism, the accumulation of phenolic compounds, and browning of whole and fresh-cut 'iceberg' lettuce, *Postharvest Biology and Technology.* **34**, 75–80.

30. Soeno, K., et al. (2010) Auxin biosynthesis inhibitors, identified by a genomics-based approach, provide insights into auxin biosynthesis, *Plant Cell Physiology.* **51**, 524–536.

31. Bradford, K. J., Hsiao, T. C. & Yang, S. F. (1982) Inhibition of ethylene synthesis in tomato plants subjected to anaerobic root stress, *Plant Physiology.* **70**, 1503–1507.

32. Lau, O. L. & Yang, S. F. (1976) Inhibition of ethylene production by cobaltous ion, *Plant Physiology.* **58**, 114–117.

33. Satoh, S. & Esashi, Y. (1980) Alpha-aminoisobutyric-acid – a probable competitive inhibitor of conversion of 1-aminocyclopropane-1-carboxylic acid to ethylene, *Plant Cell Physiology.* **21**, 939–949.

34. Bufler, G. & Streif, J. (1986) Ethylene biosynthesis of golden delicious apples stored in different mixtures of carbon-dioxide and oxygen, *Scientia Horticulturae.* **30**, 177–185.

35. Kerbel, E. L., Kader, A. A. & Romani, R. J. (1988) Effects of elevated CO_2 concentrations on glycolysis in intact Bartlett pear fruit, *Plant Physiology.* **86**, 1205–1209.

36. Burg, S. P. & Burg, E. A. (1967) Molecular requirements for biological activity of ethylene, *Plant Physiology.* **42**, 144–152.

37. Gorny, J. R. & Kader, A. A. (1996) Regulation of ethylene biosynthesis in climacteric apple fruit by elevated CO_2 and reduced O_2 atmospheres, *Postharvest Biology and Technology.* **9**, 311–323.

38. Rothan, C., et al. (1997) Suppression of ripening-associated gene expression in tomato fruits subjected to a high CO_2 concentration, *Plant Physiology.* **114**, 255–263.

39. Mathooko, F. M., et al. (2001) Regulation of genes encoding ethylene biosynthetic enzymes in peach (*Prunus persica* L.) fruit by carbon dioxide and 1-methylcyclopropene, *Postharvest Biology and Technology*. **21**, 265–281.
40. de Wild, H. P. J., Woltering, E. J. & Peppelenbos, H. W. (1999) Carbon dioxide and 1-MCP inhibit ethylene production and respiration of pear fruit by different mechanisms, *Journal of Experimental Botany.* **50**, 837–844.
41. de Wild, H. P. J., Otma, E. C. & Peppelenbos, H. W. (2003) Carbon dioxide action on ethylene biosynthesis of preclimacteric and climacteric pear fruit, *Journal of Experimental Botany.* **54**, 1537–1544.
42. Abeles, F. B., Morgan, P. W. & Saltveit, M. E. (1992) *Ethylene in plant biology*, 2nd edn, Academic Press, San Diego, CA.
43. Szyjewicz, E., Rosner, N. & Kliewer, W. M. (1984) Ethephon ((2-chloroethyl)phosphonic acid, ethrel,CEPA) in viticulture – a review, *American Journal of Enology and Viticulture.* **35**, 117–123.
44. Lawton, K. A., et al. (1994) Acquired-resistance signal-transduction in Arabidopsis is ethylene independent, *Plant Cell.* **6**, 581–588.
45. Sisler, E. C. & Serek, M. (1997) Inhibitors of ethylene responses in plants at the receptor level: recent developments, *Physiol Plantarum.* **100**, 577–582.
46. Sexton, R., et al. (1995) Effects of diazocyclopentadiene (Dacp) and silver thiosulfate (Sts) on ethylene regulated abscission of sweet-pea flowers (*Lathyrus odoratus* L.), *Annals of Botany*. **75**, 337–342.
47. Dong, C. H., et al. (2008) Subcellular co-localization of Arabidopsis RTE1 and ETR1 supports a regulatory role for RTE1 in ETR1 ethylene signaling, *Plant Journal.* **53**, 275–286.
48. Rodriguez, F. I., et al. (1999) A copper cofactor for the ethylene receptor ETR1 from Arabidopsis, *Science.* **283**, 996–998.
49. Wang, W. Y., et al. (2006) Identification of important regions for ethylene binding and signaling in the transmembrane domain of the ETR1 ethylene receptor of Arabidopsis, *Plant Cell.* **18**, 3429–3442.
50. Bayer E. M. (1976) Silver ion: a potent anti-ethylene agent in cucumber and tomato, *HortScience.* **11**, 195–196.
51. Zhao, X. C., et al. (2002) Effect of ethylene pathway mutations upon expression of the ethylene receptor ETR1 from Arabidopsis, *Plant Physiology.* **130**, 1983–1991.
52. Binder, B. M., et al. (2007) The effects of Group 11 transition metals, including gold, on ethylene binding to the ETR1 receptor and growth of *Arabidopsis thaliana*, *FEBS Letters.* **581**, 5105–5109.
53. Strader, L. C., Beisner, E. R. & Bartel, B. (2009) Silver ions increase auxin efflux independently of effects on ethylene response, *The Plant Cell Online.* **21**, 3585–3590.
54. Sisler, E. C., Blankenship, S. M. & Guest, M. (1990) Competition of cyclooctenes and cyclooctadienes for ethylene binding and activity in plants, *Plant Growth Regulation.* **9**, 157–164.
55. Hirayama, T., et al. (1999) Responsive-to-antagonist1, a Menkes/Wilson disease-related copper transporter, is required for ethylene signaling in Arabidopsis, *Cell.* **97**, 383–393.
56. Sisler, E. C. (2006) The discovery and development of compounds counteracting ethylene at the receptor level, *Biotechnology Advances.* **24**, 357–367.

57. Chao, Q. M., et al. (1997) Activation of the ethylene gas response pathway in Arabidopsis by the nuclear protein ETHYLENE-INSENSITIVE3 and related proteins, *Cell.* **89**, 1133–1144.

58. Potuschak, T., et al. (2003) EIN3-dependent regulation of plant ethylene hormone signaling by two Arabidopsis F box proteins: EBF1 and EBF2, *Cell.* **115**, 679–689.

59. Robles, L. M., et al. (2012) A loss-of-function mutation in the nucleoporin AtNUP160 indicates that normal auxin signalling is required for a proper ethylene response in Arabidopsis, *Journal of Experimental Botany.* **63**, 2231–2241.

60. Chang, C., et al. (1993) Arabidopsis ethylene-response gene ETR1: similarity of product to two-component regulators, *Science.* **262**, 539–544.

61. Alonso, J. M. (1999) EIN2, a bifunctional transducer of ethylene and stress responses in Arabidopsis, *Science.* **284**, 2148–2152.

62. Alonso, J. M., et al. (2003) Five components of the ethylene-response pathway identified in a screen for weak ethylene-insensitive mutants in Arabidopsis, *Proceedings of the National Academy of Sciences of the United States of America.* **100**, 2992–2997.

63. Stepanova, A. N., et al. (2007) Multilevel interactions between ethylene and auxin in Arabidopsis roots, *The Plant Cell Online.* **19**, 2169–2185.

64. Surpin, M., et al. (2005) The power of chemical genomics to study the link between endomembrane system components and the gravitropic response, *Proceedings of the National Academy of Sciences of the United States of America.* **102**, 4902–4907.

65. Savaldi-Goldstein, S., et al. (2008) New auxin analogs with growth-promoting effects in intact plants reveal a chemical strategy to improve hormone delivery, *Proceedings of the National Academy of Sciences of the United States of America.* **105**, 15190–15195.

66. Kim, T. H., et al. (2011) Chemical genetics reveals negative regulation of abscisic acid signaling by a plant immune response pathway, *Current Biology.* **21**, 990–997.

67. Tsuchiya, Y., et al. (2010) A small-molecule screen identifies new functions for the plant hormone strigolactone, *Nature Chemical Biology.* **6**, 741–749.

68. Grozinger, C. M., et al. (2001) Identification of a class of small molecule inhibitors of the sirtuin family of NAD-dependent deacetylases by phenotypic screening, *Journal of Biological Chemistry.* **276**, 38837–38843.

69. De Rybel, B., et al. (2009) Chemical inhibition of a subset of *Arabidopsis thaliana* GSK3-like kinases activates brassinosteroid signaling, *Chemistry & Biology.* **16**, 594–604.

70. Kieber, J. J., et al. (1993) CTR1, a negative regulator of the ethylene response pathway in Arabidopsis, encodes a member of the Raf family of protein-kinases, *Cell.* **72**, 427–441.

71. Boller, T., Herner, R. C. & Kende, H. (1979) Assay for and enzymatic formation of an ethylene precursor, 1-aminocyclopropane-1-carboxylic acid, *Planta.* **145**, 293–303.

72. Tsang, D. L., et al. (2011) Cell wall integrity controls root elongation via a general 1-aminocyclopropane-1-carboxylic acid-dependent, ethylene-independent pathway, *Plant Physiology.* **156**, 596–604.

73. Mhaske, S. B. & Argade, N. P. (2006) The chemistry of recently isolated naturally occurring quinazolinone alkaloids, *Tetrahedron.* **62**, 9787–9826.

74. Dugardeyn, J. & Van Der Straeten, D. (2008) Ethylene: fine-tuning plant growth and development by stimulation and inhibition of elongation, *Plant Science.* **175**, 59–70.

75. Ulmasov, T., Hagen, G. & Guilfoyle, T. J. (1997) AuxREs and AuxRE transcription factors: a model for auxin-responsive gene expression, *Plant Physiology.* **114**, 1255.

76. Zhao, Z., et al. (2010) Hormonal control of the shoot stem-cell niche, *Nature.* **465**, 1089–1092.
77. Mathur, J., et al. (1998) Transcription of the Arabidopsis CPD gene, encoding a steroidogenic cytochrome P450, is negatively controlled by brassinosteroids, *Plant Journal.* **14**, 593–602.
78. Yamaguchi-Shinozaki, K. & Shinozaki, K. (1994) A novel cis-acting element in an Arabidopsis gene is involved in responsiveness to drought, low-temperature, or high-salt stress, *Plant Cell.* **6**, 251–264.
79. Xie, D. X., et al. (1998) COI1: an Arabidopsis gene required for jasmonate-regulated defense and fertility, *Science.* **280**, 1091–1094.
80. Rahman, A., et al. (2002) Auxin and ethylene response interactions during Arabidopsis root hair development dissected by auxin influx modulators, *Plant Physiology.* **130**, 1908–1917.
81. Swarup, R., et al. (2007) Ethylene upregulates auxin biosynthesis in Arabidopsis seedlings to enhance inhibition of root cell elongation, *The Plant Cell Online.* **19**, 2186–2196.
82. Brown, K. M. (1997) Ethylene and abscission, *Physiol Plantarum.* **100**, 567–576.
83. Stepanova, A. N. (2005) A link between ethylene and auxin uncovered by the characterization of two root-specific ethylene-insensitive mutants in Arabidopsis, *The Plant Cell Online.* **17**, 2230–2242.
84. Stepanova, A. N., et al. (2008) TAA1-mediated auxin biosynthesis is essential for hormone crosstalk and plant development, *Cell.* **133**, 177–191.
85. Stepanova, A. N., et al. (2011) The Arabidopsis YUCCA1 flavin monooxygenase functions in the indole-3-pyruvic acid branch of auxin biosynthesis, *Plant Cell.* **23**, 3961–3973.
86. Bennett, M. J., et al. (1996) Arabidopsis AUX1 gene: a permease-like regulator of root gravitropism, *Science.* **273**, 948–950.
87. Luschnig, C., et al. (1998) EIR1, a root-specific protein involved in auxin transport, is required for gravitropism in *Arabidopsis thaliana*, *Genes & Development.* **12**, 2175–2187.
88. Muller, A., et al. (1998) AtPIN2 defines a locus of Arabidopsis for root gravitropism control, *EMBO Journal.* **17**, 6903–6911.
89. Toth, R. & van der Hoorn, R. A. (2010) Emerging principles in plant chemical genetics, *Trends in Plant Science.* **15**, 81–88.
90. Katoh, A. & Hashimoto, T. (2004) Molecular biology of pyridine nucleotide and nicotine biosynthesis, *Frontiers in Bioscience.* **9**, 1577–1586.
91. Gendron, J. M., et al. (2008) Chemical genetic dissection of brassinosteroid–ethylene interaction, *Molecular Plant.* **1**, 368–379.
92. De Grauwe, L., et al. (2005) Auxin, ethylene and brassinosteroids: tripartite control of growth in the Arabidopsis hypocotyl, *Plant Cell Physiology.* **46**, 827–836.
93. Schreiber, S. L. (2000) Target-oriented and diversity-oriented organic synthesis in drug discovery, *Science.* **287**, 1964–1969.
94. Hicks, G. R. & Raikhel, N. V. (2009) Opportunities and challenges in plant chemical biology, *Nature Chemical Biology.* **5**, 268–272.
95. Haggarty, S. J., et al. (2000) Dissecting cellular processes using small molecules: identification of colchicine-like, taxol-like and other small molecules that perturb mitosis, *Chemistry & Biology.* **7**, 275–286.

96. Cong, F., Cheung, A. K. & Huang, S. M. (2012) Chemical genetics-based target identification in drug discovery, *Annual Review of Pharmacology and Toxicology.* **52**, 57–78.

97. Girke, T., Cheng, L. C. & Raikhel, N. (2005) ChemMine. A compound mining database for chemical genomics, *Plant Physiology.* **138**, 573–577.

98. Zheng, X. F. S., Chan, T. F. & Zhou, H. H. (2004) Genetic and genomic approaches to identify and study the targets of bioactive small molecules, *Chemistry & Biology.* **11**, 609–618.

99. Huang, J., et al. (2004) Finding new components of the target of rapamycin (TOR) signaling network through chemical genetics and proteome chips, *Proceedings of the National Academy of Sciences of the United States of America.* **101**, 16594–16599.

100. Kley, N. (2004) Chemical dimerizers and three-hybrid systems: scanning the proteome for targets of organic small molecules, *Chemistry & Biology.* **11**, 599–608.

101. Kim, H. K., Choi, Y. H. & Verpoorte, R. (2010) NMR-based metabolomic analysis of plants, *Nature Protocols.* **5**, 536–549.

102. Kolb, H. C. & Sharpless, K. B. (2003) The growing impact of click chemistry on drug discovery. *Drug Discovery Today.* **8**(24), 1128–1137.

103. Kaschani, F., et al. (2009) Minitags for small molecules: detecting targets of reactive small molecules in living plant tissues using 'click chemistry', *Plant Journal.* **57**, 373–385.

104. Tan, X., et al. (2007) Mechanism of auxin perception by the TIR1 ubiquitin ligase, *Nature.* **446**, 640–645.

PART FIVE

USE OF CHEMICAL BIOLOGY TO STUDY PLANT CELLULAR PROCESSES

5.1

THE USE OF SMALL MOLECULES TO DISSECT CELL WALL BIOSYNTHESIS AND MANIPULATE THE CORTICAL CYTOSKELETON

Darby Harris and Seth DeBolt

5.1.1 INTRODUCTION

The plant cell wall is a complex and dynamic extracellular matrix that surrounds every plant cell. The wall provides a number of different yet simultaneous functions to allow for plant survival including mechanical support for upright growth, facilitation of cell shape and morphogenesis, as well as roles in signaling such as pathogen defense and abiotic stress response [1]. The primary cell wall in particular has a necessary rigidity that maintains cell growth anisotropy while also remaining highly dynamic and metabolically active for its crucial roles in cell division and differentiation [2]. The polymer networks present within cell walls that allow for these various processes to occur consist principally of a cellulose/hemicellulose network, a coextensive heterogeneous pectic polysaccharide network, some glycoproteins, and, in mature nongrowing cells with a secondary cell wall, a network of polyphenolic compounds such as lignin [3]. Due to this complexity, the biosynthesis of plant cell wall polymers involves numerous enzymes that catalyze a wide range of reactions. One example of an important family of enzymes is the processive glycosyltransferases (synthases) of which there are several hundred genes putatively annotated [4]; however, the function of only a limited number has been thoroughly characterized. This is also the case for many other genes that encode important enzymes for wall synthesis

Plant Chemical Biology, First Edition. Edited by Dominique Audenaert and Paul Overvoorde.

such as non-processive transferases, transglycosylases, and enzymes involved in the interconversion of nucleotide sugars and nucleotide-sugar transporters. The completion of the *Arabidopsis thaliana* genome sequence led to the prediction that about 15% of plant genes may be involved in the biosynthesis and metabolism of the plant cell wall [5]. Therefore, because of the large number of currently uncharacterized enzymes, understanding the details of cell wall polysaccharide biosynthesis represents a major challenge in plant biology [6].

5.1.2 THE POWER OF MOLECULAR GENETICS IN THE STUDY OF CELL WALLS

One of the investigative techniques that have provided the most insight into the molecular players and mechanisms involved in cell wall biosynthesis has been the identification and characterization of mutants with altered wall structures. Through the use of forward and reverse genetic approaches, a number of cell wall mutants have been revealed, especially in *Arabidopsis* [6]. Successful forward genetic screens include primary wall sugar compositional analyses that have led to the identification of mutants with altered pectic polysaccharides and xyloglucans [7–9], as well as screens targeting the secondary cell wall that have identified mutants with an irregular xylem phenotype due to reduced cellulose biosynthesis [10] and abnormal glucuronoarabinoxylan content [11]. A number of reverse genetic approaches have taken advantage of the public availability of *Arabidopsis* lines with T-DNA insertions in specific genes that has also allowed for the identification of specific proteins involved in wall biosynthesis and metabolism [12].

Although molecular genetics is a powerful approach, problems such as loss-of-function lethality and gene redundancy have remained a challenge in plant systems. Altering the function of an essential gene can often result in embryonic or postembryonic lethality, making the study of the gene function very difficult [13]. Traditional genetic studies have often relied on conditional mutations to circumvent the problem of lethality; however, a conditional mutant, when available for a gene of interest, often requires a substantial change in the environment of the organism to generate a phenotype. Such an example can be found with temperature-sensitive alleles, where an increase in temperature results in the destabilization of the corresponding protein and a resulting phenotype. A good example in plant cell wall research is the temperature-sensitive mutant *rsw1* that harbors a point mutation in the essential primary cell wall cellulose synthase-1 (CESA1) enzyme and was selected by a root radial swelling phenotype at the restrictive temperature [14]. The identification of CESA1 was the first strong genetic evidence to suggest that this particular subfamily of processive glycosyltransferases is involved in cellulose biosynthesis in plants. In fact, the identification of other conditional mutants has proved quite useful in plant cell wall research, as many forward genetic screens have used a conditional approach. However, changes in the growth environment invariably alter the expression of many other genes as was shown in yeast when a temperature shift from 25°C to 37°C resulted in a change in expression of 854 heat-responsive genes, 50% of which had unidentified function [15]. Therefore, care must be taken when evaluating phenotypes resulting from a screen of this nature, so as not to inaccurately assign a phenotypic role to a mutation under investigation. Another challenge with the traditional genetic approach is gene redundancy, as is quite common in plants, that can mask the phenotype of an altered gene by causing the upregulation of other genes with analogous function, thereby rescuing the loss of the altered gene [16]. A common example in plant cell wall research is found in the redundancy of genes involved in the biosynthesis of hemicelluloses such as glucuronoxylans (GXs) [17, 18]. The GXs are a critical component required for cross-linking cellulose microfibrils in the secondary walls of plant like *Arabidopsis*. Gene redundancy likely represents an evolutionary advantage to safeguard the proper making of secondary walls, which are the key structure of vessels essential for water transport in the plant body. However, overcoming redundancy of gene function and the proteins for which they encode in the cell wall biosynthetic process has posed a significant challenge.

5.1.3 WHY USE CHEMICAL GENETICS TO STUDY PLANT CELL WALLS?

The chemical genetics approach is similar to traditional genetic studies where, instead of relying on genetic mutations to disrupt protein activities within a signaling network, the use of small cell-permeable molecules is employed that directly bind and functionally alter a protein. There are several advantages of chemical genetics over the traditional genetic approach. Small molecules can be used to address the loss-of-function lethality problem by essentially mimicking a conditional mutation [19]. Important with respect to the inhibition of biosynthetic enzymes of the plant cell wall, temporal knockouts can be created at different points in development due to the relative flexibility of the inhibition process, which is rapid and reversible in most cases [20]. Often a small molecule that binds to and alters the function of one protein will also bind to closely related members of that protein family, which can frequently alleviate the problem of gene redundancy. In addition, small-molecule effects are often tunable allowing gradient phenotypes to be observed by varying the concentration of the small molecule, which can add confidence to the apparent biological effect of the small-molecule probe [21]. The identification of the small-molecule target has traditionally proceeded through a biochemical approach where a traceable derivative of the small molecule is followed during fractionation of cell extracts or the molecule is immobilized via a chemical linker on a solid-phase support, followed by affinity purification [22]. A disadvantage of these techniques is that they require chemical derivatization of the small molecule, which can be time-consuming and in some instances impossible to do and still maintain bioactivity. Therefore, target identification approaches that do not require the small molecule to be derivatized are inherently more attractive [21]. Two alternative techniques that provide possible target candidates and that can be used separately or in conjunction with each other are transcript or protein abundance profiling and map-based cloning (MBC). In the former, it is quite likely that application of the bioactive small molecule causes changes in gene expression that may reveal clues to the mechanism of action of the molecule. Additionally, this pattern of gene expression can be used as a fingerprint and matched with transcriptional profiles for that of a specific gene deletion that has been created by a T-DNA or transposon insertion [21]. The MBC approach is employed when a mutant, either resistant or hypersensitive to the small molecule, is produced through the use of chemicals like ethyl methanesulfonate (EMS) or radiation. This approach is more labor intensive and uses genetic markers linked to the mutated gene to delimit the region containing the gene of interest [23]. Using the aforementioned methods in combination might allow the investigator to better preselect putative targets of the small molecule based on changes in gene expression as the genetic window for the mutation is slowly closed using MBC. Both of these techniques require the availability of considerable genetic resources in the system being studied, and fortunately for the plant biology research community, there are a variety of genetic tools available for numerous plants, especially *Arabidopsis* [23]. However, as next-generation sequencing (NGS) technologies continually improve and become more affordable, it is quite likely that whole-genome sequencing of resistant mutants will provide the quickest route to resistant locus identification. The most recent example of NGS technology being used to directly map mutations responsible for phenotypes of interest is the Next-Gen Mapping (NGM) method [24]. Next-Gen Mapping allows for the rapid and robust mapping of the physical location of EMS mutations by sequencing a small pooled F2 population and then using a chastity-like statistic that determines the relative contribution of the mutant and mapping parental lines to each single nucleotide polymorphism in the pooled F2 population. This information can then be used to objectively delineate the candidate mutation based on its exclusive segregation with the mutant parental line ([24]; http://bar.utoronto.ca/NGM).

5.1.4 THE USE OF CHEMICAL GENETICS IN CELL WALL RESEARCH

Some of the first examples of the use of small molecules as tools to study cell wall assembly were through the use of previously characterized chemicals and herbicides that either inhibit cellulose biosynthesis or affect the cortical cytoskeleton by inhibiting microtubule (MT) or actin dynamics (see Fig. 5.1.1). Isoxaben (*N*-[3(1-ethyl-1-methylpropyl)-5-isoxazolyl]; Fig. 5.1.2**a**) is a preemergence, broadleaf herbicide that was shown to inhibit cellulose biosynthesis [25]. Following a forward genetic screen for mutant *Arabidopsis* plants' resistance to the herbicide [26, 27], the loci conferring resistance to isoxaben were identified by MBC and found to be *cesa3*^{ixr1} and *cesa6*$^{ixr2\text{-}1}$ [28, 29]. This result is significant because at the time of this finding, the only known components of the large hexameric cellulose synthase complex (CSC), also referred to as a "rosette," were the CESA enzymes, although conclusive biochemical analysis was still lacking. Genetic evidence has since revealed that three distinct CESA proteins are required to form the CSC, CESA4, CESA7, and CESA8 in the secondary cell wall [30] and CESA1, CESA3, and CESA6 or the CESA6-related proteins CESA2, CESA5, and CESA9 in the primary cell wall [31]. The predicted membrane topology of a typical

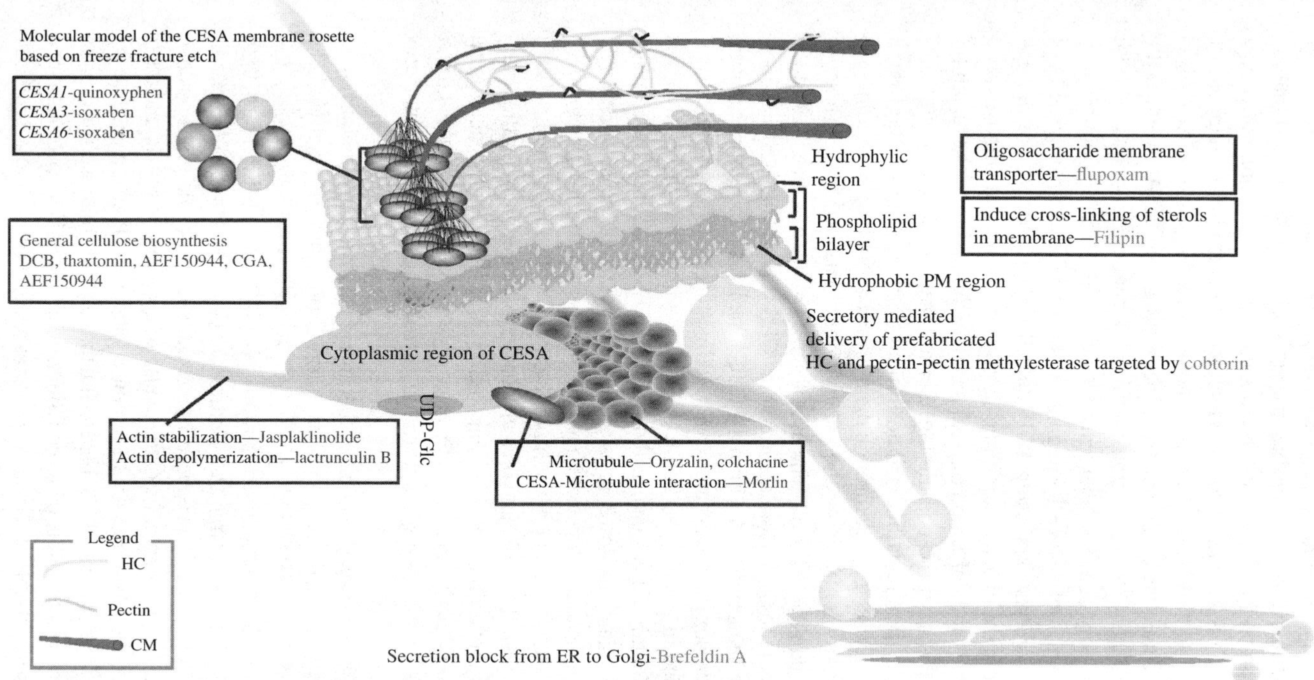

FIGURE 5.1.1 Chemical toolbox for dissecting cellulose biosynthesis. Numerous cellulose synthesis inhibitors are classified as being mapped to a target gene for forward chemical genetics such as isoxaben (*CESA3* and *CESA6*) or quinoxyphen (*CESA1*). Furthermore, DCB, CGA, thaxtomin, and AE F150944 have not been mapped to a target gene, although DCB is proposed to interact with a MAP [47]. Morlin putatively targets the interaction between MTs and cellulose synthase. Oryzalin and colchicine are MT-targeting compounds and have been used to illustrate the molecular rail hypothesis for guidance of cellulose synthase [55]. Actin depolymerization by latrunculin B has been used to show the requirement for actin-mediated trafficking in placing cellulose synthase uniformly at the plasma membrane [57]. Stabilization of actin by jasplakinolide reveals functional association between MTs and actin, which has yet to be examined in context of cellulose biosynthesis [59]. Cobtorin, a cellulose inhibitor [61] was examined genetically using FOX lines, and resistance was gained via mutations in a lectin family protein, a pectin methylesterase (AtPME1), and a putative polygalacturonase [62]. Furthermore, examining the plasma membrane cell wall continuum in relation to cellulose synthase may utilize tools such as filipin or the new drug flupoxam, which appears to target an oligosaccharide membrane transporter [24]. For color detail, please see color plate section.

plant CESA suggests a cytoplasmic N-terminal region containing zinc-finger domains followed by two transmembrane domains (TMDs), then a large cytoplasmic domain containing the catalytic motifs, and finally a cluster of six more TMDs at the C-terminus [32]. Hypothetical three-dimensional models based on this topology all suggest that the eight TMDs anchor the protein in the plasma membrane and form a pore, either as a single polypeptide or as a CESA dimer, through which the growing glucan chain passes to reach the cell wall [32, 33]. Interestingly, the mutations conferring resistance to isoxaben are not found near the putative active site for either CESA3 or CESA6. Rather, the mutations are located in the C-terminal transmembrane-spanning domains of both enzymes. In addition, not only does isoxaben cause a reduction in the amount of acid-insoluble cell wall material, but live-cell imaging of the rapidly elongating cells of *Arabidopsis* etiolated hypocotyls expressing YFP-labeled CESA6 has shown that the YFP signal disappears from the plasma membrane shortly after treatment with isoxaben [34]. This indicates that the mode of action of isoxaben in cellulose biosynthesis inhibition could be to prevent CESAs from coming together to form the CESA rosette complex that appears necessary for crystalline cellulose microfibril assembly. Therefore, it is possible that isoxaben may act directly through the physical disruption of protein–protein interactions or indirectly by blocking the putative pore that these TMDs form to extrude the glucan chain into the cell wall. The use of isoxaben has also been helpful in supporting the premise that the CESA6-related proteins CESA2 and CESA5 both compete for the same binding site as CESA6 within the CSC. The isoxaben-resistant mutant $cesa6^{ixr2\text{-}1}$ displays a lower isoxaben resistance compared with $cesa3^{ixr1}$; however, isoxaben resistance is increased in $cesa6^{ixr2\text{-}1}$ when it is crossed with *cesa2* or *cesa5* and is even higher in the triple mutant. This suggests that all three CESA6-related isoforms are isoxaben targets, which explains the lower resistance of $cesa6^{ixr2\text{-}1}$ compared to $cesa3^{ixr1}$, because CESA3 does not appear to compete with other isoforms for its positions in the CSC [35]. Therefore, isoxaben has been a very useful tool to help evaluate gene redundancy issues among primary cell wall CESA enzymes and has provided a basis for the selection of a non-conditional mutation in the essential *CESA3* gene. However, concerning a resistance mechanism, ultimately, the absence of a crystal structure for the CESA enzymes or an *in vitro* assay makes it difficult to fully resolve how isoxaben interferes with cellulose biosynthesis.

There are other cellulose biosynthesis inhibitors that appear to cause similar cellular phenotypes to that of isoxaben. *N*2-(1-ethyl-3-phenylpropyl)-6-(1-fluoro-1-methylethyl)-1,3,5-triazine-2,4-diamine (AE F150944) (Fig. 5.1.2**b**) and 1-cyclohexyl-5-(2,3,4,5,6-pentafluorophenoxyl)-1λ4,2,4,6-thiatriazin-3-amine (CGA) (Fig. 5.1.2**c**) are both experimental herbicides, while thaxtomin A (4-nitroindol-3-yl-containing 2,5-dioxopiperazine) (Fig. 5.1.2**d**) is a phytotoxin produced by several species of the gram-positive filamentous bacteria in the genus *Streptomyces* that cause scab disease in potato and other taproot crops [36–38]. AE F150944 appears to inhibit crystalline cellulose synthesis by destabilizing plasma membrane rosettes [37]. Freeze-fracture electron microscopy showed that the plasma membrane below the patterned thickenings of AE F150944-treated tracheary elements in mesophyll cells of *Zinnia elegans* was depleted of cellulose-synthase-containing rosettes, which appeared to be inserted intact into the plasma membrane followed by their rapid disaggregation [37]. Likewise, CGA and thaxtomin A were also shown to inhibit the synthesis of crystalline cellulose, and live-cell imaging shows that the GFP-labeled CESA3 signal in etiolated *Arabidopsis* hypocotyls disappears from the plasma membrane shortly after treatment with either chemical [39, 40]. Interestingly, while there has not been a resistant mutant identified for AE F150944 or CGA, a forward resistance screen to thaxtomin A in *Arabidopsis* identified a gene *TXR1* that encodes a novel small protein most likely involved in the regulation of a transport mechanism and thus may provide resistance by reducing the uptake of thaxtomin A by the plant [38]. The identifications of mutants of this nature are good examples of how resistance to a small molecule does not always require the mutation of a molecular target of the molecule, but may occur through mutations to mechanisms that metabolize, modify, or, in this instance, cause reduced uptake of the molecule. In the future, if forward resistance screens are successful toward AE F150944 or CGA, it will be interesting to learn whether the loci map to CESA or to new molecular players in cellulose biosynthesis.

2,6-Dichlorobenzonitrile (DCB) (Fig. 5.1.2**i**) is another synthetic herbicide that is known to inhibit cellulose biosynthesis, but its mode of action appears to be different than that of isoxaben. Although DCB has been shown to interfere with the assembly of the linear terminal complexes implicated in cellulose biosynthesis in the alga *Vaucheria hamata* [41] and to cause changes in the number of intact rosettes at the plasma membranes of moss (*Funaria hygrometrica*) and wheat (*Triticum aestivum*) [42, 43], more recent live-cell imaging results have shown DCB to cause simultaneous accumulation and cessation of CESA mobility within localized regions at the plasma membrane of *Arabidopsis* [34]. This suggests that DCB does not block the interaction of CESA proteins during CSC formation; however, the mechanism by which this small molecule prevents β-1,4-glucan polymerization and thus CESA mobility is still uncertain. An early clue toward the molecular function of DCB was that it was shown to bind to small proteins of 12–18 kD, the amount of which seemed to increase significantly at the onset of secondary cell wall synthesis in different plants [44]. In addition, DCB induces changes in the cortical MT networks in the fucoid alga *Pelvetia compressa* and *Arabidopsis* roots [45], which suggests that the DCB-binding protein might be associated with MTs [46]. Recently, a target for DCB was identified in hybrid aspen (*Populus tremula*×*tremuloides*) using a biochemical approach and found to be MAP20, a protein that was also shown to bind with MTs [47]. This may be one of the first examples of a direct link between a specific microtubule-associated protein (MAP) and cellulose biosynthesis and provides a possible explanation for the action of DCB.

Understanding the association between cellulose deposition and cortical MTs has been investigated for a number of years [48, 49]. In the typical plant cell, MTs are positioned just under and parallel with the plasma membrane and constitute the majority of the plant interphase cortical cytoskeletal array [50]. In fact, the discovery of interphase MTs was stimulated in part by the observation that the mitotic spindle-disrupting drug colchicine (Fig. 5.1.2**e**), a small molecule that traditionally has been used as a medicine for humans, also caused giant algal cells to swell in a radial pattern [48]. It was then observed in subsequent studies that cellulose microfibril deposition in the walls of growing plant cells tended to be parallel to, and were often coincident with, the subjacent MTs [50]. Other groups of anti-MT compounds seem to be specific to plant MTs and are ineffective against vertebrate MTs, making them effective herbicides with widespread use [51].

The dinitroaniline classes of small molecules such as oryzalin (Fig. 5.1.2**f**) are agriculturally important herbicides commonly used to control the emergence of annual grasses and certain broadleaf weeds. Oryzalin has been shown to bind to the maize tubulin dimer to form a tubulin–oryzalin complex, which is then thought to copolymerize with unliganded tubulin and to slow further MT assembly [52]. Interestingly, as the different hypotheses on the alignment of cellulose microfibrils by cortical MTs have developed [53], the use of both anti-cellulose and anti-MT small molecules has indicated that the relationship between the cortical MTs and cell morphogenesis is more complicated than early models predicted. For example, it is possible to observe uncoupling of MT and cellulose orientation and to see order appear in one array when the other array is experimentally disrupted [54–56]. Microtubule depolymerization via oryzalin was also used to dissect trafficking of cellulose synthase to the plasma membrane [40, 57]. Here, a cellulose-synthase-rich vesicle tracks on the depolymerizing end of cortical MTs. It was also shown that effective trafficking of cellulose synthase required an intact actin cytoskeleton using the small-molecule actin-depolymerizing drug latrunculin B [57] (Fig. 5.1.2**g**). Pharmacological stabilization of actin by jasplakinolide (Fig. 5.1.2**h**), a cyclo-depsipeptide produced by an Indo-Pacific sponge, *Jaspis johnstoni* [58], has more recently been used to show interdependence between the MT and actin cytoskeletons [59]. It will be intriguing to use jasplakinolide to dissect the role of actin and, more generally, actin-mediated trafficking in cellulose synthase delivery to the plasma membrane in plant cells.

Several other forward chemical genetic screens for compounds affecting cell wall synthesis and morphology have been conducted. To identify new chemical inhibitors of cell wall synthesis, a screen utilizing a library of 20,000 compounds from the DIVERSet collection (ChemBridge, San Diego, CA) was conducted, and the coumarin derivative morlin (7-ethoxy-4-methyl chromen-2-one, Fig. 5.1.2**j**) was identified by a swollen root phenotype in *Arabidopsis* [60]. Further analysis using live-cell imaging of fluorescently labeled microtubule-associated protein-4 (MAP4) and CESA revealed that morlin caused a defect in cytoskeleton organization and its functional interaction with CESA. The identification of the target(s) for morlin may provide further insight into the mechanisms of cortical MT interaction with the cell membrane and cellulose biosynthesis. Likewise, in a smaller but similar screen looking for a swollen cell phenotype in tobacco BY-2 cells, the compound cobtorin (4-[(2-chlorophenyl)-methoxy]-1-ntirobenzene) was identified [61] (Fig. 5.1.2**k**). Further analysis showed that the effect of cobtorin was to perturb the parallel alignment of preexisting cortical MTs and nascent cellulose microfibrils, and thus the target for this molecule is likely to have an important role in the relationship between MTs and microfibrils [61]. Recently, as a method for screening for genetic-resistant mutants to cobtorin, these same authors employed the Arabidopsis FOX hunting library, an activation-tagging technology that makes use of full-length cDNAs that create gain-of-function mutants [62]. From approximately 13,000 FOX lines, three cobtorin-resistant lines were identified. Interestingly, the genes that were overexpressed in the three FOX lines corresponded to a lectin family protein, a pectin methylesterase (AtPME1), and a putative polygalacturonase. This study goes on to show some important features of pectin in relation to the formation and orientation of cellulose microfibrils, which depend on the methylation ratio of pectin and its distribution [62], which has recently been experimentally characterized by magic angle spinning NMR [63].

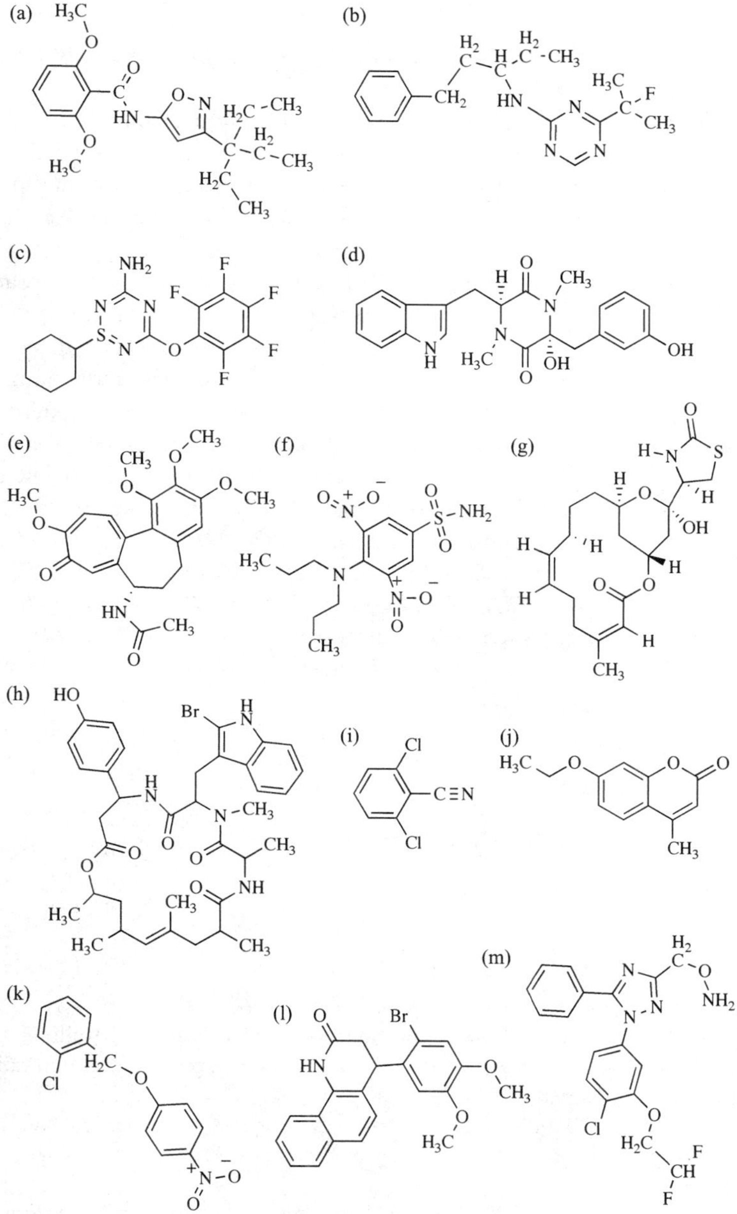

FIGURE 5.1.2 Chemical structures of molecules described in the text
(a) Isoxaben, N-[3(1-ethyl-1-methylpropyl)-5-isoxazolyl], (b) AE F150944, N2-(1-ethyl-3-phenylpropyl)-6-(1-fluoro-1-methylethyl)-1,3,5-triazine-2,4-diamine, (c) CGA 325'615, 1-cyclohexyl-5-(2,3,4,5,6-pentafluorophenoxyl)-1λ4,2,4,6-thiatriazin-3-amine, (d) Thaxtomin A, 4-nitroindol-3-yl-containing 2,5-dioxopiperazine, (e) Colchicine, N-[(7*S*)-1,2,3,10-tetramethoxy-9-oxo-5,6,7,9-tetrahydrobenzo[*a*]heptalen-7-yl]acetamid, (f) Oryzalin, 4-(dipropylamino)-3,5-dinitrobenzenesulfonamide, (g) Latrunculin-B, 4R-[(1R, 4Z,8Z,10S,13R,15R)-15-hydroxy-5,10-dimethyl-3-oxo-2,14-dioxabicyclo[11.3.1]heptadeca-4,8-dien-15-yl]-2-thiazolidinone, (h) Jasplakinolide, 7-[(2-bromo-1H-indol-3-yl)methyl]-4-(4-hydroxyphenyl)-8,10,13,15,17,19-hexamethyl-1-oxa-5,8,11-triazacyclononadec-15-ene-2,6,9,12-tetrone, (i) DCB, 2,6, dichlorobenzonitrile, (j) morlin, 7-ethoxy-4-methyl chromen-2-one, (k) Cobtorin, 4-[(2-chlorophenyl)-methoxy]-1-ntirobenzene, (l) Quinoxyphen, 4-(2-bromo-4,5-dimethooxy-phenyl)-3,4-dihydro-1H-benzo-quinolin-2(1H)-one, (m) Flupoxam, 1-[4-chloro-3-[(2,2,3,3,3-pentafluoro-propoxy)methyl]phenyl]-5-phenyl-1H-1,2,4-triazole-3-carboximide.

As additional chemical screens are completed and new molecules are identified that target the cell wall (see Fig. 5.1.1), it is certain that many will be followed up with successful forward resistance or hypersensitive screening and new molecular players in cell wall biosynthesis will be identified at an ever-increasing efficiency. An example of the former is a drug classified as a quinoline derivative, named quinoxyphen (Fig. 5.1.2**l**), that seems to mimic isoxaben in its cellular phenotype by causing the disappearance of YFP::CESA6 at the plasma membrane; however, the resistant locus for this drug was determined to be *CESA1* (Harris and DeBolt, unpublished). This mutant shows a growth phenotype only slightly reduced to that of wild type, thus representing the first fully viable, non-conditional mutant in CESA1. An example of a recent chemical genetic hypersensitivity screen was that of an EMS-mutagenized population of *Arabidopsis* hypersensitive to the cellulose biosynthesis inhibitor flupoxam [24]. Two mutants were identified, *flupoxam hypersensitive 1 and 2* (*fph1*, *fph2*), and through the use of NGM technology, the loci were identified as *ECTOPIC ROOT HAIR3* (*ERH3*) for the *fph1* locus and *OLIGOSACCHARIDE TRANSMEMBRANE TRANSPORTER* (*OST3/OST6)* for the *fph2* locus. Neither *ERH3/FPH1* nor *OST3/OST6/FPH2* encodes cell wall biosynthetic enzymes, and consequently, this screen did not identify new biosynthetic enzymes involved in *de novo* synthesis of cell walls, but rather for regulators of cell wall composition [24] (Fig. 5.1.2**m**).

ACKNOWLEDGMENTS

This work was supported by the National Science Foundation grants EFRI-0937657 and NSF-IOS-0922947.

REFERENCES

1. Carpita NC, Gibeaut DM (1993) Structural models of primary cell walls in flowering plants: consistency of molecular structure with the physical properties of the walls during growth. The Plant Journal **3**: 1–30.
2. Farrokhi N, et al. (2006) Plant cell wall biosynthesis: genetic, biochemical and functional genomics approaches to the identification of key genes. Plant Biotechnology Journal **4**: 145–167.
3. Gille S, Haensel U, Ziemann M, Pauly M 2009, Identification of plant cell wall mutants by means of a forward chemical genetic approach using hydrolases. Proceeding of the National Academy of Sciences of the United States of America **106**(34): 14699–14704.
4. Coutinho PM, Henrissat B (1999) Life with no sugars? Journal of Molecular Microbiology and Biotechnology **1**: 307–308.
5. Carpita NC, Tierney M, Campbell M 2001a. Molecular biology of the plant cell wall: searching for the genes that define structure, architecture and dynamics. Plant Molecular Biology **47**, 1–5.
6. Somerville C, et al. (2004) Toward a systems approach to understanding plant cell walls. Science **306**: 2206–2211.
7. Zablackis E, et al. (1996) Substitution of L-fucose by L-galactose in cell walls of Arabidopsis mur1. Science **272**: 1808–1810.

8. Reiter W-D, Chapple C, Somerville CR (1993) Altered growth and cell walls in a fucose-deficient mutant of Arabidopsis. Science **261**: 1032–1035.
9. Reiter W-D, Chapple C, Somerville CR (1997) Mutants of *Arabidopsis thaliana* with altered cell wall polysaccharide composition. The Plant Journal **12**: 335–345.
10. Turner SR, Somerville CR (1997) Collapsed xylem phenotype of Arabidopsis identifies mutants deficient in cellulose deposition in the secondary cell wall. The Plant Cell **9**: 689–701.
11. York WS, O'Neill MA (2008) Biochemical control of xylan biosynthesis – which end is up? Current Opinion in Plant Biology **11**: 258–265.
12. Alonso JM, et al. (2003) Genome-wide insertional mutagenesis of *Arabidopsis thaliana*. Science **301**: 653–657.
13. Meinke DW (1985) Embryo-lethal mutants of *Arabidopsis thaliana*: analysis of mutants with a wide range of lethal phases. Theoretical and Applied Genetics **69**: 543–552.
14. Arioli T, et al. (1998) Molecular analysis of cellulose biosynthesis in Arabidopsis. Science **279**: 717–720.
15. Causton HC, et al. (2001) Remodeling of yeast genome expression in response to environmental changes. Molecular Biology of the Cell **12**: 323–337.
16. Byrne ME, Simorowski J, Martienessen RA (2002) ASYMMETRIC LEAVES1 reveals knox gene redundancy in Arabidopsis. Development **129**: 1957–1965.
17. Lee C, et al. (2009) The F8H glycosyltransferase is a functional paralog of FRA8 involved in glucuronoxylan biosynthesis in Arabidopsis. Plant and Cell Physiology **50**: 812–827.
18. Brown DM, et al. (2009) Characterization of IRX10 and IRX10-like reveals an essential role in glucuronoxylan biosynthesis in Arabidopsis. The Plant Journal **57**: 732–746.
19. Robert S, Raikhel N, Hicks G (2008) Powerful partners: Arabidopsis and chemical genomics. *In* The Arabidopsis Book. The American Society of Plant Biologists, Rockville, MD, pp 1–16.
20. Zabotina O, et al. (2008) Identification and preliminary characterization of a new chemical affecting glucosyltransferase activities involved in plant cell wall biosynthesis. Molecular Plant **1**: 977–989.
21. Spring DR (2005) Chemical genetics to chemical genomics: small molecules offer big insights. Chemical Society Reviews **34**: 472–482.
22. Mayer TU (2003) Chemical genetics: tailoring tools for cell biology. Trends in Cell Biology **13**: 270–277.
23. Peters JL, Cnudde F, Gerats T (2003) Forward genetics and map-based cloning approaches. Trends in Plant Science **8**: 484–491.
24. Austin RS, et al. (2011) Next-generation mapping of Arabidopsis genes. The Plant Journal **67**:715–725.
25. Heim DR, et al. (1990) Isoxaben inhibits the synthesis of acid insoluble cell wall materials in *Arabidopsis thaliana*. Plant Physiology **93**: 695–700.
26. Heim DR, et al. (1989) Mutation of a locus of *Arabidopsis thaliana* confers resistance to the herbicide isoxaben. Plant Physiology **90**: 146–150.
27. Heim DR, et al. (1990) A second locus, *Ixr B1* in *Arabidopsis thaliana*, that confers resistance to the herbicide isoxaben. Plant Physiology **92**: 858–861.

28. Scheible W-F, et al. (2001) Modifications of cellulose synthase confer resistance to isoxaben and thiazolidinone herbicides in Arabidopsis *Ixr1* mutants. Proceedings of the National Academy of Sciences of the United States of America **98**: 10079–10084.
29. Desprez T, et al. (2002) Resistance against herbicide isoxaben and cellulose deficiency caused by distinct mutations in same cellulose synthase isoform CESA6. Plant Physiology **128**: 482–490.
30. Taylor NG, et al. (2003) Interactions among three distinct CesA proteins essential for cellulose synthesis. Proceedings of the National Academy of Sciences of the United States of America **100**: 1450–1455.
31. Persson S, et al. (2007) Genetic evidence for three unique components in primary cell-wall cellulose synthase complexes in Arabidopsis. Proceedings of the National Academy of Sciences of the United States of America **104**: 15566–15571.
32. Delmer DP (1999) Cellulose biosynthesis: exciting times for a difficult field of study. Annual Review of Plant Physiology and Plant Molecular Biology **50**: 245–276.
33. Carpita NC (2011) Update on mechanisms of plant cell wall biosynthesis: How plants make cellulose and other $(1\rightarrow4)$-β-D-glycans. Plant Physiology **155**: 171–184.
34. DeBolt S, et al. (2007) Nonmotile cellulose synthase subunits repeatedly accumulate within localized regions at the plasma membrane in Arabidopsis hypocotyl cells following 2,6-dichlorobenzonitrile treatment. Plant Physiology **145**: 334–338.
35. Desprez T, et al. (2007) Organization of cellulose synthase complexes involved in primary cell wall synthesis in *Arabidopsis thaliana*. Proceedings of the National Academy of Sciences USA **104**: 15572–15577.
36. Peng L, et al. (2001) The experimental herbicide CGA 325′615 inhibits synthesis of crystalline cellulose and causes accumulation of non-crystalline β-1,4-glucan associated with CesA protein. Plant Physiology **126**: 981–992.
37. Kiedaisch B, Blanton R, Haigler C (2003) Characterization of a novel cellulose synthesis inhibitor. Planta **217**: 922–930.
38. Scheible W-R, et al. (2003) An Arabidopsis mutant resistant to thaxtomin A, a cellulose synthesis inhibitor from Streptomyces species. The Plant Cell **15**: 1781–1794.
39. Bischoff V, et al. (2009) Thaxtomin A affects CESA-complex density, expression of cell wall genes, cell wall composition, and causes ectopic lignification in *Arabidopsis thaliana* seedlings. Journal of Experimental Botany **60**: 955–965.
40. Crowell EF, et al. (2009) Pausing of golgi bodies on microtubules regulates secretion of cellulose synthase complexes in Arabidopsis. The Plant Cell **21**: 1141–1154.
41. Mizuta S, Brown RM (1992) Effects of 2,6-dichlorobenzonitrile and Tinopal LPW on the structure of the cellulose synthesizing complexes of *Vaucheria hamata*. Protoplasma **166**: 200–207.
42. Herth W (1987) Effects of 2,6-DCB on plasma membrane rosettes of wheat root cells. Naturwissenschaften **74**: 556–557.
43. Rudolph U, Gross H, Schnepf E (1989) Investigations of the turnover of the putative cellulose-synthesizing particle "rosettes" within the plasma membrane of *Funaria hygrometrica* protonema cells. Protoplasma **148**: 57–69.
44. Delmer DP, Read SM, Cooper G (1987) Identification of a receptor protein in cotton fibers for the herbicide 2,6-dichlorobenzonitrile. Plant Physiology **84**: 415–420.
45. Himmelspach R, Williamson RE, Wasteneys GO (2003) Cellulose microfibril alignment recovers from DCB-induced disruption despite microtubule disorganization. The Plant Journal **36**: 565–575.

46. Bisgrove SR, Kropf DL (2001) Cell wall deposition during morphogenesis in fucoid algae. Planta **212**: 648–658.
47. Rajangam AS, et al. (2008) MAP20, a microtubule-associated protein in the secondary cell walls of hybrid aspen, is a target of the cellulose synthesis inhibitor 2,6-dichlorobenzonitrile. Plant Physiology **148**: 1283–1294.
48. Green PB (1962) Mechanism for plant cellular morphogenesis. Science **138**: 1404–1405.
49. Ledbetter MC, Porter KR (1963) A "microtubule" in plant cell fine structure. The Journal of Cell Biology **19**: 239–250.
50. Ehrhardt DW, Shaw SL (2006) Microtubule dynamics and organization in the plant cortical array. Annual Review of Plant Biology **57**: 859–875.
51. Morejohn LC (1991) The molecular pharmacology of plant tubulin and microtubules. *In* The Cytoskeletal Basis of Plant Growth and Form, CW Lloyd, ed. Academic Press, London, pp 29–55.
52. Hugdahl JD, Morejohn LC (1993) Rapid and reversible high-affinity binding of the dinitroaniline herbicide oryzalin to tubulin from *Zea mays* L. Plant Physiology **102**: 725–740.
53. Baskin TI (2001) On the alignment of cellulose microfibrils by cortical microtubules: a review and a model. Protoplasma **215**: 150–171.
54. Wasteneys GO, Galway ME (2003) Remodeling the cytoskeleton for growth and form: an overview with some new views. Annual Review of Plant Biology **54**: 691–722.
55. Paredez AR, Somerville CR, Ehrhardt DW (2006) Visualization of cellulose synthase demonstrates functional association with microtubules. Science **312**: 1491–1495.
56. Paredez AR, et al. (2008) Genetic evidence that cellulose synthase activity influences microtubule cortical array organization. Plant Physiology **147**: 1723–1734.
57. Gutierrez R, et al. (2009) Arabidopsis cortical microtubules position cellulose synthase delivery to the plasma membrane and interact with cellulose synthase trafficking compartments. Nature Cell Biology **11**: 797–806.
58. Senderowicz AM, et al. (1995). Jasplakinolide's inhibition of the growth of prostate carcinoma cells in vitro with disruption of the actin cytoskeleton. Journal of the National Cancer Institute **87**: 46–51.
59. Sampathkumar A, et al. (2011) Live cell imaging reveals structural associations between the actin and microtubule cytoskeleton in Arabidopsis. Plant Cell **23**: 2302–2313.
60. DeBolt S, et al. (2007) Morlin, an inhibitor of cortical microtubule dynamics and cellulose synthase movement. Proceedings of the National Academy of Sciences of the United States of America **104**: 5854–5859.
61. Yoneda A, et al. (2007) Chemical genetic screening identifies a novel inhibitor of parallel alignment of cortical microtubules and cellulose microfibrils. Plant Cell Physiology **48**: 1393–1403.
62. Yoneda A, et al. (2010) Cobtorin target analysis reveals that pectin functions in the deposition of cellulose microfibrils in parallel with cortical microtubules. The Plant Journal **64**: 657–667.
63. Dick-Pérez M, et al. (2011) Structure and interactions of plant cell-wall polysaccharides by two- and three-dimensional magic-angle-spinning solid-state NMR. Biochemistry **15**:989–1000.

5.2

THE USE OF CHEMICAL BIOLOGY TO STUDY PLANT CELLULAR PROCESSES: SUBCELLULAR TRAFFICKING

ASH HAEGER, MALGORZATA ŁANGOWSKA,
AND STÉPHANIE ROBERT

5.2.1 THE PLANT ENDOMEMBRANE SYSTEM, A COMPLEX NETWORK ESSENTIAL FOR DIVERSE DEVELOPMENTAL FUNCTIONS

The complexity of developmental processes necessitates the functional compartmentalization of the cell. Material secretion, transport, and exchange between cellular compartments rely on highly regulated trafficking pathways along the network of the endomembrane system. Their proper organization and regulation are essential since within a multicellular organism, a given cell may have different developmental tasks from its closest neighbor and may require a particular signaling machinery. The components of the endomembrane system are partially conserved among the eukaryotic kingdoms, and despite the simplicity of yeasts, the key functions are similar in this unicellular organism to those found in animals and plants [1, 2]. However, major players in the plant trafficking network have evolved during the rise of multicellular organisms [1]. For example, the syntaxin protein family is evolutionarily conserved in its function in vesicle fusion, but the yeast genome contains 8 members while 24 are present in *Arabidopsis thaliana*. This implies a differential development of the trafficking machinery in lower and higher eukaryotes, reflecting the existence of processes such as the establishment of organ polarity or vacuolar protein storage [3, 4].

Plant Chemical Biology, First Edition. Edited by Dominique Audenaert and Paul Overvoorde.

Within the plant endomembrane system, the secretory pathway is involved in the synthesis, maturation, and sorting of cellular components prior to delivery to their final destination [5] (Fig. 5.2.1). The secretory pathway is composed of the endoplasmic reticulum (ER), the Golgi apparatus, and the *trans*-Golgi network (TGN). From the TGN, cargo molecules are targeted either to the plasma membrane (PM), from where they can be recycled to the TGN [6], or to the vacuole via different intermediate compartments such as multivesicular bodies (MVBs) for storage or degradation [7, 8]. The dynamics of the endomembrane system permits rapid modifications of the plant membrane composition in response to developmental or environmental stimuli [9, 10]. The importance of these processes is illustrated by the dynamic polar localization of the auxin efflux carrier PIN-FORMED (PIN) family proteins and their contribution to directional auxin transport, linking the endomembrane system to fundamental developmental processes. Additionally, the trafficking network has a role in the perception, integration, and transduction of both inter- and intracellular signals such as brassinosteroids and in plant immune responses [11–13]. For instance, the endocytosis of flagellin-sensing 2 (FLS2) receptor after the recognition of bacterial flagellin is necessary to trigger downstream immune responses [14, 15].

Overall, the endomembrane network is involved in the maturation and targeting of proteins to their site of activity and permits a sessile lifestyle by redirecting cellular material in response to developmental and environmental stimuli. The elucidation of these cellular processes represents a major challenge to gain a better understanding of plant developmental biology.

5.2.2 CHEMICAL GENOMICS STRATEGIES TO STUDY THE PLANT TRAFFICKING NETWORK

Chemical genomics is a valuable approach to address the different problems inherent to trafficking network characterization, such as lethality and gene redundancy. Recently, endomembrane trafficking studies have been facilitated by advances in combinatorial chemical synthesis techniques and establishment of large screening collections of small drug-like molecules (for review see Robert et al. [16]). We will focus on a subset of chemical genomics strategies used in the understanding of endomembrane trafficking mechanisms. We will consecutively discuss (i) the use of small molecules to phenocopy the effect of genetic mutations affecting vesicular trafficking, (ii) the screening for compounds affecting evolutionarily conserved mechanisms, and (iii) the activity of some natural toxins targeting the vital function of the endomembrane system.

5.2.3 THE IDENTIFICATION OF NOVEL COMPOUNDS TARGETING THE PLANT TRAFFICKING NETWORK

The screening of large chemical libraries to identify new molecules that induce changes in the plant endomembrane trafficking offers a targeted way to probe the plant endomembrane system. For example, the *shoot gravitropism* (*sgr*) or *zig* mutants impaired in

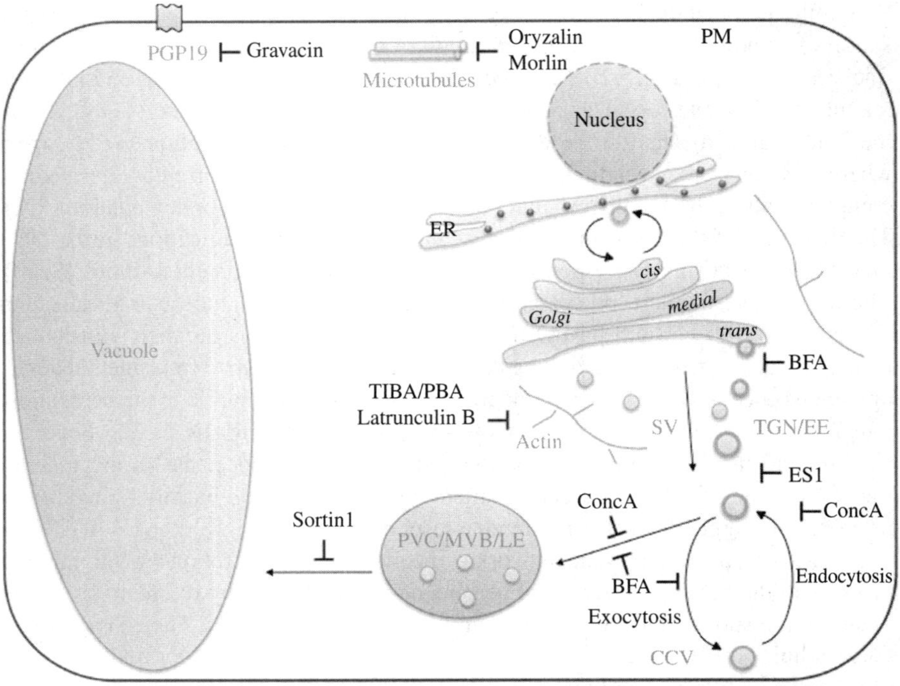

FIGURE 5.2.1 The plant endomembrane system. The trafficking machinery consists of several major compartments: the endoplasmic reticulum (ER), the Golgi complex, the *trans*-Golgi network/early endosome (TGN/EE), the pre-vacuolar compartment (PVC), the vacuole, and the plasma membrane (PM). Proteins are transported from the ER via Golgi/TGN to the PM or vacuole. Newly synthesized proteins are transported from the ER to the Golgi where they pass sequentially through the *cis*-, *medial*-, and *trans*-cisterna as they mature and finally onto the TGN. In the TGN, proteins are sorted into pathways leading either to the vacuole via the PVC (also known as a late endosome (LE) or the multivesicular body (MVB)) or to the PM. Proteins targeted to the PM are alternatively sorted in secretory vesicles (SV), uncoated vesicles or clathrin-coated vesicles (CCV). Proteins destined for the vacuole are sorted into CCV and transported to the PVC. Also illustrated are the points of action for several chemical agents used in subcellular trafficking research. Brefeldin A (BFA) inhibits exocytosis by preventing vesicle formation, resulting in the formation of large compartments. Concanamycin A (ConcA) inhibits vesicular trafficking between Golgi and TGN. Sortin1 affects the biogenesis of vacuoles and vacuolar transport. Gravacin interacts with PGP19, a membrane-localized auxin transporter. Oryzalin causes microtubule depolymerization, affecting vesicle trafficking. Similarly, latrunculin B and the auxin transport inhibitors *N*-1-naphthylphthalamic acid (NPA), PBA, and TIBA disrupt polymerization of actin and vesicle trafficking. Endosidin 1 (ES1) disrupts endocytosis via interaction with an unknown target. For color detail, please see color plate section.

the function of syntaxin proteins exhibit a reduced gravitropic response [17]. Based on this, the reduction of gravity response was taken as a suitable phenotype correlating with vesicular trafficking in a screen of 10,000 molecules from the DIVERSet library (ChemBridge Corporation, San Diego, CA) [18]. A total of 219 hit compounds were found to affect gravitropic responses in roots or hypocotyls of wild-type *Arabidopsis* seedlings. Based upon structural analysis, 69 of the 219 hit compounds were selected to perform dose–response analyses on gravitropism and 34 were confirmed to affect gravitropism. Subsequently, these 34 compounds were applied to seedlings expressing the tonoplast-localized GFP:δ-TIP marker protein to select chemicals affecting tonoplast morphology. Five molecules were shown to affect the morphology of the endomembrane system or the localization and distribution of GFP:δ-TIP. One of the five molecules that were selected for further analysis was gravacin [18, 19]. One of the targets of gravacin, the ABC type B transport protein phosphoglycoprotein 19 (PGP19), previously identified as an auxin efflux carrier [20], was identified through a screen for resistant mutants. Phosphoglycoprotein 19 is known to be active in the basipetal transport of auxin, playing a role in the maintenance of long-range auxin transport, in partnership with PGP1 and in synergy with PIN proteins [21, 22]. Auxin plays an essential role in regulating responses to gravity. The perception of a change in the gravity vector results in a redirection of the auxin transport and modification of the auxin gradient. This triggers differential growth processes resulting in organ curvature and realignment with the gravity vector. Through its interaction with PGP19, gravacin was found to be a strong inhibitor of polar auxin efflux and gravitropic responses in hypocotyls. This finding defines PGP19 as an important component in the gravity response [19]. Additionally, gravacin causes the mislocalization of PGP19 to FM4–64-negative endomembrane compartments, suggesting that its activity might prevent the interaction of PGP19 with other auxin efflux transporters [21, 23]. Interestingly, the analysis of eight structural analogs of gravacin established that different structural domains of the molecule were responsible for the inhibition of gravitropic response and the disruption of GFP:δ-TIP trafficking, uncoupling these two phenotypes [19]. The discovery of gravacin led to a better understanding of the regulation and activity of PGP19 in auxin efflux in response to gravity and specifically the importance of the second nucleotide-binding domain mutated in *pgp19-4* (E1174K) in the interaction with PIN1 [19]. The screening of chemicals inducing endomembrane-related phenotypes can efficiently lead to the identification of new proteins, their function, and their interactions.

Another example comes from the investigation of pollen germination to identify chemicals disturbing vesicular trafficking [11]. Pollen tube growth in *Arabidopsis* and tobacco is highly dependent on secretion and recycling of proteins, membrane material, and cell wall components at the PM [24]. It was used as a model system for subcellular trafficking studies allowing a high-throughput chemical genomics screening (2016 molecules from MicroSource's Spectrum Collection) supported by automated image-based analysis [11]. To specifically characterize polar targeting processes, a secondary screen was performed to identify compounds impairing the polar localization of the pollen tip-localized GFP-ROP-INTERACTING PROTEIN1 (RIP1, also known as Interactor of Constitutively Active ROP1, ICR1; [25]). Compounds causing mislocalization of GFP-RIP1 at the tip of the pollen tube were retained for further analysis. The identification of several compounds that impair trafficking pathways in tobacco pollen tubes and *Arabidopsis* roots showed the existence of conserved mechanisms between plant species and similarities between polar-tip and non-tip growth mechanisms. Among them, endosidin 1 (ES1) impairs the trafficking of the PM-localized brassinosteroid receptor BRASSINOSTEROID-INSENSITVE 1 (BRI1) and the auxin carriers PIN2 and AUX1, resulting in agglomeration of endomembrane called "ES1 bodies." The syntaxin SYP61 and V-ATPase subunit VHA-a1, previously identified as TGN and endosomal proteins, are also sensitive to ES1. The sequestration of BRI1 with TGN-derived endosomal bodies caused an attenuation of the signaling response to brassinosteroids. Taken together, the application of ES1 demonstrated the importance of the endomembrane network in the transduction of hormonal signaling mechanisms.

5.2.4 CHEMICAL GENOMICS CAN TARGET EVOLUTIONARILY CONSERVED PATHWAYS IN EUKARYOTES

Yeast and *Arabidopsis* share a large number of homologous genes and display similarities in the molecular machinery such as the soluble *N*-ethylmaleimide-sensitive factor adaptor protein receptor (SNARE) components of vesicular trafficking and proteins involved in membrane fusion [1, 2]. However, in contrast to *Arabidopsis*, the vacuole is not essential in yeast [26, 27]. In addition, the yeast endomembrane system is relatively simple and better understood than in plants, which makes it an ideal organism for the discovery of chemicals that are likely to possess activity against vacuolar trafficking. Yeast carboxypeptidase Y (CPY) is normally sorted from the Golgi apparatus to the vacuole [28]. However, if this pathway is altered, CPY is secreted to the outside of the cell. Part of a structurally diverse library of chemicals (4800 compounds from the DIVERSetE library, ChemBridge, San Diego) was screened for compounds impairing CPY trafficking to the vacuole and thus directing CPY to the cell exterior. The screen resulted in the identification of 14 chemicals [29]. Two of these compounds, Sortin1 and Sortin2, were shown to reversibly alter vacuole morphology and root development in *Arabidopsis*.

Recently, a screen for Sortin1-hypersensitive *Arabidopsis* mutants led to the isolation of lines that displayed a deficiency in flavonoid accumulation in the seed coat [30]. These mutants were shown to be impaired in flavonoid accumulation or biosynthesis pathways, leading to the hypothesis that Sortin1 activity might target one of these two processes. Physiological effects induced by Sortin1 treatment are similar to the one observed after inhibition of flavonoid transport by buthionine sulfoximine (BSO), an ABC transporter inhibitor. In addition, microarray analyses provided evidence that Sortin1 treatment led to an upregulation in the expression of glutathione-*S*-transferases (GSTs) among other indicators of oxidative stress. This suggested that Sortin1 might act as a causative agent of oxidative stress, depleting the pool of GST enzymes available for the formation of GST–flavonoid complexes. Overall, this study revealed a cross talk between vacuolar targeting pathways and flavonoid biosynthesis and indicated that flavonoids might regulate cargo sorting to the vacuole but could also have a function in the maintenance of vacuolar integrity under oxidative stress [30]. In a complementary study, the chemical Sortin2 was used to screen a haploid yeast deletion library to identify mutant strains possessing differential sensitivity to the drug with respect to wild-type yeast [31]. A total of 217 hypersensitive mutants among the 4800 screened were identified, and some of them were previously known as vacuolar protein sorting (*vps*) mutants or displaying a *vps* phenotype upon exposure to the drug. Bioinformatic approaches revealed enrichment of protein-trafficking-related genes, reflecting the specific activity of Sortin2 toward the endomembrane system. Among the 217 hypersensitive mutants, 30 were impaired in genes of unknown function. Further investigation allowed the assignment of a functional location to the products of 15 of these genes of unknown function, as being within the ER-to-vacuole trafficking pathways, opening up new gene targets for future studies [31].

The knowledge acquired in different organisms has resulted in a better understanding of the mechanisms facilitating and regulating subcellular trafficking in plant development. Additionally, the high degree of conservation of the endomembrane machinery reflects its importance in cellular functions and illustrates the power of lower eukaryotes such as yeast as a model system to study vesicular trafficking in plants.

5.2.5 BROADLY USED BIOACTIVE NATURAL COMPOUNDS TO INVESTIGATE VESICULAR TRAFFICKING IN PLANTS

As a defense strategy, several toxins synthesized and secreted by microorganisms affect the plant trafficking network, which due to its complexity represents a broad range of potential targets for microbial toxins. For example, brefeldin A (BFA), a fungal macrocyclic lactone inhibits vesicle budding via action on the protein family of exchange factors for ARF-GTPases (ARF-GEF) with a SEC7 domain [32]. Brefeldin A was originally characterized in mammalian cell studies [33, 34] and was soon defined as an inhibitor of anterograde trafficking from the ER to the Golgi [35]. Interestingly, BFA activity in plants is similar albeit not identical, demonstrating the complex evolutionarily conserved nature of the molecular mechanisms regulating vesicle trafficking. In tobacco Bright Yellow-2 (BY-2) suspension cultures, BFA prevents the release of vesicles from the Golgi apparatus within 5 min [36], while prolonged treatment leads to the complete loss of Golgi architecture resulting in the formation of ER/Golgi hybrid structures termed "BFA compartments" [37]. Remarkably, in the presence of BFA, the GFP-tagged Golgi marker α-mannosidase 1 (Man1-GFP, [38]), which is usually present on the *cis*-Golgi stacks and permanently recycles from downstream cisternae, was detected in *trans*-like cisternae. This constituted evidence for the maturation of *cis*-Golgi cisternae developing into *medial*- and *trans*-Golgi over time, even in the absence of vesicular transport. This study led to the confirmation of the organelle maturation hypothesis [36] and helped further the understanding of Golgi morphology and function. Notably, ER–Golgi fusion was not systematically observed in all tissues or plant species, demonstrating some specification in the localization and function of ARF-GEF proteins [36, 39–41]. In *Arabidopsis* roots and at low concentration, BFA inhibits exocytosis to the PM by affecting the activity of GNOM, an ARF-GEF localized at the endosomes. Thus, in several studies BFA was used as a chemical tool to identify components subjected to endocytic cycling [42–44]. This led to the discovery of the importance of ARF-GEF-dependent recycling pathways in the polar localization of proteins, such as the PIN auxin efflux carriers [43, 45–47]. Overall, the analysis of the BFA-induced effects on the endomembrane system provided key discoveries for dissecting the endomembrane system and its implications in diverse developmental processes.

Small molecules, more specifically actin and microtubule depolymerizing drugs, have also been extensively used to elucidate the precise subcellular functions of the cytoskeleton in vesicular trafficking. The synthetic molecules 2,3,5-triiodobenzoic acid (TIBA) and 2-(1-pyrenoyl) benzoic acid (PBA), which interfere with actin filament dynamics, inhibit vesicle and organelle motility [48]. The application of TIBA and PBA impairs the targeting and cycling of PIN proteins at the PM and consequently the auxin flux. Another example is the natural toxin latrunculin B, whose activity inhibits the transport from the Golgi to the central vacuole [49] showing an essential function of actin filament in this particular pathway. The use of latruncunlin B also demonstrated that actin filaments represent physical tracks for secretion and endocytosis, which are essential mechanisms mediating tip-driven processes such as pollen tube growth [50]. The effect of TIBA and PBA on actin is distinct from that observed with lantrunculin B. TIBA and PBA, originally characterized as auxin transport inhibitors, cause the stabilization of the actin network [48], while lantrunculin B depolymerizes the actin proteins, disrupting the filament structure [51]. This illustrates the variety of effects that can be induced by structurally unrelated molecules on the same cellular process. Moreover, disruption of the microtubule orientation and polymerization by the coumarin-derivative morlin result in the retardation and redistribution of cellulose synthase complexes (CSCs) in *Arabidopsis* [52]. This strongly suggested that microtubules are involved in the guidance of CSCs to the PM, regulating the rate and direction of cellulose synthesis and thus plant morphology [53, 54].

The cytoskeleton is also essential for polar targeting of proteins. Interestingly, actin filaments have been shown to play a particular role in the polar targeting of proteins on the apical side of the cell. The apical targeting of the auxin influx carrier AUX1 is more sensitive to actin filament disruption by latrunculin B than the targeting of PIN1 to the basal domain [46, 55]. On the other hand, intact microtubule filaments are essential for targeting to the basal side of the cell as illustrated by the effect of the depolymerizing agent oryzalin on basally localized PIN2 proteins [46]. This demonstrates that the cytoskeleton network establishes trafficking pathways to the different sides of the cell, specifically using the network of intact actin or microtubule filaments for the correct targeting to apical or basal sides of the cell, ensuring proper polar localization of vesicle-bound materials.

These examples illustrate that natural compounds, together with synthetic chemicals, have increased our understanding of vesicular trafficking, especially with regard to the role played by the cytoskeleton. Given the importance of the endomembrane system to plant cellular processes, it is not surprising to find that non-plant organisms produce a number of compounds capable of interacting with and altering the function of its components.

5.2.6 ADVANTAGES OF A CHEMICAL GENOMICS STRATEGY TO DISSECT THE PLANT ENDOMEMBRANE SYSTEM

A primary strength of chemical genomics approaches in general and more specifically to study the plant endomembrane system is the circumvention of problems arising from gene functional redundancy. Small molecules that modify the action of every member of a class of proteins provide an opportunity to induce a detectable phenotype even in a highly redundant gene family, whereas a single mutant line may exhibit no discernable phenotype [56–58]. In this case, rather than the requirement to mutagenize each member of the gene family, the applied chemical can act across a whole class of proteins by inhibiting a particular biochemical activity. An example of this comes from the activity of BFA targeting an entire ARF-GEF protein family. Brefeldin A sensitivity depends on the dual specificity of the SEC7 domain within the ARF-GEFs and on the actual interacting ARF [59]. Therefore, the BFA concentration is critical (Fig. 5.2.2). Fifty micromolar BFA affects the recycling of PM-localized proteins and their targeting to the vacuole, whereas at 25 μM BFA, the pathway to the vacuole remains largely unaffected [60, 61].

Moreover, the compound's activity is tunable, which presents an important advantage to address the function of proteins. Brefeldin A can induce diverse effects at different concentrations (Fig. 5.2.2 and Fig. 5.2.3). This is opposite to the consequence of permanent genetic mutations affecting *GNOM* or *MIN7/BEN1* genes ([44, 47]; Fig. 5.2.3). If shutting down vital functions is important to understand molecular processes, then the reversibility of this action is an essential parameter. The washing out of BFA and the subsequent observation of the cell as it returns to equilibrium have provided insights into the physical construction of the plant Golgi apparatus and the origin of specific cytoplasmic vesicles [37] (Fig. 5.2.2). Moreover, due to the importance of vesicular transport, embryonic lethality is a common consequence of the genetic mutation impairing a component of the endomembrane system such as *VACUOLELESS1* (*vcl1*) [26] or genes coding for syntaxins [62]. Consequently, attempts to isolate mutants for genes important in the generation of vacuolar structure have been hampered by their early lethality. This is illustrated by the observation that a collection of EMS-mutagenized *Arabidopsis* vacuolar mutants exhibited an approximately 50% rate of mortality [63]. Alternatively, the identification of Sortin1 and Sortin2, small synthetic molecules that induce reversible defects in vacuole protein targeting in *Arabidopsis*, provided the opportunity to understand the activity of vacuole-related proteins that perform an essential function [29].

Interestingly, a compound can possess more than one biologically active structural domain due to the large number of possible protein binding domains. In the field of vesicular transport, at least one study was able to separate diverse effects of distinct chemical substructures. Gravacin was found to have an effect on the localization of the tonoplast marker GFP-δTIP that could be separated from its activity on the gravitropic response by means of assaying structural analogs [19]. Only gravacin and one of eight other structurally related compounds were found to inhibit the gravitropic response.

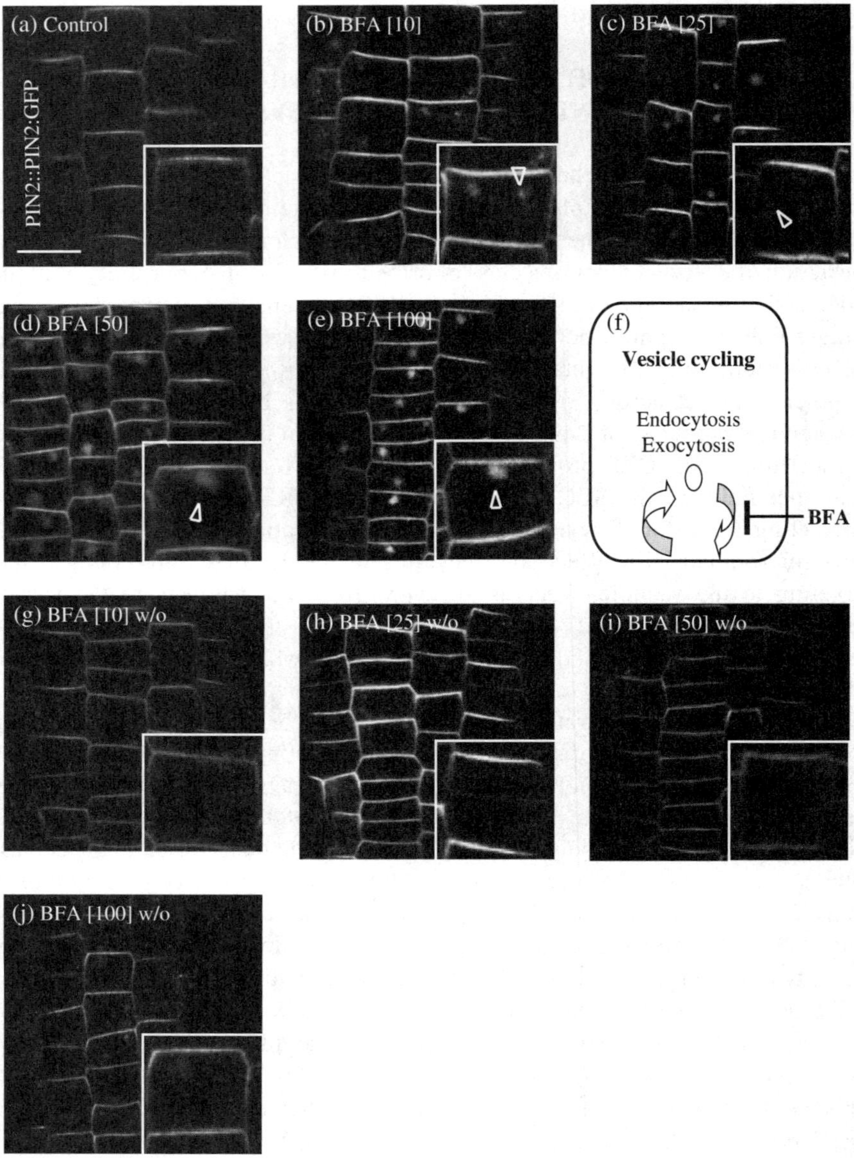

FIGURE 5.2.2 Brefeldin A, a drug with dose-dependent reversible effect on vesicular cycling. Laser scanning confocal microscopy images showing the dose-dependent effects of BFA treatment on the localization of PIN2::PIN2-GFP in the *Arabidopsis* root tip. Brefeldin A treatment blocks the action of BFA-sensitive ARF-GEFs, including GNOM, necessary for the formation of recycling endocytic vesicles that maintain the polar localization of PIN2-GFP. As a result, PIN2-GFP accumulates in clearly visible BFA compartments (arrowheads). (**a**) Untreated, (**b**) treatment with 10 μM, (**c**) 25 μM, (**d**) 50 μM, and (**e**) 100 μM of BFA for 2 h. (**f**) Schematic representation of the activity of BFA targeting exocytosis. Seedlings incubated with (**g**) 10 μM, (**h**) 25 μM, (**i**) 50 μM, and (**j**) 100 μM of BFA for 1 h and transferred to regular media AM(+) for 2 h. Note the dose-dependent difference in size of the BFA-induced agglomerations. Washing out BFA allows PIN2-GFP to return to a normal distribution w/o stands for washout. Scale bar represents 10 μm.

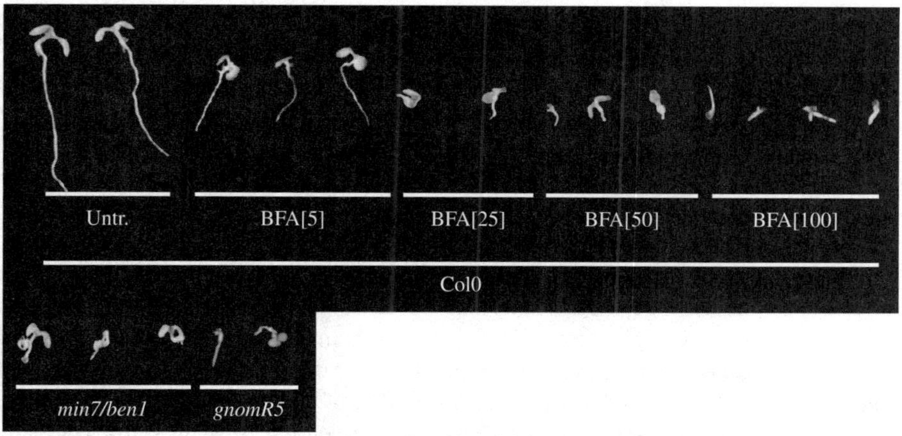

FIGURE 5.2.3 Brefeldin A activity is tunable. (**a**) Five-day-old Col-0 seedlings grown in presence of 5, 25, 50, or 100 μM BFA. (**b**) Phenotype of 5-day-old mutants (*gnomR5*, *min7/ben1*) impaired in the function of ARF-GEF protein with an SEC7 domain. For color detail, please see color plate section.

However, all of the analogs were impairing the trafficking of the tonoplast marker protein, with several compounds showing higher activity than the original gravacin. Thus, it was concluded that the structural domain of the molecules required to affect gravitropism and tonoplast trafficking was different, interacting with separate target molecules to induce the two effects.

The chemical genomics approach is beneficial to the dissection of rapid processes, such as trafficking mechanisms. After genetic perturbation, the resulting phenotype reports the status of the cells in equilibrium with a lesion. The rapid and highly dynamic nature of endomembrane transport and intracellular signaling provides only a short time window for study and therefore requires a rapid intervention [64]. Small molecules are capable of interacting with the components of these pathways in a short time frame and thus are more compatible to study dynamic processes. This fact has been exploited in the investigation of early pathogen-associated-molecular-pattern (PAMP) responses mediated by the *ARABIDOPSIS TOXICOS EN LEVADURA* (*ATL2*) gene. Previous genetic approaches had failed to recover non-ATL2 elicitor-responsive plants [65]. However, a chemical screen in which an ATL2 promoter–GUS reporter line was used led to the identification of compounds capable of modifying early ATL2 responses, from which some were able to block the endocytosis of FLS2 [66]. In this case, the rapid action of the chemicals allowed the investigation of short-lived early immune responses, essential for the efficient dissection of these pathways.

Overall, given the vital functions of the endomembrane network, the high redundancy found among its fundamental actors, and the rapid and dynamic nature of its processes, the development of additional approaches besides classical genetics is extremely valuable.

5.2.7 CONCLUDING REMARKS

The use of small molecules has improved our understanding of subcellular trafficking by helping to resolve issues related to highly complex and essential functions of the endomembrane system in *Arabidopsis*. The chemical genomics field has evolved tremendously, thanks to the input of combinatorial synthetic chemistry and, increasingly, automated microscopy [63, 67]. The growing number of GFP-tagged membrane marker expressing plant lines and the increasing power of computational techniques, such as image analysis, will maintain endomembrane trafficking studies at the forefront of chemical genomics. Automated data collection and analysis techniques will allow for more rapid screening for novel chemicals expanding the chemical toolbox for researchers studying endomembrane trafficking. Cheminformatic and bioinformatic advances will enable the correlation of chemical structures with their biological effects and will further help our understanding of plant trafficking networks and contribute in the development of targeted chemical libraries.

ACKNOWLEDGMENTS

We thank Jan Zouhar for the helpful discussion and Jan Zouhar, Glenn Hicks, Yohann Boutté, Olivier Keech, and Rozenn S.L. Lehir for the critical reading of the manuscript. Vetenskapsråde and VINNOVA supported this work. A.H. is a Postdoctoral Fellow of the Kempe Foundation.

REFERENCES

1. Sanderfoot, A., Assaad, F., Raikhel, N. (2000). The Arabidopsis genome. An abundance of soluble N-ethylmaleimide-sensitive factor adaptor protein receptors. *Plant Physiology*, *124*, 1558–1569.
2. Rojo, E., et al. (2003). The AtC-VPS protein complex is localized to the tonoplast and the prevacuolar compartment in *Arabidopsis*. *Molecular Biology of the Cell*, *14*, 361–369.
3. Mohr, H., Schopfer, P., *Plant Physiology*, Berlin: Springer-Verlag, 1995.
4. Ebine, K., et al. (2008). A SNARE complex unique to seed plants is required for protein storage vacuole biogenesis and seed development of *Arabidopsis thaliana*. *The Plant Cell*, *20*, 3006–3021.
5. Jürgens, G., Geldner, N. (2002). Protein secretion in plants: from the trans-Golgi network to the outer space. *Traffic*, *3*, 605–613.
6. Dettmer, J., et al. (2006). Vacuolar H^+-ATPase activity is required for endocytic and secretory trafficking in *Arabidopsis*. *The Plant Cell*, *18*, 715–730.
7. Gomez, L., Chrispeels, M. (1993). Tonoplast and soluble vacuolar proteins are targeted by different mechanisms. *The Plant Cell*, *5*, 1113–1124.
8. Jolliffe, N., Craddock, C., Frigerio, L. (2005). Pathways for protein transport to seed storage vacuoles. *Biochemical Society Transactions*, *33*, 1016–1018.
9. Friml, J., et al. (2002). Lateral relocation of auxin efflux regulator PIN3 mediates tropism in *Arabidopsis*. *Nature*, *415*, 806–809.

10. Kleine-Vehn, J., et al. (2010). Gravity-induced PIN transcytosis for polarization of auxin fluxes in gravity-sensing root cells. *Proceedings of the National Academy of Sciences, 107*, 22344–22349.
11. Robert, S., et al. (2008). Endosidin1 defines a compartment involved in endocytosis of the brassinosteroid receptor BRI1 and the auxin transporters PIN2 and AUX1. *Proceedings of the National Academy of Sciences, 105*, 8464–8469.
12. Geldner, N., et al. (2007). Endosomal signaling of plant steroid receptor kinase BRI1. *Genes & Development, 21*, 1598–1602.
13. Wang, Y., et al. (2002). Endosomal signaling of epidermal growth factor receptor stimulates signal transduction pathways leading to cell survival. *Molecular and Cellular Biology, 22*, 7279–7290.
14. Hoefle, C., Hückelhoven, R. (2008). Enemy at the gates: traffic at the plant cell pathogen interface. *Cellular Microbiology, 10*, 2400–2407.
15. Robatzek, S., Chinchilla, D., Boller, T. (2006). Ligand-induced endocytosis of the pattern recognition receptor FLS2 in *Arabidopsis. Genes & Development, 20*, 537–542.
16. Robert, S., Raikhel, N., Hicks, G. (2009). Powerful partners: *Arabidopsis* and chemical genomics. In *The Arabidopsis Book*. American Society of Plant Biologists, Rockville, MD, pp 1–16.
17. Yano, D., et al. (2003). A SNARE complex containing SGR3/AtVAM3 and ZIG/VTI11 in gravity-sensing cells is important for *Arabidopsis* shoot gravitropism. *Proceedings of the National Academy of Sciences, 100*, 8589–8594.
18. Surpin, M., et al. (2005). The power of chemical genomics to study the link between endomembrane system components and the gravitropic response. *Proceedings of the National Academy of Sciences, 102*, 4902–4907.
19. Rojas-Pierce, M., et al. (2007). *Arabidopsis* P-glycoprotein19 participates in the inhibition of gravitropism by gravacin. *Chemistry & Biology, 14*, 1366–1376.
20. Geisler, M., Murphy, A. (2006). The ABC of auxin transport: the role of P-glycoproteins in plant development. *FEBS Letters, 580*, 1094–1102.
21. Blakeslee, J., et al. (2007). Interactions among PIN-FORMED and P-glycoprotein auxin transporters in Arabidopsis. *The Plant cell, 19*, 131–147.
22. Lewis, D., et al. (2007). Separating the roles of acropetal and basipetal auxin transport on gravitropism with mutations in two *Arabidopsis* multidrug resistance-like ABC transporter genes. *The Plant Cell, 19*, 1838–1850.
23. Mravec, J., et al. (2008). Interaction of PIN and PGP transport mechanisms in auxin distribution-dependent development. *Development, 135*, 3345–3354.
24. Moscatelli, A., Idilli, A. (2009) Pollen tube growth: a delicate equilibrium between secretory and endocytic pathways. *Journal of Integrative Plant Biology, 51*, 727–739.
25. Lavy, M., et al. (2007). A novel ROP/RAC effector links cell polarity, root-meristem maintenance, and vesicle trafficking. *Current Biology, 17*, 947–952.
26. Rojo, E., et al. (2001). VACUOLELESS1 is an essential gene required for vacuole formation and morphogenesis in *Arabidopsis. Developmental Cell, 1*, 303–310.
27. Horazdovsky, B., Emr, S. (1993). The VPS16 gene product associates with a sedimentable protein complex and is essential for vacuolar protein sorting in yeast. *Journal of Biological Chemistry, 268*, 4953–4962.
28. Jung, G., Ueno, H., Hayashi, R. (1999). Carboxypeptidase Y: structural basis for protein sorting and catalytic triad. *Journal of Biochemistry, 126*, 1–6.

29. Zouhar, J., Hicks, G., Raikhel, N. (2004). Sorting inhibitors (Sortins): chemical compounds to study vacuolar sorting in *Arabidopsis*. *Proceedings of the National Academy of Sciences*, *101*, 9497–9501.
30. Rosado, A., et al. (2011). Sortin1-hypersensitive mutants link vacuolar-trafficking defects and flavonoid metabolism in Arabidopsis vegetative tissues. *Chemistry & Biology*, *18*, 187–197.
31. Norambuena, L., et al. (2008). Identification of cellular pathways affected by Sortin 2, a synthetic compound that affects protein targeting to the vacuole in *Saccharomyces cerevisiae*. *BMC Chemical Biology*, *8*, 1–8.
32. Peyroche, A., et al. (1999). Brefeldin A acts to stabilize an abortive ARF-GDP-Sec7 domain protein complex: involvement of specific residues of the Sec7 domain. *Molecular Cell*, *3*, 275–285.
33. Singleton, V., Bohonos, N., Ullstrup, A. (1953). Decumbin, a new compound from a species of *Penicillium*. *Nature*, *181*, 1072–1073.
34. Klausner, R., Donaldson, J., Lippincott-Schwartz, J. (1992). Brefeldin A: insights into the control of membrane traffic and organelle structure. *The Journal of Cell Biology*, *116*, 1071–1080.
35. Misumi, Y., et al. (1986). Novel blockade by brefeldin A of intracellular transport of secretory proteins in cultured rat hepatocytes. *Journal of Biological Chemistry*, *261*, 11396–11403.
36. Ritzenthaler, C., et al. (2002). Reevaluation of the effects of brefeldin A on plant cells using tobacco Bright Yellow 2 cells expressing Golgi-targeted green fluorescent protein and COPI antisera. *The Plant Cell*, *14*, 237–261.
37. Satiat-Jeunemaitre, B., Hawes, C. (1992). Redistribution of a Golgi glycoprotein in plant cells treated with brefeldin A. *Journal of Cell Science*, *103*, 1153–1166.
38. Nebenfuhr, A., et al. (1999). Stop-and-go movements of plant Golgi stacks are mediated by the acto-myosin system. *Plant Physiology*, *121*, 1127–1141.
39. Satiat-Jeunemaitre, B., et al. (1996). Brefeldin A effects in plant and fungal cells: something new about vesicle trafficking? *Journal of Microscopy*, *181*, 162–177.
40. Staehelin, L., Driouich, A. (1997). Brefeldin A effects in plants. *Plant Physiology*, *114*, 401–403.
41. Robinson, D., et al. (2008). BFA effects are tissue and not just plant specific. *Trends in Plant Science*, *13*, 405–408.
42. Baluška, F., et al. (2002). F-actin-dependent endocytosis of cell wall pectins in meristematic root cells. Insights from brefeldin A-induced compartments. *Plant Physiology*, *130*, 422–431.
43. Geldner, N., et al. (2001). Auxin transport inhibitors block PIN1 cycling and vesicle trafficking. *Nature*, *413*, 425–428.
44. Geldner, N., et al. (2003). The *Arabidopsis* GNOM ARF-GEF mediates endosomal recycling, auxin transport, and auxin-dependent plant growth. *Cell*, *112*, 219–230.
45. Kleine-Vehn, J., et al. (2006). Subcellular trafficking of the Arabidopsis auxin influx carrier AUX1 uses a novel pathway distinct from PIN1. *The Plant Cell*, *18*, 3171–3181.
46. Kleine-Vehn, J., et al. (2008). Cellular and molecular requirements for polar PIN targeting and transcytosis in plants. *Molecular Plant*, *1*, 1056–1066.
47. Tanaka, H., et al. (2009). Fluorescence imaging-based screen identifies ARF GEF component of early endosomal trafficking. *Current Biology*, *19*, 391–397.
48. Dhonukshe, P., et al. (2008). Auxin transport inhibitors impair vesicle motility and actin cytoskeleton dynamics in diverse eukaryotes. *Proceedings of the National Academy of Sciences*, *105*, 4489–4494.

49. Kim, H., et al. (2005). Actin filaments play a critical role in vacuolar trafficking at the Golgi complex in plant cells. *The Plant Cell*, *17*, 888–902.
50. Chen, T., et al. (2007). Disruption of actin filaments by latrunculin B affects cell wall construction in *Picea meyeri* pollen tube by disturbing vesicle trafficking. *Plant and Cell Physiology*, *48*, 19–30.
51. Coué, M., et al. (1987). Inhibition of actin polymerization by latrunculin A. *FEBS Letters*, *213*, 316–318.
52. DeBolt, S., et al. (2007). Morlin, an inhibitor of cortical microtubule dynamics and cellulose synthase movement. *Proceedings of the National Academy of Sciences*, *104*, 5854–5859.
53. Gutierrez, R., et al. (2009). *Arabidopsis* cortical microtubules position cellulose synthase delivery to the plasma membrane and interact with cellulose synthase trafficking compartments. *Nature Cell Biology*, *11*, 797–808.
54. Crowell, E., et al. (2009). Pausing of Golgi bodies on microtubules regulates secretion of cellulose synthase complexes in *Arabidopsis*. *The Plant Cell*, *21*, 1141–1154.
55. Boutté, Y., et al. (2006). The plasma membrane recycling pathway and cell polarity in plants: studies on PIN proteins. *Journal of Cell Science*, *119*, 1255–1265.
56. Tóth, R., van der Hoorn, R. (2010). Emerging principles in plant chemical genetics. *Trends in Plant Science*, *15*, 81–89.
57. Hicks, G., Raikhel, N. (2009). Opportunities and challenges in plant chemical biology. *Nature Chemical Biology*, *5*, 268–272.
58. Cutler, S., McCourt, P. (2005). Dude, where's my phenotype? Dealing with redundancy in signaling networks. *Plant Physiology*, *138*, 558–559.
59. Zeeh, J., et al. (2006). Dual specificity of the interfacial inhibitor brefeldin A for Arf proteins and Sec7 domains. *Journal of Biological Chemistry*, *281*, 11805–11814.
60. Kleine-Vehn, J., et al. (2008). ARF GEF-dependent transcytosis and polar delivery of PIN auxin carriers in *Arabidopsis*. *Current Biology*, *18*, 526–531.
61. Robert, S., et al. (2010). ABP1 mediates auxin inhibition of clathrin-dependent endocytosis in *Arabidopsis*. *Cell*, *143*, 111–121.
62. Sanderfoot, A., et al. (2001). Disruption of individual members of *Arabidopsis* syntaxin gene families indicates each has essential functions. *The Plant Cell*, *13*, 659–666.
63. Avila, E., et al. (2003) Tools to study plant organelle biogenesis. Point mutation lines with disrupted vacuoles and high-speed confocal screening of green fluorescent protein-tagged organelles. *Plant Physiology*, *133*, 1673–1676.
64. Drakakaki, G., et al. (2009). Chemical dissection of endosomal pathways. *Plant Signaling & Behaviour*, *4*, 57–62.
65. Serrano, M., Guzmán, P. (2004). Isolation and gene expression analysis of *Arabidopsis thaliana* mutants with constitutive expression of ATL2, an early elicitor-response RING-H2 zinc-finger gene. *Genetics*, *167*, 919–929.
66. Serrano, M., et al. (2007). Chemical interference of pathogen-associated molecular pattern-triggered immune responses in *Arabidopsis* reveals a potential role for fatty-acid synthase type II complex-derived lipid signals. *Journal of Biological Chemistry*, *282*, 6803–6811.
67. DeBlasio, S., Sylvester, A., Jackson, D. (2010). Illuminating plant biology: using fluorescent proteins for high-throughput analysis of protein localization and function in plants. *Briefings in Functional Genomics*, *9*, 129–138.

PART SIX

TARGET IDENTIFICATION

6.1

TARGET IDENTIFICATION OF BIOLOGICALLY ACTIVE SMALL MOLECULES

PAUL OVERVOORDE AND DOMINIQUE AUDENAERT

6.1.1 INTRODUCTION

Cellular activity emerges from the diffusion of molecules throughout the cell. Molecules that are less than approximately 600 daltons facilitate cellular metabolism and regulate the conformation, localization, and activity of macromolecules. The emergence of efficient diversity-oriented chemical synthesis enabled libraries of novel chemical molecules that can be screened for biological activity. The validation of potential "lead" molecules provides novel tools for exploring and manipulating a biological process of interest. While lead molecules selected via high-throughput screens provide non-endogenous compounds that affect a particular process, a thorough understanding of its effect at the cellular level requires the identification of the target protein(s). In a manner that parallels screening for mutants that affect a biological process of interest, identification of chemical probes can be approached as target-based or phenotype-based screens.

The target-based approach is comparable to reverse genetic screening and entails purification of a protein of interest, subsequently screening for compounds that interfere with the activity of the protein and finally using this compound to determine the phenotypic consequences of altering the function of this protein in a cellular context. Assays of this kind often involve biochemical assays in cell-free format that reports the activity of an enzyme, receptor–ligand binding, or protein–protein interaction. Because these are homogeneous assays, they can be adapted for high-throughput screening (e.g., 1536- or 3456-well plates) with the

Plant Chemical Biology, First Edition. Edited by Dominique Audenaert and Paul Overvoorde.

assistance of automated liquid handling systems. The lead compounds identified via target-based screenings might have limitations during later stages of the screening process, such as when applied in the context of cellular systems or whole organisms. These limitations include possible cytotoxic effects, the inability to enter cells, and off-target effects. Each of these would go unnoticed during the screening campaign itself.

Analogous to a forward genetics screen, the phenotype-based approach depends on a robust biological assay that uses a cell-based system or a small model organism to identify compounds that impact a phenotype of interest. Morphological changes or reporter gene-based analysis serves as outputs in these screens. Compounds that alter the phenotype of interest are then selected, and the target protein(s) of these hit compounds is identified. Phenotypic assays have the advantage of being performed under more relevant physiological contexts. In addition, phenotypic screening selects membrane-permeable compounds and allows the identification of cytotoxic compounds. Unlike target-based assays, phenotypic assays do not require a priori knowledge of a protein target and aim to identify compounds that alter known or novel pathways that impact a biological process rather than targeting a specific protein. As such, the phenotype-based screens allow identification of distinct chemical classes that act on different primary protein targets, which could generate similar phenotypes.

An inherent limitation for the phenotypic approach arises from the lack of knowledge of what proteins serve as primary target(s) of the compound. Identification of the protein targets represents the most challenging step during the process of phenotype-based screening. Although no single systematic solution to this challenge exists, the most widely used methods to address this challenge can be placed into two broad categories: affinity-based and genetic. In the remainder of this chapter, we provide an overview of these methods and describe specific examples where these approaches have been successfully utilized.

6.1.2 AFFINITY-BASED APPROACHES TO IDENTIFY PROTEIN TARGETS

For the affinity-based approaches, specific ligand–protein interactions form the basis of these techniques. In addition to the widely used affinity purification approach, alternative proteomic techniques that take advantage of genomic resources or that attempt to directly detect protein targets of small molecules also fall into this category. Regardless of the approach, inherent pros and cons exist when considering any given protein–target interaction.

Affinity purification based on ligand–protein interactions is a well-known technique [1], but its widespread use for identifying cellular protein targets of small molecules remains limited by a number of factors. First, affinity-based approaches require a thorough understanding of the structure–activity relationship (SAR) of a compound. In SAR analysis, the effect of modifying or removing functional groups and/or substructures on the biological activity of a compound defines the substituents that are necessary for its activity. A defined SAR provides important information about whether a compound can undergo substitutions or elaborations (e.g., addition of a linker molecule). In cases where any substitution of functional groups disrupts activity, affinity approaches become limited. An example of such a molecule that does not tolerate any modification is bikinin, an inhibitor of *Arabidopsis thaliana* GSK3-like kinases [2]. Bikinin interacts with the ATP-binding pocket of these kinases, and any alteration of the substituents of this compound drastically decreases its activity. In such cases, identification of the target must rely on prior knowledge of a pathway or require genetic approaches.

The second potential limitation of affinity purification arises from concerns about the specificity of protein interactions with the matrix or the molecule of interest. After the molecule has been immobilized, it is mixed with cell lysates to allow the compound to form a complex with its protein target. After extensive washing to remove nonspecific bound proteins, the bound proteins are eluted, separated by electrophoresis, and identified by mass spectrometry. In order to avoid false-positive identification, several approaches can be employed.

One approach to enhance the likelihood of finding specific targets involves using negative controls that consist of a matrix coupled to an inactive analog. Alternatively, active compared to inactive compounds can be used as a basis for binding competition experiments. Comparison of proteins isolated from the active compound with the proteins isolated from a close compound analog that lacks biological activity can help pinpoint target candidates. A study that examined the selectivity of the cyclin-dependent kinase (CDK) inhibitor purvalanol B highlights the benefits of this approach [3]. Structure–activity relationship analysis showed that the addition of a methyl group greatly reduced the CDK inhibitory activity of purvalanol, which could serve as a negative control. Based upon the CDK2-purvalanol crystal structure, variants of the active purvalanol and the inactive methylated purvalanol were made and immobilized on agarose beads. Comparative affinity chromatography revealed seven protein kinases that bind selectively to purvalanol, but not to methylated purvalanol. For several of these recovered kinases, which were not previously known to be targets of purvalanol, biochemical assays indeed showed inhibition of kinase activity by purvalanol [3].

An additional consideration is that the biological inactivity of the negative control molecule should be the consequence of its inability to bind the target. For example, if a compound's activity is decreased due to reduced solubility or membrane permeability, this type of compound will still be able to bind the target proteins during the purification steps and no differences in protein binding will be observed. This concern could be overcome by immobilizing the active compound at a site that is crucial for its activity, implying impaired interaction with its target proteins, and using this compound as a negative control. Ding et al. [4] used this approach to identify the target of TWS119, a synthetic molecule that can induce neurogenesis in murine embryonic stem cells (SCs). Structure–activity relationship data indicated that TWS119 could be linked to an affinity matrix via its anilino group without losing biological activity. An active analog of the compound was linked via a nonpermissive site for activity and used as a negative control. Via SDS/PAGE, two bands were identified that were specifically bound to the affinity matrix containing the active compound. These two bands were identified to be two splicing isoforms of GSK3β via LC/MS. Subsequent bimolecular interaction analysis and biochemical assays independently validated these interactions.

An alternative strategy that does not require an inactive variant of the compound of interest involves including a sample that includes an excess of free compound in the lysate-binding suspension. The free compound will compete with the immobilized compound and reduce the amount of target protein available for pulldown. Proteins that show a reduced recovery in the competition experiment are likely to be genuine targets. An example of this approach comes from the work of Yang et al. [5], who identified the target of quinostatin, a modulator of the mammalian target-of-rapamycin (mTOR) signaling network. The affinity approach yielded dozens of silver-stained bands. When the binding solution contained 100 μM of free quinostatin, the binding of a single band decreased. This protein was then identified as the catalytic subunit of class Ia phosphatidylinositol 3-kinase.

The third potential limitation of the affinity purification comes from the affinity of a target protein for the small molecule of interest. Proteins with a low K_D for a compound are considered to be well suited for target identification. Such high-affinity interactions allow for more extensive washing steps and reduce nonspecific protein retention. In contrast, lower-affinity interactions would be more transient and could lead to their loss during the washing procedure. These concerns are magnified if the target protein is to remain part of a complex in order to retain its ability to interact with the compound of interest. The inclusion of a photoaffinity moiety in the modified compound can alleviate this concern [6]. Upon radiation, the photoreactive group forms a covalent bond with proteins in close proximity and as such stabilizes the compound–target complex during the washing steps. Another concern related to affinity is the abundance of the target. For example, if a protein target is of low abundance or expressed in a limited subset of cells, even if it has a high affinity for a compound, the underrepresentation of the target in the extract presents technical challenges. On the other hand, identifying an abundant protein target that has low affinity for a compound of interest remains possible. For example, Sato et al. [7] identified glyoxylase 1 (GLO1), an abundant metabolic enzyme, as a target of indomethacin despite having a low affinity for the enzyme.

The fourth potential limitation of the affinity purification involves consideration of the column elution strategy. Applying strong denaturation steps for elution results in many nonspecific proteins being released into the eluate. This might mask the separation of proteins from the genuine target, especially if the target is of limited abundance. An alternative approach involves incorporating a cleavable linker between the compound of interest and the affinity tag such as a disulfide bridge [8]. Addition of a reducing agent during the elution step releases the compound and its associated proteins from the matrix without releasing nonspecific binders to the column. In addition, more selective elution procedures that involve adding the free, unlabeled compound in the eluent can be used. Emami et al. [9] used selective elution to successfully identify the target of ICG-001, a compound that downregulates signaling by catenin/T cell factor. Elution with a denaturation agent resulted in dozens of nonspecific proteins, whereas selective elution with free ICG-001 released a single protein that was identified as CREB-binding protein. Because CREB-binding protein is a low-abundant transcription factor, this study illustrates the strength of selective elution for identifying rare proteins.

If a molecule becomes covalently attached to a target protein *in vivo*, an alternative strategy for identifying these proteins has been described. In order for this approach to work, the compound must either contain a reactive group (e.g., epoxide, vinyl sulphone) or retain biological activity when such a group is introduced. Kaschani et al. [10] demonstrated that the two-step labeling procedure developed by Speers and Cravatt [11] for target identification in mice also works in plants. This approach involves synthesizing bioactive compounds with a terminal, biologically inert azide or alkyne, which is used in the second step to attach biotin via "click chemistry". These molecules are then fed to either detached leaves, seedlings, or cell cultures where they become covalently attached to their targets. The second step of the labeling involves coupling biotin to the labeled molecules through the azide or alkyne in a reaction that is catalyzed by copper. These biotin-labeled proteins are now ready to be detected or purified through the use of streptavidin-conjugated reagents. Using this approach, Kaschani et al. [10] were able to identify targets of E-64, an inhibitor of cysteine proteins, and MVA178, a proteasome inhibitor.

6.1.3 VARIATION ON THE AFFINITY APPROACH TO TARGET IDENTIFICATION

In addition to using biological tissues as sources for target proteins, other approaches take advantage of heterologous expression of proteins to identify compound targets. The first of these is a molecular genetic technique referred to as the yeast three-hybrid (Y3H) system, which represents a variation of the widely known and used yeast two-hybrid (Y2H) system [12]. The Y2H reveals the interaction of two proteins through the formation of a binary complex that contains an activation domain (AD) and a DNA-binding domain (BD). The small molecule Y3H system extends this by involving the formation of a ternary complex composed of a protein fused to the AD, a protein fused to the BD, and a molecule that serves as a "chemical inducer of dimerization" (CID). Unlike the Y2H system, it is not the direct interaction of two proteins that leads to the formation of an active transcription factor, but rather the interaction of fusion proteins with components of the bifunctional CID.

Several variations of the small molecule Y3H system have been developed (Table 6.1.1). The iterations involve changes in the ligand-binding proteins used, which alter the nature of the CID, and alternative DNA-BDs that interact with distinct DNA targets to control the expression of various reporter genes. The initial report of the Y3H made use of a mutant version of the rat glucocorticoid receptor (rGR) that had been selected for higher affinity and protein expression in yeast fused to the B42 bacterial AD and the lexA DNA-binding domain (lexA-BD) fused to the FK506-binding protein [13]. The CID used in these experiments contained dexamethasone (DEX) connected to FK506 by a 10-carbon linker. In addition to showing that a ternary complex forms to activate reporter gene expression, Licitra and Liu demonstrated

TABLE 6.1.1 Combinations of fusion proteins used in Y3H systems.

Ligand-binding protein fusion 1	Ligand-binding protein fusion 2	Chemical inducer of dimerization (CID)	Reporter genes	References
lexBD-rGR	B42-FKBP12	MTX-FK506	LEU2 β-galactosidase	[13]
lexBD-eDHFR	Gal4AD-rGR	MTX-DEX	HIS3	[17]
lexBD-eDHFR	B42-rGR	MTX-DEX	β-galactosidase	[18]
lexBD-eDHFR	B42-FKBP12 + variants	MTX-SLF	β-galactosidase LEU2 HIS3	[19]
lexBD-eDHFR	B42-rGR	MTX-DEX	β-galactosidase LEU2	[16]
lexBD-eDHFR	B42-rGR	MTX-DEX TMP-DEX	β-galactosidase	[20]
Gal4BD-rGR	Gal4AD-eDHFR + variants	MTX-DEX	β-galactosidase HIS3	[15]
Gal4BD-rGR	Gal4AD-FKBP12	DEX-FK506	β-galactosidase HIS3	[15]

lexBD, DNA-binding domain from lexA protein; rGR, rat glucocorticoid receptor hormone-binding domain with F620S and C656G mutations; B42, bacterial-derived transcriptional AD [21]; FKBP12, FK506-binding protein; MTX, methotrexate; LEU2, required for leucine biosynthesis in yeast; eDHFR, *E. coli*-derived dihydrofolate reductase; Gal4AD, transcriptional AD of the yeast Gal4 protein; DEX, dexamethasone; HIS3, required for histidine biosynthesis in yeast; SLF, synthetic ligand for FKBP12; TMP, trimethoprim.

that a human FK506-binding protein could be isolated from a Jurkat cell cDNA library when introduced to yeast cells expressing the lexBD-rGR and grown in the presence of DEX-FK506. This result indicates that the Y3H system has the potential to identify protein targets of small molecules. After this initial report, Lin et al. reported that the original system could be modified by substituting an *Escherichia coli*-derived dihydrofolate reducatase (DHFR) fused to the lexA-BD for the FK506-binding protein and using a DEX-methotrexate (MTX) as the CID [14]. Subsequently, Henthorn et al. [15] adopted the use of a GAL4 promoter system with the DHFR and FK506-binding protein and further showed the utility of this approach by identifying four DHFR clones from a mouse cDNA library screen for proteins able to interact with MTX. The Gal4-based system was also used to optimize a DEX-MTX system [16].

Evidence that the Y3H system could be used to identify targets of a compound from a cDNA library was established in a screen to characterize multiple ATP-competitive active site kinase inhibitors [17]. Using MTX as an "anchor" and three or more ethyleneglycol units as a linker, they attached either purvalanol B or roscovitine to create novel CIDs. They then introduced cDNA libraries prepared from a number of human tissues into a yeast strain expressing lexA-BD fused to the *E. coli* DHFR and selected for the activation of reporter gene expression. Through the use of MTX-purvalanol B in this screen, they validated 35 kinase interactions, 15 of which represented novel targets of this compound. Of these targets, several were in the low micromolar range, suggesting that low-affinity interactions could be identified using this approach.

Two other three-hybrid systems complement the Y3H system by not relying on nuclear translocation of the ternary complex. The first is a modification of the split ubiquitin system [22]. This system relies on the association of amino- (N_{ub}) and carboxy- (C_{ub}) halves of the ubiquitin protein. The reassembly of a quasi-native ubiquitin is recognized by ubiquitin-specific proteases, which leads to the removal of Ura3p. The removal of Ura3p can be monitored by lack of yeast growth on media that lack uracil or the ability of yeast to grow in the presence of 5-fluoroorotic acid, which is converted to a toxic compound by the enzyme activity of Ura3p. Dirnberger et al. [23] adapted this system by using a DEX-MTX molecule to induce the assembly of an N_{ub}-DHFR with a GR-C_{ub}-URA3p. This system broadens the types of proteins (e.g., membrane proteins) that might be detected using this type of approach. The second system addresses an additional potential limitation of the yeast system: in order to maintain a high-enough affinity for the compound, a target protein needs to be correctly folded or might require posttranslational modifications or assembly in a multimeric complex. In order to circumvent these issues, Caligiuri et al. [24] adapted the mammalian protein–protein interaction trap (MAPPIT) for use as a three-hybrid system. Upon a protein–protein interaction, the MAPPIT system restores ligand-dependent activation of the JAK/STAT pathway to a signaling-deficient mutant cytokine receptor leading to changes in gene expression. The *ma*mmalian *s*mall-molecule–*p*rotein *i*nteraction *t*rap (MASPIT) system takes advantage of the MAPPIT concept and depends on the interactions of MTX-compound hybrid molecules to activate reporter gene expression. This system works with a number of small-molecule chemotypes and can be used to identify protein targets of these compounds [24].

Phage display is another strategy in which expression libraries are screened to identify the target of a compound. In this approach, a phage display library is constructed by packing a cDNA library within phages. The protein encoded by the cDNA is displayed as a fusion protein with the coat protein at the surface of the phage. The phage population is passed over a column containing immobilized compound and phage clones that display the target protein are retained onto the column. After washing, retained phages are eluted, rescued by immediate infection in *E. coli*, and amplified. The identity of the target protein can be determined by isolation and sequencing the phage DNA. An advantage is that, prior to sequencing, the amplified phage can be used in additional selection rounds to enrich the target proteins. Jin et al. [25] used a T7 phage library derived from human liver cDNA to identify binding proteins of the anticancer drug doxorubicin. After four rounds of elution, the nucleolar phosphoprotein hNopp140 was identified as a direct target of the compound. The interaction between doxorubicin and hNopp140 was independently validated via surface plasmon resonance analysis [25]. This approach has also been used to identify FK506 and taxol-binding proteins [26, 27].

In the format of proteome microarrays, direct target identification of a compound can be achieved by immobilizing the proteome on a chip. In this method, the compound of interest is labeled (e.g., with a fluorescent tag) and added to the chip Subsequent washing and scanning of the array reveals the physical location of the compound on the chip and allows identification of the protein target. Huang et al. [28] used this approach to identify the targets of compounds that modify the cellular effects of rapamycin, which is a polyketide macrolide with antiproliferative activity. Yeast proteome chips were probed with biotinylated small-molecule inhibitors of rapamycin (SMIRs). Several cellular targets were identified including Tep1p, a homolog of the mammalian tumor suppressor phosphatase and tensin homolog deleted on chromosome 10 (PTEN) and Ybr077cp (Nir1p). The function of Nir1p was previously unknown but was shown to be a component of the target-of-rapamycin signaling network. The targets of the SMIRs were validated by analyzing sensitivity and resistance of yeast deletion mutant strains [28].

6.1.4 LABEL-FREE APPROACHES

The previously described methods all require labeling of the compound with a tag. Modification of the compound might affect the interaction with its target. In addition, the target must be displayed in a correctly folded confirmation, which might be an issue with expression libraries and proteome arrays. To overcome these limitations, Lomenick et al. [29] developed a method that they named *d*rug *a*ffinity *r*esponsive *t*arget *s*tability (DARTS). It is a universal target identification approach that is based upon the observation that a protein becomes less susceptible to proteolysis when it is drug bound than when it is drug-free [30]. Target proteins are stabilized upon binding to the drug in a global or local fashion, for example, in a specific conformation or by masking proteolytic cleavage sites, which makes them less susceptible to proteases [29]. Comparing proteolytic degradation of the proteome both in the absence and in the presence of the compound allows the protein target to be revealed. Compared to affinity-based target identification, a major difference is the way by which enrichment of the target proteins occurs. In affinity chromatography, washing away unbound proteins and selectively pulling out target proteins achieve positive enrichment. In DARTS, negative enrichment occurs by digesting away nontarget proteins and leaving behind the protease-resistant target proteins. In addition, unlike affinity-based approaches, no washing steps are required for DARTS, which also allows identifying lower-affinity interactions [31].

A similar approach to DARTS is *s*tability of *p*roteins from *r*ates of *ox*idation (SPROX) [32]. SPROX is based upon changes in folding and thermodynamic stability of proteins upon interaction with small molecules. Oxidation rates of methionine residues are utilized as a readout to report the thermodynamic properties of global and/or local unfolding/refolding reactions [32]. However, unlike affinity-based approaches and DARTS, a major limitation of the SPROX methodology is the lack of enrichments steps during the procedure. This makes it difficult to detect low-abundant proteins, unless extensive up-front fractionation of the sample has been performed [31].

A final category of label-free approaches includes the use of information gained by exploring patterns of gene expression when organisms are treated with a molecule. This approach was pioneered in yeast, where primary and secondary drug targets have been described [33]. Online resources that compute and display correlations between gene expression patterns from different experiments have been developed for *Arabidopsis* data sets [34]. While this approach is indirect, it can inform subsequent experiments that reveal the primary protein target. In addition, responses to a combination of chemical molecules hold the potential to uncover functional connections between components of a system that might not be revealed by examining individual pieces. Such an approach has been referred to as "combination chemical genetics" (CCG) [35]. CCG has been most well developed experimentally using yeast where changes in the landscape of a cell upon treatment with one or more compounds have enabled drug target identification [36].

6.1.5 GENETIC APPROACHES FOR IDENTIFYING TARGETS OR PATHWAYS OF NOVEL COMPOUNDS

For plant chemical biologists, the primary mode of target identification to date involves the identification of *e*thyl *m*ethane*s*ulfonate (EMS)-induced mutants that do not respond to a compound of interest. While this approach has yielded several success stories—see chapters on ABA, auxin, and the endomembrane system for detailed examples—the fact that the targets of many molecules with bioactivity in plants remain to be identified highlights this approach as a bottleneck. Furthermore, working through mutants in a "one by one" manner limits this approach and highlights the need to develop additional strategies. Recent advances in NexGen sequencing and associated statistical methods have accelerated the pace of cloning by reducing the number of recombinants required to identify a mutant's map position [37]. In addition, the development of community resources for reference organisms such as *A. thaliana* and rice allows additional alleles to be identified and function tested. In *Arabidopsis*, additional loss-of-function analysis is facilitated by the large database of T-DNA insertional mutants [38]. In a complementary manner, the *Arabidopsis* full-length cDNA overexpressing gene (FOX) hunting system and RiceFOX represent additional gain-of-function resources [39, 40].

The large collection of cataloged ecotypes of *Arabidopsis* has been useful for defining and studying many processes in plant development, physiology, and adaptation [41]. With the growing availability of whole-genome sequences for multiple accessions [42–44], the use of these resources represents a venue for identifying plants that display distinct responses to treatment with endogenous or synthetic compounds. For example, by exploring the response of 20 *Arabidopsis* accessions upon treatment with the naturally occurring auxin (indole-3-acetic acid) and two synthetic auxins (2,4-dinitrophenol and 1-napthaleneacetic acid), Delker et al. [45] observed large phenotypic variation. To further explore the molecular basis for the phenotypic variation, they used microarray analysis to document the transcriptional changes in seven accessions. These data revealed that the distinct phenotypes were not due to single or few polymorphisms giving rise to novel protein isoforms, but rather resulted from changes in the relative steady-state levels of signaling components [45]. A second example of the use of natural variation to describe the mode of action for the small molecule hypostatin, which inhibits hypocotyl elongation in some ecotypes but not others, is described in Chapter 5.1.

REFERENCES

1. Cuatrecasas, P., M. Wilchek, and C.B. Anfinsen, *Selective enzyme purification by affinity chromatography*. Proceedings of the National Academy of Sciences of the United States of America, 1968. **61**(2): p. 636–43.
2. De Rybel, B., et al., *Chemical inhibition of a subset of* Arabidopsis thaliana *GSK3-like kinases activates brassinosteroid signaling*. Chemistry & Biology, 2009. **16**(6): p. 594–604.
3. Knockaert, M, et al., *Intracellular targets of cyclin-dependent kinase inhibitors: identification by affinity chromatography using immobilised inhibitors*. Chemistry & Biology, 2000. **7**(6): p. 411–422.
4. Ding, S., et al., *Synthetic small molecules that control stem cell fate*. Proceedings of the National Academy of Sciences of the United States of America, 2003. **100**(13): p. 7632–7.
5. Yang, J., et al., *Identification of a small-molecule inhibitor of class Ia PI3Ks with cell-based screening*. Chemistry & Biology, 2007. **14**(4): p. 371–7.
6. Mackinnon, A.L. and J. Taunton, *Target identification by diazirine photo-cross-linking and click chemistry*. Current Protocols in Chemical Biology, 2009. **1**: p. 55–73.
7. Sato, S., et al., *Polyproline-rod approach to isolating protein targets of bioactive small molecules: isolation of a new target of indomethacin*. Journal of the American Chemical Society, 2007. **129**(4): p. 873–80.
8. Mano, N., K. Sato and J. Goto, *Specific affinity extraction method for small molecule-binding proteins*. Analytical Chemistry, 2006. **78**(13): p. 4668–4675.
9. Emami, K.H., et al., *A small molecule inhibitor of beta-catenin/CREB-binding protein transcription [corrected]*. Proceedings of the National Academy of Sciences of the United States of America, 2004. **101**(34): p. 12682–7.
10. Kaschani, F., et al., *Minitags for small molecules: detecting targets of reactive small molecules in living plant tissues using 'click chemistry.'* The Plant Journal: For Cell and Molecular Biology, 2009. **57**(2): p. 373–85.
11. Speers, A.E. and B.F. Cravatt, *Profiling enzyme activities in vivo using click chemistry methods*. Chemistry & Biology, 2004. **11**(4): p. 535–46.
12. Fields, S. and O. Song, *A novel genetic system to detect protein–protein interactions*. Nature, 1989. **340**(6230): p. 245–6.
13. Licitra, E.J. and J.O. Liu, *A three-hybrid system for detecting small ligand–protein receptor interactions*. Proceedings of the National Academy of Sciences of the United States of America, 1996. **93**(23): p. 12817–21.
14. Lin, H.N., et al., *Dexamethasone–methotrexate: an efficient chemical inducer of protein dimerization in vivo*. Journal of the American Chemical Society, 2000. **122**(17): p. 4247–8.
15. Henthorn, D.C., A.A. Jaxa-Chamiec, and E. Meldrum, *A GAL4-based yeast three-hybrid system for the identification of small molecule–target protein interactions*. Biochemical Pharmacology, 2002. **63**(9): p. 1619–28.
16. Baker, K., et al., *An optimized dexamethasone–methotrexate yeast 3-hybrid system for high-throughput screening of small molecule–protein interactions*. Analytical Biochemistry, 2003. **315**(1): p. 134–7.

17. Becker, F., et al., *A three-hybrid approach to scanning the proteome for targets of small molecule kinase inhibitors*. Chemistry & Biology, 2004. **11**(2): p. 211–23.
18. Abida, W.M., et al., *Receptor-dependence of the transcription read-out in a small-molecule three-hybrid system*. Chembiochem, 2002. **3**(9): p. 887–95.
19. de Felipe, K.S., et al., *Correlation between ligand–receptor affinity and the transcription readout in a yeast three-hybrid system*. Biochemistry, 2004. **43**(32): p. 10353–63.
20. Gallagher, S.S., L.W. Miller, and V.W. Cornish, *An orthogonal dexamethasone-trimethoprim yeast three-hybrid system*. Analytical Biochemistry, 2007. **363**(1): p. 160–2.
21. Ma, J. and M. Ptashne, *A new class of yeast transcriptional activators*. Cell, 1987. **51**(1): p. 113–9.
22. Johnsson, N. and A. Varshavsky, *Split ubiquitin as a sensor of protein interactions in vivo*. Proceedings of the National Academy of Sciences of the United States of America, 1994. **91**(22): p. 10340–4.
23. Dirnberger, D., et al., *A small-molecule–protein interaction system with split-ubiquitin as sensor*. Chembiochem, 2006. **7**(6): p. 936–42.
24. Caligiuri, M., et al., *MASPIT: Three-hybrid trap for quantitative proteome fingerprinting of small molecule–protein interactions in mammalian cells*. Chemistry & Biology, 2006. **13**(7): p. 711–22.
25. Jin, Y., J. Yu, and Y.G. Yu, *Identification of hNopp140 as a binding partner for doxorubicin with a phage display cloning method*. Chemistry & Biology, 2002. **9**(2): p. 157–62.
26. Rodi, D.J., et al., *Screening of a library of phage-displayed peptides identifies human Bcl-2 as a taxol binding protein*. Journal of Molecular Biology, 1999. **285**(1): p. 197–203.
27. Sche, P.P., et al., *Display cloning: functional identification of natural product receptors using cDNA-phage display*. Chemistry & Biology, 1999. **6**(10): p. 707–16.
28. Huang, J., et al., *Finding new components of the target of rapamycin (TOR) signaling network through chemical genetics and proteome chips*. Proceedings of the National Academy of Sciences of the United States of America, 2004. **101**(47): p. 16594–9.
29. Lomenick, B., et al., *Target identification using drug affinity responsive target stability (DARTS)*. Proceedings of the National Academy of Sciences, 2009. **106**(51): p. 21984–9.
30. Park, C. and S. Marqusee, *Pulse proteolysis: a simple method for quantitative determination of protein stability and ligand binding*. Nature Methods, 2005. **2**(3): p. 207–12.
31. Lomenick, B., R.W. Olsen, and J. Huang, *Identification of direct protein targets of small molecules*. ACS Chemical Biology, 2011. **6**(1): p. 34–46.
32. West, G.M., et al., *Quantitative proteomics approach for identifying protein–drug interactions in complex mixtures using protein stability measurements*. Proceedings of the National Academy of Sciences of the United States of America, 2010. **107**(20): p. 9078–82.
33. Marton, M.J., et al., *Drug target validation and identification of secondary drug target effects using DNA microarrays*. Nature Medicine, 1998. **4**(11): p. 1293–301.
34. Toufighi, K., et al., *The Botany array resource: e-Northerns, expression angling, and promoter analyses*. The Plant Journal: For Cell and Molecular Biology, 2005. **43**(1): p. 153–63.
35. Lehar, J., et al., *Combination chemical genetics*. Nature Chemical Biology, 2008. **4**(11): p. 674–81.

36. Costanzo, M., et al., *The genetic landscape of a cell*. Science, 2010. **327**(5964): p. 425–31.
37. Schneeberger, K., et al., *SHOREmap: simultaneous mapping and mutation identification by deep sequencing*. Nature Methods, 2009. **6**(8): p. 550–1.
38. Alonso, J.M., et al., *Genome-wide insertional mutagenesis of* Arabidopsis thaliana. Science, 2003. **301**(5633): p. 653–7.
39. Sakurai, T., et al., *RiceFOX: a database of Arabidopsis mutant lines overexpressing rice full-length cDNA that contains a wide range of trait information to facilitate analysis of gene function*. Plant & Cell Physiology, 2011. **52**(2): p. 265–73.
40. Ichikawa, T., et al., *The FOX hunting system: an alternative gain-of-function gene hunting technique*. The Plant Journal: For Cell and Molecular Biology, 2006. **48**(6): p. 974–85.
41. Alonso-Blanco, C., et al., *What has natural variation taught us about plant development, physiology, and adaptation?* Plant Cell, 2009. **21**(7): p. 1877–96.
42. Gan, X., et al., *Multiple reference genomes and transcriptomes for* Arabidopsis thaliana. Nature, 2011. **477**(7365): p. 419–23.
43. Weigel, D. and R. Mott, *The 1001 genomes project for* Arabidopsis thaliana. Genome Biology, 2009. **10**(5): p. 107.
44. Cao, J., et al., *Whole-genome sequencing of multiple* Arabidopsis thaliana *populations*. Nature Genetics, 2011. **43**(10): p. 956–63.
45. Delker, C., et al., *Natural variation of transcriptional auxin response networks in* Arabidopsis thaliana. Plant Cell, 2010. **22**(7): p. 2184–200.

PART SEVEN

TRANSLATION OF PLANT CHEMICAL BIOLOGY FROM THE LAB TO THE FIELD

7.1

PROSPECTS AND CHALLENGES FOR TRANSLATING EMERGING INSIGHTS IN PLANT CHEMICAL BIOLOGY INTO NEW AGROCHEMICALS

TERENCE A. WALSH

7.1.1 INTRODUCTION

The global agrochemical industry is largely based on the application of small-molecule active ingredients to crops to protect them from pests and competing weed species or to act as fertilizers to improve growth and crop productivity. These chemical compounds have provided enormous societal benefits in sustaining and improving crop yields over decades and are a salient reminder of the dynamic role that innovations produced from "plant chemical biology" in its broadest sense have contributed to societal needs. With the challenges of a growing global population, shrinking cultivatable land area, growing constraints on water, changing global environments, and food security issues, the pressures to deliver further gains in crop productivity are acute. Thus, there is a continuing and urgent need for further innovations within the agrochemical industry to meet these challenges.

The current market for agrochemicals is large but has remained fairly static in recent years. The global crop protection chemicals segment had estimated sales of about $38 billion in 2010, of which $17 billion (45%) was for herbicides, $10 billion (26%) for fungicides, $10 billion (26%) for insecticides, and $1 billion (3%) in other uses [1]. The research efforts to develop the wide variety of efficacious compounds produced

Plant Chemical Biology, First Edition. Edited by Dominique Audenaert and Paul Overvoorde.

by the crop protection industry have resulted in significant insights into the types of exogenously applied compounds that are bioactive in plants and in-depth understanding of their effects on plants and metabolic fate in plant tissues. Thus, many of the tenets and principles of plant chemical biology [2, 3], especially in the search for new active molecules, have common foundations in the principles of agrochemical discovery and development and the use of agrochemicals as plant perturbing agents.

This shared foundation is well illustrated by the compound 2,4-dichlorophenoxyacetic acid (2,4-D), a herbicide first developed and commercialized during World War II and still in extensive use today [4, 5]. Indeed, it is finding a potential rejuvenation in new applications via plant biotechnological approaches [6]. As well as its commercial uses, 2,4-D has also been a mainstay for plant biologists as a convenient auxin surrogate, as it mimics the natural auxin indole-3-acetic acid [7] but is not subject to the fine metabolic control that the native hormone has. Its attractive properties as a herbicide that include ready uptake by tissues and cells and subsequent *in planta* mobility and cellular longevity are the same desirable properties that are required for a robust chemical probe. The use of 2,4-D as an auxin surrogate led to the identification of auxin-resistant *Arabidopsis* mutants [8] that were keys to the eventual identification of the TIR1/AFB class of auxin receptors [9, 10]. This dissection of auxin action using the small molecule 2,4-D led to the discovery of a signaling mechanism and pathway that was new to science, over 40 years after the original development of the compound [11]. The journey from agrochemical to informative chemical probe to discovery of a novel plant process may yet complete a full circle as new insights into the auxin signaling mechanism revealed by 2,4-D could offer novel sites to perturb plant growth and development and spur new rounds of agrochemical innovation [12, 13].

The reinvigoration of plant chemical biology in recent years [3, 14–16] has created a potential opportunity to translate newly discovered chemical effectors and their associated insights into plant biology into commercially viable agrochemicals. New knowledge on how small molecules interact with and influence plant processes is expanding, together with increasing use of small-molecule screens with diverse inputs comprised of synthetic xenobiotic compound libraries or natural products derived from diverse sources [16–18]. These all act to provide information on how small molecules interact and exert their effects in plants, as well as to potentially supply chemical leads and targets for new agrochemical development. A familiar observation in whole-organism screening for chemical leads is that better hit rates and lead outcomes are obtained from compounds that have been designed or prescreened for some indication of bioactivity [19, 20], for example, in cellular assays that are a central theme of plant chemical biology studies. This is in contrast to hits from *in vitro* biochemical assays and high-throughput screenings (HTS) of enzyme targets, for example, that can have a poor translation rate into bioactivity.

The journey from molecular mode-of-action and *in vitro* chemically induced phenotypic observations (equivalent to early-stage discovery in the agrochemical development process) to the successful launch of a new compound for commercial use is long, arduous, and expensive. Current costs for the discovery and development costs for a new agrochemical are estimated to average around $256 million over 10 or more years, with about $146 million of these costs being in post-discovery development [21].

The costs associated with regulatory submissions are continuing to escalate, and there is little reason to assume that compounds that are applied to crops and so can potentially enter the food chain will avoid these mounting costs without significant shifts in regional and global regulatory environments. Consequently, a significant barrier for the new approaches using plant chemical biology to impact agrochemical discovery and deliver new actives is the identification of those commercial opportunities that are sufficiently robust and viable to sustain the significant anticipated cost of chemical development and registration. Clearly, commercial crop protection opportunities are well established for herbicides, fungicides, and insecticides, but other opportunities beyond these are considerably less prescribed from either technical or commercial perspectives.

Plant biotechnology has generated genetic solutions to enhance crop productivity that have reduced or in some cases eliminated the need for many crop chemical applications. Discovery and development costs for commercial biotechnology traits are also escalating with current estimates of around $136 million to discover and launch a new trait, of which about $105 million is in post-discovery optimization, development, and deregulation costs [22]. Nevertheless, these are below the average cost of development of an agrochemical. However, leveraging a beneficial trait across different crops and geographies requires additional R&D investments for transformation of each crop species and subsequent deregulation of each transgenic event prior to commercial launch. This does create a potential "portability" advantage for a chemical reagent that can induce a desirable crop attribute because the active ingredient has the potential to be more easily leveraged into additional crops and geographies, including minor crops. These may not be commercially addressable via a biotechnology trait as the R&D investment required to develop a trait in a minor crop may be too high, or there may be lack of acceptance of plant biotechnology-derived traits in certain geographies.

Despite the high cost barriers to market entry for agrochemicals, the new approaches from plant chemical biology can inform and enrich the agrochemical discovery pipeline through either the discovery of new bioactive compounds or the elucidation of novel target sites and plant process for chemical intervention [23, 24]. Clearly, the most straightforward applications from plant biology studies using small-molecule interventions and screens are toward herbicide discovery. Uncovering of plant processes that are particularly sensitive to chemical perturbation and that also could have lethal downstream consequences at the whole-plant level is potentially exploitable as novel herbicidal mechanisms [23, 25]. The use of low-barrier model plant systems (at the intact plant or organ level) in many plant chemical biology studies offers a level of sensitivity to the effect of small molecules that may be missed in agrochemical screens using larger intact plants [12, 18, 26]. Clearly, translation of such "weak" hits to robust whole-plant activity will be a challenge to the agrochemist (as is discussed later in this chapter), but the appeal of novelty in either mode of action or chemical structure may provide sufficient incentives to pursue such leads.

Other potential agrochemical applications from plant chemical biology beyond conventional pesticides are more challenging to discern and predict as the commercial concepts and markets for many of these have not be completely defined or have little precedence. Many current plant growth regulators have fairly small niche markets, and there are relatively few new entries with novel active ingredients, perhaps because of the R&D and commercial barriers discussed previously. Additional breakthrough products or concepts could further invigorate the potential for new agrochemical concepts, perhaps stimulated by recent developments in plant chemical biology. Clearly there are still significant new areas of plant/small-molecule interactions to be elucidated, illustrated by the recent discoveries of the role of strigolactones in plant architecture and parasitic plant relationships [27–29]. These and other discoveries may offer a suite of novel chemical effectors that await more thorough structural exploration and further exploitation. The next three sections will review three areas of potential for application, and translation of plant chemical biology studies toward agrochemical discovery will be discussed, followed by a discussion on what attributes may be required for a chemical biology probe to be translated into an agrochemical lead.

7.1.2 PLANT CHEMICAL BIOLOGY TOWARD HERBICIDE DISCOVERY

Conventional herbicide lead discovery involves screening compounds from various sources to find novel compounds that exert phytotoxic or lethal effects on target weed species. *In vivo* HTS using miniaturized assay systems with small model plant species such as *Arabidopsis*, duckweed, or bentgrass are often deployed to minimize the amount of compound required for primary screening [30]. Hit compounds are promoted to greenhouse spray tests on target weed species, and novel compounds with desirable phytotoxic and structural attributes can be further optimized via additional bioassay-directed synthesis around the areas of lead chemistry. Despite the early emphasis on whole-organism activity (in contrast to typical pharmaceutical discovery pipelines that place a larger emphasis on target site activity at early lead discovery stages), the shortfall in productivity of this "conventional" approach in recent years has been well documented [31]. Although there has been a slow but steady introduction of new active ingredients, these have predominantly comprised additional members of known classes of chemistry and modes of action. No chemistries with new herbicidal modes of action have been introduced into the marketplace since the 1980s [31] with the introduction of the hydroxyphenylpyruvate dioxygenase inhibitors that block carotenoid biosynthesis. This may be partially attributed to the impact and success of herbicide-resistant crop technology that resulted in the shifting of the discovery and technological R&D resources of agrochemical companies from traditional synthetic chemical lead discovery toward the development of herbicide-tolerant traits. However, the broad adoption of herbicide-resistance technology, particularly resistance to glyphosate, is now leading to new opportunities in weed control due to the emergence of glyphosate-resistant weeds and weed shifts associated with the widespread deployment of glyphosate-resistance crop technology [32].

In the 1990s and early 2000s, following research paradigms established in the pharmaceutical industry, great hope was placed in the identification of new herbicidal target sites via genetic methods such as screening mutant collections of plants (principally *Arabidopsis*) that contained knockout or downregulated genes. These methods did uncover many new genes and their encoded proteins that were important for plant growth and development. These protein targets could then be used for HTS of chemical libraries (comprised of $>10^4$–10^5 compounds) [33] or targeted elaboration of chemical leads using structure-based design and molecular modeling techniques if a protein/ligand co-crystal structure could be procured [34]. The impact from these efforts appears to have been modest as no major herbicidal leads have been reported to date. A weakness of these approaches is that target site validation that relies solely on genetic techniques does not provide a high degree of confidence that the target can effectively bind or interact with a small molecule, particularly those that would constitute a good agrochemical lead. In the pharmaceutical arena, this target attribute is often referred to as protein "druggability" [35]. There are a variety of other reasons why nonchemical target validation via genetic knockouts or downregulation may be difficult [36] or relatively uninformative for agrochemical lead generation, as discussed in [23] and [37].

After target identification, the efforts for HTS of targets and triage of *in vitro* hits can be time-consuming and confounded by "frequent-hitter" compounds [38]. Perhaps the most significant empirical barrier to the further development of hits from *in vitro* biochemical HTS and/or structure-based design campaigns (at least from unpublished experiences at Dow AgroSciences) is the poor and relatively inconsistent translation of compounds with high *in vitro* target affinity into corresponding predictable bioactivity at the intact plant cell level. There are many possible reasons for this; nevertheless, it remains a significant practical barrier to *in vitro* screening. These barriers can be largely circumvented by the "chemistry-first" paradigm of plant chemical biology and chemical genetics that deploy chemical inputs to elicit a desired cellular phenotype [23, 39]. Thus, any molecular target site that is subsequently elucidated from these studies will by definition have a cognate starting ligand associated with it. This target–ligand pairing associated with a desired biological activity can serve as a powerful combination for agrochemical lead generation as it allows dissection of barriers to enhanced efficacy (compound delivery to the site of action, target affinity, downstream effects from target interaction, etc.) that are otherwise difficult to ascertain when the biological target is unknown [23].

The paucity in the discovery of new herbicidal modes of action despite many years of direct and intensive compound screening on plants by large agrochemical companies indicates that commercially viable mechanisms to kill weeds are quite rare [31]. This suggests that further intensive exploitation of those mechanisms that are effective in killing plants may be worthwhile using the tools of plant chemical biology. A similar approach has been proposed in the pharmaceutical area by researchers at GlaxoSmithKline after an analysis of their discovery processes [40]. Consequently, targeted biological screens that focus on these chemically well-validated sensitive plant processes may help uncover new sites that are susceptible to interference or modulation by small molecules. For example, the unraveling of the complex biological processes in the auxin signaling pathway may lead to new sites of chemical intervention and ways of interfering with this sensitive process [12]. Similarly, microtubule and cytoskeletal assembly [41] and cellulose biosynthesis [42, 43] are additional complex cellular processes in plants that are the targets for commercial herbicides and appear to involve many sites of chemical interaction.

The primary emphasis of many plant chemical biology studies has understandably focused on the use of the model plant system *Arabidopsis*, a dicotyledonous plant with a vast amount of genetic resources available. However, there are many significant unmet needs for new methods of control of grass weeds (*Poaceae*) and sedges (*Cyperaceae*). Consequently, a fruitful area for plant chemical biology to inform and have a future impact on herbicide discovery is in deeper investigation of grass systems, both in dissecting processes and chemistries that are unique to (or preferentially active in) grasses and in the translation and comparative study of processes characterized in *Arabidopsis* to grass systems (e.g., in the model grass *Brachypodium*). An illustrative example could be the comparison of the mechanism of auxin action and signaling in dicots and monocots where there are key differences in auxin-related plant responses and development between broadleaf and grass plants [44], as well as several diverse auxin herbicide chemistries that are available as chemical probes for comparative studies [4].

7.1.3 PLANT CHEMICAL BIOLOGY TOWARD AGROCHEMICALS FOR PLANT PROTECTION

Most commercial pesticides that are applied to crops to control pests are effective by having a direct lethal effect on the target fungi or insect species, interfering with some vital function of the pest to kill it or prevent it from feeding or establishing on or in the plant. Typical screens and assays for the discovery and optimization of these agents involve application or exposure of the compound to the target pest (or a suitable surrogate), often in the absence of plant material. However, divergence from this paradigm is illustrated by compounds that can be applied to crops to "prime" them so that the crop plant responds more effectively to subsequent pathogen or pest attacks and thus provide enhanced crop protection either via foliar application or as seed treatments with the compound [45–47]. This approach has been commercially validated by compounds discovered via both targeted and serendipitous means. The compound acibenzolar-*S*-methyl (or benzothiadiazole) has been developed and marketed by Syngenta (as Actigard or Bion) to mimic the effect of the plant signaling compound salicylic acid (as do other known but less effective chemical priming agents such as 2,6-dichloroisonicotinic acid) [48]. The compound primes plant defense responses via the systemic acquired resistance pathways to elicit multilayered protection against fungal and bacterial plant pathogens. In a more serendipitous example, investigations into the field-observed "plant-protective" side effects after crop applications of the insecticide imidacloprid (that targets neural nicotinic receptors in pest insects) indicate that the compound is metabolized *in planta* to 6-chloronicotinic acid that can also act as a priming agent to boost protection against biotic and abiotic stresses [49]. A further serendipitous example of the protective effects of an agrochemical involves the strobilurin family of fungicides that targets the QoI component of the fungal mitochondrial respiratory pathway. These have also been shown to elicit increased resistance to viral and bacterial pathogens and to have beneficial effects by a "stay-green" leaf effect to extend plant productivity and enhance yield [47, 50]. Induced resistance to insect herbivory is triggered by the jasmonate response pathway (and can be antagonistic to the salicylic acid-mediated systemic acquired resistance pathway for pathogen protection) [51]. Treatment of seeds with jasmonate can confer increased resistance to insects (and necrotrophic fungi) providing another example of a small-molecule application that can elicit some degree of plant protection via the plant's induced resistance responses [46].

None of these compounds have as yet achieved widespread adoption as mainstream crop protection agents that can effectively compete with classical highly efficacious curative fungicides, for example. This may be because in the field, crop plants are already in a state of heightened alert depending on the environment and exposure to biotic and abiotic stresses, relative to a greenhouse testing system [52], or that there is a growth penalty for mounting the defenses relatively indiscriminately [53]. This may mean that activation of native plant defense responses by exogenous application of a compound will be inherently variable and "attack specific," rather than giving a broad spectrum of control [54]. However, it is intriguing to note that there appears to be both a suite of structurally diverse compounds that can be exogenously applied and a set of native plant small-molecule effectors that are able to elicit a variety of plant-protective responses. Continuing plant chemical biological studies and screens may provide additional insights to reveal new effector compounds in these areas. These could provide agrochemical leads that are more effective in the field or can stimulate cross talk and synergism (or avoid antagonism) between the various plant responses to provide higher levels of efficacy and reliability for protection to pathogens or herbivorous insects.

7.1.4 PLANT CHEMICAL BIOLOGY TOWARD AGROCHEMICALS FOR PLANT IMPROVEMENT AND YIELD

Perhaps the most intriguing concepts for new agrochemicals that could emerge from plant chemical biology studies and associated screens are for compounds that induce commercially desirable plant traits. Chief among these is the ability to reliably improve crop yields by the application of a compound at an appropriate stage in the crop growth cycle. This was a "holy grail" of many agrochemical companies with "plant growth regulator" discovery programs during the heyday of agrochemical innovation in the 1970s and 1980s. Most of these industry research programs were later terminated or shelved after it proved difficult to find and expensive to test compounds that could consistently and significantly increase yields in the greenhouse and moreover to translate those effects into the field across different environments and crop genotypes. This still remains a significant challenge for research in this area today. However, the increasing body of genetic, physiological, and metabolic knowledge around crop yield determinants and subcomponents together with the improved analytical toolsets in plant chemical biology is creating a more fertile discovery environment for "chemically induced traits."

Other chapters in this volume deal with the intricate plant chemical biology of plant hormones, and some of these compounds and structurally related molecules have been commercially exploited as agrochemical plant growth regulators. An interesting agrochemical compound that is finding increasing commercial applications is 1-methylcyclopropene (1-MCP) marketed as SmartFresh™ by AgroFresh. This gaseous compound blocks the perception of ethylene at plant ethylene receptors and so can interfere with a variety of ethylene-induced plant processes. The compound has found many useful applications in regulating postharvest quality in a wide variety of fruits and vegetables by preventing or controlling ripening and delaying spoilage [55]. More recently, its use is being explored in row crop applications to ameliorate the effects of stress by blocking ethylene-modulated stress responses in the field [56–58]. The bioactivity of the compound has been known since the mid-1990s, but a key technological breakthrough for the use of 1-MCP gas was the discovery of an appropriate formulation vehicle using γ-cyclodextrin to encapsulate the gas and form a powder that allows subsequent aqueous release on application to fruits and vegetables. 1-MCP is an interesting model for future discovery efforts involving plant chemical biology as it demonstrates that a small molecule interacting specifically with a key plant metabolic process can be translated into a new and useful (non-pest control) commercial opportunity.

Following this model, chemical screens might be envisaged to identify compounds (and their corresponding targets) that induce other phenotypes other than ethylene-induced ripening. For example, compounds that affect key processes in plant growth, development, and morphology could provide beneficial yield attributes. The recent elegant elucidation of the ABA signaling pathway [59], in part catalyzed by the use of the chemical probe compound pyrabactin [60], opens up the possibility of modulating water use efficiency and drought response in crops via application of small molecules that could provide leads for agrochemical design around this concept. Screens for compounds that alter partitioning of metabolites could identify chemical leads for modifying harvest quality traits such as seed composition.

7.1.5 CHEMICAL PROBES VERSUS AGROCHEMICAL LEAD

The use of chemical probes to dissect plant processes is a key principle of plant chemical genetics [14, 39]. A well-characterized chemical probe (either xenobiotic or naturally derived) can be useful to explore the function and interactions of its cognate target(s) in different biological processes and systems [61, 62]. However, a robust and useful chemical probe is not necessarily a good lead for agrochemical development. Most useful probes are active at the cellular level and so have the ability to penetrate plant cells and exert a biological effect, providing a distinct advantage over hits obtained from a biochemical *in vitro* assay. It is naïve to expect that a chemical probe discovered in a cellular screen will be immediately effective in a typical whole-plant agrochemical screen or greenhouse assay. There may be a rapid drop-off in efficacy or potency of a probe compound that has been characterized for use in a specific assay system when the compound is deployed in more complex testing systems. For example, many early-stage assay systems are tested in bathing medium that provides a constant (and sometimes saturating) source of the compound and are also conducted over a relatively short window of time. The compound therefore can have a relatively direct route to the target tissues that are often actively dividing or growing tissues such as a root or young leaf. As the size and complexity of a plant system is increased, effective leads have to be able to move through tissue barriers to the site of biological action and be effective after considerable dilution in larger amounts of tissue. They also have to survive exposure to a wider variety of plant metabolic processes that may degrade or sequester the compound. Similar considerations can be applied to the robustness of target sites to influence phenotype at the whole-plant level. Processes that can have profound effects at the cellular level, perhaps in a certain tissue type or developmental stage or observed at the subcellular level, may have limited effects on the intact plant. Those target sites worthy of agrochemical targeting must have a profound influence on the intact plant to manifest chemically induced phenotypic traits that will be of eventual commercial value.

For agrochemical discovery, the eventual preferred delivery method for the compound in a field situation has to be considered during compound selection and development. For example, an effective modern herbicide lead must have the ability to move to and intoxicate the meristematic growing points of the plant to ensure sustained control of weed growth. A compound for a seed treatment has to be stable in the soil environment, be able to penetrate tissues upon germination, and maintain its effect for a substantial amount of time while the plant rapidly grows and expands tissues. Most plant chemical biology labs obtain their chemical inputs for screens from vendors (or academic colleagues) that have tuned their inputs for the pharmaceutical lead generation industry. While the intrinsic characteristics of compounds that allow effective interaction with protein targets are not likely to differ across plant, microbial, and animal systems ("proteins are proteins"), the route that compounds have to traverse to reach the protein target and exert their biological effect is rather different, as is their method of delivery. Plants have lipophilic cuticles, cellulosic cell walls, large vacuoles, and a relatively high abundance and diversity of genes associated with xenobiotic metabolism (e.g., CytP450s, glutathione-*S*-transferases) and pumps (e.g., ABC-type pumps). Thus, compounds that worked quite well in a low-barrier system may fail in larger whole-plant systems without considering and addressing these barriers at the chemical structure level.

Several analyses have concluded that a majority of commercial agrochemical products and lead compounds have similar properties to pharmaceutical drugs [20, 63] and generally follow the Lipinski Rule of Five [64] for drug-likeness. This perhaps reflects some general overall chemical constraints for "bioactive" molecules (although there are many exceptions, especially in the structures of natural product actives). Herbicides and fungicides in particular tend to have more pharmaceutical "lead-like" properties [63] that could be attributed to the greater need for higher solubility, moderate lipophilicity, and reduction in the number of potential of "metabolic handles" to manifest robust bioactivity [20]. This is particularly true for herbicides, compounds that are applied to plant tissue, translocate to a molecular target site, and exert a strong (and, in the case of herbicides, lethal) biological effect. These are all desirable attributes of a robust chemical probe and so are worth paying a high degree of attention to in chemical biology screens to deliver probes useful in plant biology that will have better potential for translation into an agrochemical lead.

Most active modern herbicides are anionic in character (in their active form) to facilitate translocation via the phloem to plant meristematic regions. This is of importance in translating bioactivity from test systems that directly expose the target tissue to the compounds of interest (e.g., in a bathing solution) to more challenging applications that use lower volumes and require the compound to translocate to target cells and tissues. To effectively move out of a treated leaf to growing points such as the root and meristem, a compound must be mobile in the phloem. The physicochemical attributes that allow mobility of compounds in the phloem have been well characterized and indicate that compounds that are weakly acidic with moderate lipophilicity are trapped and translocated in the phloem stream [65, 66]. In contrast, the xylem stream (with a much higher solute volume) will move leaf-applied compounds toward the leaf margins and root-applied compounds up into the aerial parts of the plant. However, there is generally no significant basipetal translocation of uncharged lipophilic compounds from a leaf to the roots [67]. Consideration of these physical properties may greatly enhance the probability of a probe compound becoming an agrochemical lead and for translation of bioactivity in low-barrier systems to the greenhouse and field.

7.1.6 CONCLUSION

Plant chemical biology is undergoing a renaissance with a renewed emphasis on chemical probes to interrogate plant biological systems and increased accessibility to a larger variety of chemical compound resources, coupled with a rapidly expanding array of complementary genetic, genomic, and analytical technology. In addition to generating novel insights into basic questions of plant biology, the field of plant chemical biology also has the potential to provide additional societal and commercial benefits by generating new opportunities in the form of new agrochemical leads or commercially viable target processes amenable to chemical intervention.

REFERENCES

1. Anonymous. (2011) Industry Overview – 2010 Market. Phillips McDougall, Midlothian, UK.
2. Tóth, R., and van der Hoorn, R. A. L. (2010) Emerging principles in plant chemical genetics, *Trends in Plant Science 15*, 81–88.
3. Kaschani, F., and van der Hoorn, R. (2007) Small molecule approaches in plants, *Current Opinion in Chemical Biology 11*, 88–98.

4. Sterling, T. M., and Hall, J. C. (1997) Mechanism of action of natural auxins and the auxinic herbicides, In *Herbicide Activity: Toxicology, Biochemistry and Molecular Biology* (Roe, R. M., Burton, J. D., and Kuhr, R. J., Eds.), pp 111–141, IOS Press, Amsterdam.
5. Mithila, J., et al. (2011) Evolution of resistance to auxinic herbicides: historical perspectives, mechanisms of resistance, and implications for Broadleaf weed management in agronomic crops, *Weed Science 59*, 445–457.
6. Wright, T. R., et al. (2010) Robust crop resistance to broadleaf and grass herbicides provided by aryloxyalkanoate dioxygenase transgenes, *Proceedings of the National Academy of Sciences of the United States of America 107*, 20240–20245.
7. Estelle, M. A., and Somerville, C. (1987) Auxin-resistant mutants of *Arabidopsis thaliana* with an altered morphology, *Molecular and General Genetics 206*, 200–206.
8. Hobbie, L., and Estelle, M. (1994) Genetic approaches to auxin action, *Plant, Cell and Environment 17*, 525–540.
9. Dharmasiri, N., Dharmasiri, S., and Estelle, M. (2005) The F-box protein TIR1 is an auxin receptor, *Nature 435*, 441–445.
10. Kepinski, S., and Leyser, O. (2005) The Arabidopsis F-box protein TIR1 is an auxin receptor, *Nature 435*, 446–451.
11. Mockaitis, K., and Estelle, M. (2008) Auxin receptors and plant development: a new signaling paradigm, *Annual Review of Cell Developmental Biology 24*, 55–80.
12. De Rybel, B., et al. (2009) The past, present, and future of chemical biology in auxin research, *ACS Chemical Biology 4*, 987–998.
13. Kelley, K. B., and Riechers, D. E. (2007) Recent developments in auxin biology and new opportunities for auxinic herbicide research, *Pesticide Biochemistry and Physiology 89*, 1–11.
14. Robert, S., Raikhel, N. V., and Hicks, G. R. (2009) Powerful partners: Arabidopsis and chemical genomics, *The Arabidopsis Book 7*, e0109.
15. Raikhel, N. (2010) Seeing is believing: opportunities, challenges, and solutions in plant cell biology, *Molecular Biology of the Cell 21*, 3816–3817.
16. Hicks, G. R., and Raikhel, N. V. (2009) Opportunities and challenges in plant chemical biology, *Nature Chemical Biology 5*, 268–272.
17. Kolukisaoglu, U., and Thurow, K. (2010) Future and frontiers of automated screening in plant sciences, *Plant Science 178*, 476–484.
18. Hicks, G. R., and Raikhel, N. V. (2010) Advances in dissecting endomembrane trafficking with small molecules, *Current Opinion in Plant Biology 13*, 706–713.
19. Drakakaki, G., et al. (2011) Clusters of bioactive compounds target dynamic endomembrane networks in vivo, *Proceedings of the National Academy of Sciences of the United States of America 108*, 17850–17855.
20. Delaney, J., et al. (2006) Modern agrochemical research: a missed opportunity for drug discovery? *Drug Discovery Today 11*, 839–845.
21. Anonymous. (2010) The cost of new agrochemical product discovery, development and registration and research and development predictions for the future. Phillips MacDougall, Midlothian, UK.
22. Anonymous. (2011) The cost and time involved in the discovery, development and authorisation of a new plant biotechnology derived trait. A consultancy study for Crop Life International. Crop Life International, Pathhead, Midlothian, UK.
23. Walsh, T. A. (2007) The emerging field of chemical genetics: potential applications for pesticide discovery, *Pest Management Science 63*, 1165–1171.

24. Armstrong, J. I. (2007) Chemical genetics: catalysing pathway exploration and new target discovery, *Journal of the Science of Food and Agriculture 87*, 1985–1990.
25. Dayan, F. E., Duke, S. O., and Grossmann, K. (2010) Herbicides as probes in plant biology, *Weed Science 58*, 340–350.
26. Grossmann, K. (2005) What it takes to get a herbicide's mode of action. Physionomics, a classical approach in a new complexion, *Pest Management Science 61*, 423–431.
27. Tsuchiya, Y., and McCourt, P. (2012) Strigolactones as small molecule communicators, *Molecular Biosystems 8*, 464–469.
28. Tsuchiya, Y., et al. (2010) A small-molecule screen identifies new functions for the plant hormone strigolactone, *Nature Chemical Biology 6*, 741–749.
29. Beveridge, C. A., and Kyozuka, J. (2010) New genes in the strigolactone-related shoot branching pathway, *Current Opinion in Plant Biology 13*, 34–39.
30. Tietjen, K., Drewes, M., and Stenzel, K. (2005) High throughput screening in agrochemical research, *Combinatorial Chemistry & High Throughput Screening 8*, 589–594.
31. Duke, S. O. (2012) Why have no new herbicide modes of action appeared in recent years? *Pest Management Science 68*, 505–512.
32. Green, J. M., and Owen, M. D. K. (2011) Herbicide-resistant crops: utilities and limitations for herbicide-resistant weed management, *Journal of Agricultural and Food Chemistry 59*, 5819–5829.
33. Lein, W.,et al. (2004) Target-based discovery of novel herbicides, *Current Opinion in Plant Biology 7*, 219–225.
34. Walter, M. W. (2002) Structure-based design of agrochemicals, *Natural Product Reports 19*, 278–291.
35. Russ, A. P., and Lampel, S. (2005) The druggable genome: an update, *Drug Discovery Today 10*, 1607–1610.
36. Cutler, S., and McCourt, P. (2005) Dude, where's my phenotype? Dealing with redundancy in signaling networks, *Plant Physiology 138*, 558–559.
37. Akers, A. (1996) Molecular biology as virtual biology: limitations of molecular biology in pesticide discovery, *Pesticide Science 46*, 85–91.
38. Thorne, N., Auld, D. S., and Inglese, J. (2010) Apparent activity in high-throughput screening: origins of compound-dependent assay interference, *Current Opinion in Chemical Biology 14*, 315–324.
39. McCourt, P., and Desveaux, D. (2010) Plant chemical genetics, *New Phytologist 185*, 15–26.
40. Payne, D. J., et al. (2007) Drugs for bad bugs: confronting the challenges of antibacterial discovery, *Nature Reviews Drug Discovery 6*, 29–40.
41. Mayer, U., and Jurgens, G. (2002) Microtubule cytoskeleton: a track record, *Current Opinion in Plant Biology 5*, 494–501.
42. Carpita, N. C. (2011) Update on mechanisms of plant cell wall biosynthesis: how plants make cellulose and other (1→4)-beta-D-glycans, *Plant Physiology 155*, 171–184.
43. DeBolt, S., et al. (2007) Morlin, an inhibitor of cortical microtubule dynamics and cellulose synthase movement, *Proceedings of the National Academy of Sciences of the United States of America 104*, 5854–5859.
44. McSteen, P. (2010) Auxin and monocot development, *Cold Spring Harbor Perspectives in Biology 2*, a001479.

45. Conrath, U. (2011) Molecular aspects of defence priming, *Trends in Plant Science 16*, 524–531.
46. Worrall, D., et al. (2012) Treating seeds with activators of plant defence generates long-lasting priming of resistance to pests and pathogens, *New Phytologist 193*, 770–778.
47. Beckers, G. J. M., and Conrath, U. (2007) Priming for stress resistance: from the lab to the field, *Current Opinion in Plant Biology 10*, 425–431.
48. Gorlach, J., et al. (1996) Benzothiadiazole, a novel class of inducers of systemic acquired resistance, activates gene expression and disease resistance in wheat, *Plant Cell 8*, 629–643.
49. Ford, K. A., et al. (2010) Neonicotinoid insecticides induce salicylate-associated plant defense responses, *Proceedings of the National Academy of Sciences of the United States of America 107*, 17527–17532.
50. Bartlett, D. W., et al. (2002) The strobilurin fungicides, *Pest Management Science 58*, 649–662.
51. Smith, J. L., De Moraes, C. M., and Mescher, M. C. (2009) Jasmonate- and salicylate-mediated plant defense responses to insect herbivores, pathogens and parasitic plants, *Pest Management Science 65*, 497–503.
52. Walters, D. R. (2009) Are plants in the field already induced? Implications for practical disease control, *Crop Protection 28*, 459–465.
53. Heil, M., et al. (2000) Reduced growth and seed set following chemical induction of pathogen defence: does systemic acquired resistance (SAR) incur allocation costs? *Journal of Ecology 88*, 645–654.
54. Walters, D. R., and Fountaine, J. M. (2009) Practical application of induced resistance to plant diseases: an appraisal of effectiveness under field conditions, *The Journal of Agricultural Science 147*, 523–535.
55. Chris B, W. (2006) The use of 1-methylcyclopropene (1-MCP) on fruits and vegetables, *Biotechnology Advances 24*, 389–409.
56. da Costa, V. A., Cothren, J. T., and Bynum, J. B. (2011) Abiotic stress effects on plant growth and yield components of 1-MCP treated cotton plants, *Agronomy Journal 103*, 1591–1596.
57. Djanaguiraman, M., Prasad, P. V. V., and Al-Khatib, K. (2011) Ethylene perception inhibitor 1-MCP decreases oxidative damage of leaves through enhanced antioxidant defense mechanisms in soybean plants grown under high temperature stress, *Environmental and Experimental Botany 71*, 215–223.
58. Below, F. E., and Uribelarrea, M. (August 26, 2009) Stress control for achieving high yields: the quest for 300 bushel per acre corn, *Proceedings – Plant Growth Regulation Society of America 36*, Asheville, NC, 45–52.
59. Weiner, J. J., et al. (2010) Structural and functional insights into core ABA signaling, *Current Opinion in Plant Biology 13*, 495–502.
60. Nishimura, N., et al. (2009) Structural mechanism of abscisic acid binding and signaling by dimeric PYR1, *Science 326*, 1373–1379.
61. Frye, S. V. (2010) The art of the chemical probe, *Nature Chemical Biology 6*, 159–161.
62. Garcia-Serna, R., and Mestres, J. (2011) Chemical probes for biological systems, *Drug Discovery Today 16*, 99–106.

63. Clarke, E. D., and Delaney, J. S. (2003) Physical and molecular properties of agrochemicals: an analysis of screen inputs, hits, leads, and products, *CHIMIA 57*, 731–734.
64. Tice, C. M. (2001) Selecting the right compounds for screening: does Lipinski's Rule of 5 for pharmaceuticals apply to agrochemicals? *Pest Management Science 57*, 3–16.
65. Hsu, F. C., and Kleier, D. A. (1996) Phloem mobility of xenobiotics. 8. A short review, *Journal of Experimental Botany 47*, 1265–1271.
66. Lichtner, F. (2000) Phloem mobility of crop protection products, *Australian Journal of Plant Physiology 27*, 609–614.
67. Liu, Z. Q. (2006) Leaf epidermal cells: a trap for lipophilic xenobiotics, *Journal of Integrative Plant Biology 48*, 1063–1068.

7.2

IN VITRO PROPAGATION

Hans Motte, Stefaan Werbrouck, and Danny Geelen

7.2.1 PLANT TISSUE CULTURE AS A HISTORICAL BASIS FOR THE DISCOVERY OF PLANT GROWTH REGULATORS

Plant tissue culture was truly on its way when phytohormones or plant growth regulators were used to control growth and development. In 1926 Went was the first to isolate such a plant growth regulator from oat coleoptiles: the auxin indole-3-acetic acid (IAA) [1, 2]. Previously, Haberlandt [3] suggested already that a plant growth regulator might be responsible for cell division, but it took until 1955 before the first cytokinin, kinetin, was discovered as a degradation product from DNA [4]. During more than 40 years, kinetin was considered as a synthetic phytohormone, but more recently it is reported to be endogenously present in coconut and Australian pine [5, 6]. The first cytokinin that was extracted from plants was zeatin [7]. Skoog and Miller [8] showed that organ differentiation could be manipulated by changing the relative concentrations of auxin and cytokinin. This concept of hormonal regulation of organogenesis is now applicable to a lot of plant species and made plant tissue culture a widely applied technique for propagation and genetic modification. Although *in vitro* propagation still mainly uses auxins and cytokinins, other plant growth regulators are implemented for specific applications. By screening collections of small molecules, the toolbox of plant growth regulators available for plant tissue culture has been steadily growing since the early days of plant growth regulator discovery. The recent discovery of pyrabactin as an abscisic acid (ABA) agonist [9] and the novel plant hormone strigolactone [10] is exemplary for the prospect of identifying additional compounds with growth regulatory activity in tissue culture. Still more molecules are being discovered by screening commercial and custom-made

Plant Chemical Biology, First Edition. Edited by Dominique Audenaert and Paul Overvoorde.

chemical libraries. This approach, adopted from cancer drug screening, generates great opportunities for the plant tissue culturist who is trying to develop protocols for plants that are difficult to cultivate and for the commercial grower who seeks the most cost-effective method for plant reproduction and propagation.

7.2.2 CYTOKININS USED IN TISSUE CULTURE

Cytokinins play a central role in the regulation of proliferation and differentiation of plant cells. Exogenous cytokinins allow to control the adventitious shoot/root balance, degree of apical dominance, and delay of senescence [11]. All natural cytokinins are N^6-substituted adenine derivatives (Fig. 7.2.1) that are classified, according to the configuration of their N^6-side chain, as isoprenoid or aromatic compounds. In the early years of cytokinin research, only cytokinins with an isoprenoid side chain were thought to be endogenous. However, Horgan et al. [12] isolated and identified *ortho*-topolin (oT), a benzyladenine derivative in mature poplar leaves. Later on, other aromatic cytokinins were extracted from plants [5, 13, 14]. Some synthetic molecules with a cytokinin activity will be discussed in the next paragraphs. An overview of the natural cytokinins can be found in Table 7.2.1.

The active form of an adenine-type cytokinin is the free base, but different conjugates also occur in plants (reviewed in Reference [15]). Nucleosides (cytokinin ribosides) and nucleotides (cytokinin riboside 5′-mono-/di-/triphosphates) (Fig. 7.2.1) are important transport forms and precursors of the free bases. Some of these conjugates are also able to activate certain cytokinin receptors [16] and hence are sometimes applied in *in vitro* tissue culture (see Section 7.2.4). Inactivation of the free bases can occur by *N*-glucosylation on the N^7- or N^9-position [17] or, dependent on the side chain, by *O*-glucosylation [18].

7.2.2.1 Urea-Type Cytokinins

Other types of cytokinins that, at first sight, have no structural relationship with the natural cytokinins are phenylurea compounds. The discovery of phenylureas with cytokinin-like activity was an accidental finding. Coconut milk, which is a rich source of cytokinins, was used by Shantz and Steward [19] to identify *N,N'*-diphenylurea (DPU) as a highly active substance. DPU is however not naturally

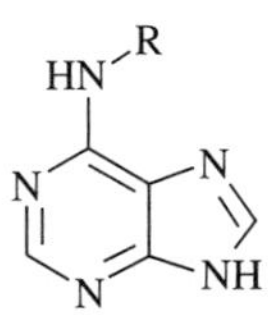

FIGURE 7.2.1 N^6-substituted adenine. R represents the substituted structure. The natural occurring substitutions are presented in Table 7.2.1. Nucleosides and nucleotides include a β-D-ribose or β-D-ribose 5′-mono-/di-/triphosphate at the N^9 position, respectively.

TABLE 7.2.1 Natural cytokinins

Trivial name	Class	Abbreviation	IUPAC name	Substituted structure (R)	Molecular formula	Molecular weight (g/mol)
2-Isopentenyladenine	Isoprenoid	2-iP; iP	*N*-(3-methylbut-2-enyl)-7H-purin-6-amine	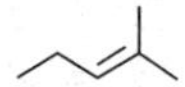	$C_{10}H_{13}N_5$	203.24
***trans*-Zeatin**	Isoprenoid	tZ	(E)-2-methyl-4-(7H-purin-6-ylamino) but-2-en-1-ol	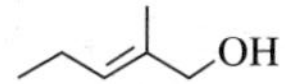	$C_{10}H_{13}N_5O$	219.24
***cis*-Zeatin**	Isoprenoid	cZ	(Z)-2-methyl-4-(7H-purin-6-ylamino) but-2-en-1-ol	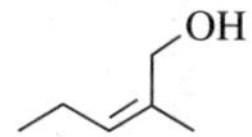	$C_{10}H_{13}N_5O$	219.24
Dihydrozeatin	Isoprenoid	DHZ	2-methyl-4-(7H-purin-6-ylamino) butan-1-ol	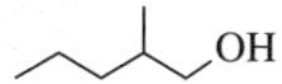	$C_{10}H_{15}N_5O$	221.26
***N*⁶-Benzyladenine**	Aromatic	BA	*N*-benzyl-7H-purin-6-amine	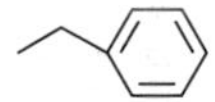	$C_{12}H_{11}N_5$	225.25
***ortho*-Topolin**	Aromatic	oT	*N*-(2-hydroxybenzyl)-7H-purin-6-amine	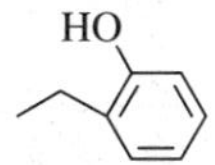	$C_{12}H_{11}N_5O$	241.25
***meta*-Topolin**	Aromatic	mT	*N*-(3-hydroxybenzyl)-7H-purin-6-amine	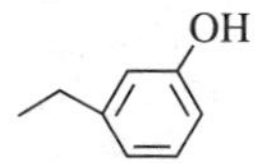	$C_{12}H_{11}N_5O$	241.25
***ortho*-Methoxytopolin**	Aromatic	MeoT	*N*-(2-methoxybenzyl)-7H-purin-6-amine	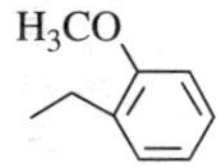	$C_{13}H_{13}N_5O$	255.28
***meta*-Methoxytopolin**	Aromatic	MemT	*N*-(3-methoxybenzyl)-7H-purin-6-amine	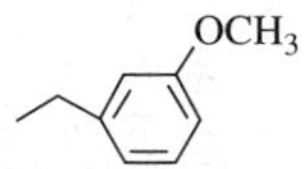	$C_{13}H_{13}N_5O$	255.28
Kinetin	Aromatic	Kin	*N*-(furan-2-ylmethyl)-7H-purin-6-amine	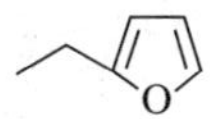	$C_{10}H_9N_5O$	215.21

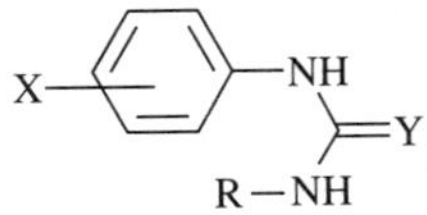

FIGURE 7.2.2 Phenylurea. Derivatives carry different R and X side chains.

present in coconut milk, but it was in fact a contaminant derived from equipment used in experiments with chemically synthesized herbicides. Nevertheless, the fortunate mistake led to the discovery of a group of synthetic compounds with some derivatives having a very strong cytokinin-like activity (reviewed in Reference [20]).

The common structure of the urea-type cytokinins is phenylurea, with variations on the phenyl moiety or on *N′* (Fig. 7.2.2). As there are many variations, the urea derivatives provide a large potential to develop chemical libraries for screening compounds with putative cytokinin-like activity. With this purpose, a group of about 500 urea and thiourea derivatives were evaluated by testing the tobacco pith cell division, senescence retardation in radish, lettuce seed germination, and pea lateral bud development, leading to relationships between chemical structure and biological activity [21, 22]. Similar screens were repeated and enlarged by other labs, and hence the relationship between structure and biological activity was refined, expanded, and sometimes adjusted. Additional urea derivatives with activity in the standard tobacco callus assay were found to carry specific substitutions on the phenyl ring (X, Fig. 7.2.2) and unsubstituted or substituted pyridyl rings on *N′* (R, Fig. 7.2.2) [23, 24]. The highly complex phenylurea derivative *N*-phenyl-*N′*-(1,2,3-thiadiazol-5-yl)urea (thidiazuron (TDZ)) was originally developed as a cotton defoliant [25], named Dropp®. Its cytokinin-like activity was compared with other urea and thiadazolylurea derivatives by Mok et al. [26] in the *Phaseolus* callus growth bioassay. Others have synthesized and tested a wide range of urea derivatives [27–30]. Based on such screens, multiple bioactive phenylurea derivatives have been identified, with a diversity of physiological effects. For example, *N*-(2-chloro-4-pyridyl)-*N′*-phenylurea (CPPU) is highly effective to promote bud break [31–33] and parthenocarpic fruit development ([18] and refs. therein), whereas TDZ is effective for promoting shoot regeneration, even for some recalcitrant species, and somatic embryogenesis ([34] and refs. therein). *N*-Phenyl-*N′*-benzothiazol-6-ylurea (PBU) was shown to have shoot morphogenetic activity [28, 35]. In contrast to the isoprenoid-type cytokinins, some diarylurea derivatives have been shown to stimulate root formation while they lack other cytokinin- or auxin-like activities [36].

There is no clear structural similarity between DPU derivatives and the adenine-type cytokinins. Miller [7] proposed that DPU was a precursor of unknown natural cytokinins, but Mok et al. [37] suggested that DPU enhanced endogenous cytokinin biosynthesis as this compound induced cytokinin autonomy. This idea was rejected by Chatfield and Armstrong [38] and Laloue and Fox [39], who hypothesized that phenylurea derivatives inhibit the cytokinin-degrading cytokinin oxidase/

dehydrogenase (CKX) enzymes and thus mimic cytokinin action by raising the endogenous cytokinin levels. This mode of action was recently demonstrated by structural data analysis [40], although some derivatives also act as cytokinin signal transduction molecules by directly interacting with cytokinin receptors. For example, TDZ can activate the *Arabidopsis* histidine kinase 3 (AHK3) and AHK4/cytokinin response 1 (CRE1) cytokinin receptors similar to adenine-type cytokinins [16, 41, 42], while DPU does not activate the AHKs [16] but only influences the endogenous cytokinin level via CKX inhibition.

7.2.2.2 Novel Cytokinins

Since the discovery of kinetin, a lot of chemically related compounds were synthesized and tested for cytokinin activity. For instance, Skoog et al. [43] tested 69 compounds related to the natural cytokinins and found that variations in N^6-monosubstituted aminopurines could have a positive effect on the cytokinin-like activity. The research group of Miroslav Strnad synthesized different libraries by adding side groups or modifying substructures of cytokinins, which resulted in a total collection of more than 1000 cytokinin-like molecules. For example, 84 aromatic cytokinin derivatives were synthesized and compared with BA for their cytokinin activity [44, 45]. The majority of these compounds exhibited a high activity in the tobacco callus, wheat senescence, and *Amaranthus* bioassays. Concerning the substitutions on the aromatic ring, the results suggested a general trend of cytokinin activity being *meta* ≥ *ortho* > *para*. The *meta* hydroxy-substituted compounds were already previously noted to be more active than the *ortho* and *para* isomers [46]. Generally, the fluoro derivatives were recognized as the most active compounds. Doležal et al. [44] concluded that as there were strong differences in activities of the same cytokinin compounds in the different bioassays, it may be possible to design specific compounds that can be used to target particular cytokinin-dependent processes and to eliminate undesirable side effects.

Some more complex variants of compounds with adenine as substructure have been tested as well. Szüčová et al. [47] tested 6-benzylamino-9-tetrahydropyran-2-yl and 9-tetrahydrofuran-2-ylpurine derivatives and found a higher resistance to enzymatic degradation if the benzyl ring contains a hydroxy or methoxy group in the *meta* position. Experiments with additional structural variants of adenine revealed compounds enhancing the cytokinin response indirectly, by inhibiting CKX [48]. Other research groups synthesized and tested cytokinin variants, mainly starting from the isoprenoid cytokinins (e.g., References [49, 50]), and found inhibitors of CKX that, in contrast to the phenylurea derivatives, irreversibly bind the cytokinin-degrading enzyme [51]. An alternative way to indirectly increase cytokinin activity is inhibition of glycosylation. A screen for such inhibitors led to the identification of a number of BA-related structures [52].

Cytokinin antagonists are another group of growth regulators that have been identified from screens that could be very useful to study the mode of action of cytokinins. The chemical structure of some of these antagonists is similar to cytokinins. Spíchal et al. [53] screened BA-related structures that do not have cytokinin activity and were shown not to activate the cytokinin receptors AHK4 and AHK3, for their ability to compete for cytokinin receptor binding. In their experiments, recombinant *Escherichia coli* expressing *AHK4* or *AHK3* receptor genes were used to determine radiolabeled tZ binding capacity in the presence of cytokinin derivatives. This screening method allowed the identification of the cytokinin antagonist 6-(2-hydroxy-3-methylb enzylamino)purine (PI-55). Based on the structure of PI-55, a second library of compounds was synthesized [54] that led to the identification of a stronger antagonist, 6-(2,5-dihydroxybenzylamino)purine (LGR-991). Arata et al. [55] screened for potential cytokinin antagonists by means of a yeast complementation assay. The yeast strain Δ*sln1*[*CRE1*] lacks an essential histidine kinase that is complemented by AHK4 expressed via the yeast CYC promoter. In the presence of BA or by introducing *Arabidopsis thaliana isopentenyltransferase 4* (*AtIPT4*) or *AtIPT5*, the modified yeast strain will grow. Cytokinin-like compounds that inhibit the growth of the yeast strain Δ*sln1*[*CRE1*] are potential cytokinin antagonists. Two compounds, sharing a 4-phenylquinazoline structure, were identified using this yeast selective growth system [55]. There are also several earlier reports of cytokinin antagonists [56–60], but these seemed to inhibit the cell cycle progression and not necessarily targeted the cytokinin receptors [61].

Synthesis of novel cytokinin-like molecules has also lead to other unexpected applications. For example, BA derivatives have been found to specifically inhibit several protein kinases [62]. Because some of these kinases are important for tumor cell divisions, the cytokinin-like compound library from Miroslav Strnad was screened for cytotoxic activity and strong anticancer properties (reviewed in Reference [63]). Olomoucin (I and II) and roscovitine are two examples that have been identified from these screens [64–66].

Microorganisms, mainly pathogens, could be a source for novel cytokinins. An example is the actinomycete *Rhodococcus fascians* that produces, besides some common plant cytokinins, also methylaminopurin, 2-methylthioisopentenyladenine (2MeSiP), 2-methylthio-*cis*-zeatin (2MeScZ), and their respective ribosides, which are tRNA degradation products in plants [67, 68]. Bacterial infection can be used *in vitro* and gives some advantages compared to the classical micropropagation technique, such as improvement of shoot regeneration for a wide range of species [69]. Also *Streptomyces turgidiscabies* possesses homolog cytokinin biosynthesis genes [70]. Some other examples of novel bacterial cytokinins are 5′-deoxyisopentenyladenosine produced by *Pantoea agglomerans* [71], 2′-deoxyzeatin riboside produced by *Pseudomonas amygdali* [72], and 1″-methyl zeatin riboside produced by *Pseudomonas syringae* ssp. *savastanoi* [73], but to our knowledge, these have not been applied in plant tissue culture. The fungus *Cladosporium* sp. *501-7W* produces two molecules that show no structural relationship with cytokinins, but do have cytokinin-like activity. As they were found in a cytokinin bioassay regarding cotyledon growth, these were called cotylenin A and B [74, 75]. Remarkably, cotylenin A is also tested for its anticancer properties and was found to affect the differentiation of leukemia cells [76].

7.2.3 AUXINS USED IN TISSUE CULTURE

7.2.3.1 A Plethora of Auxins

The term auxin refers to a group of chemicals that have the capacity to stimulate plant growth. In addition to the stimulation of cell division, auxin has a role in cell differentiation and elongation. The cumulative effect of division and elongation controls physiological processes including tropism, apical dominance, and root growth [77, 78]. In contradiction to other growth regulators, auxins are less unambiguously defined and can regulate a large diversity of morphological or developmental processes [77]. A more clearly identified property is that they, independently from other exogenously applied hormones, stimulate adventitious root formation. This property has led to their general application in plant clonal reproduction through stem cuttings and in *in vitro* shoot micropropagation protocols [79].

Auxins were initially defined as substances that have the ability to promote growth in the *Avena* coleoptile bioassay [1]. A quite diverse group of chemicals have been shown to exert auxin-like growth effects. As early as the 1930s, about 50 chemicals were tested that share some structural relationships to IAA [80]. Many more auxin-like compounds that are similar to IAA in structure or function have been identified by testing derivatives of molecules with auxin-like activity [81–83]. Because of the wide diversity of molecules belonging to this hormone group, a structural classification is not evident. However, in general, the most common auxins could be classified according to their basic structure: indole, benzene, naphthalene, or phenol. The most important auxins are listed in Table 7.2.2.

Surprisingly, there is no correlation between the structural class and the kind of physiological response, suggesting that the signal-to-response correlation follows a complex path. Indeed, several auxin receptors have been demonstrated to occur in a single species. The first receptor identified is the transport inhibitor response 1 (TIR1) protein that is required for the changes in gene expression patterns provoked by exogenous auxin [84–86]. Members of the same family as TIR1, auxin signaling F-box1–5 (AFB1–AFB5) also function as auxin receptors and are partially redundant to TIR1 [85, 87–89]. In addition to TIR1, the auxin binding protein 1 (ABP1) is another receptor shown to be involved in cell division and expansion [90]. Although the molecular function of ABP1 is not known, it has been shown to regulate at least two auxin-sensitive processes: the clathrin-dependent endocytosis of pin-formed 1 (PIN1) [91] and the interdigitated pavement cell expansion [92]. Although IAA causes an auxin effect via binding on ABP1, the response of certain other auxins, like naphthalene-1-acetic acid (NAA), is caused via an ABP1-independent pathway [93]. More recently, the S-phase kinase-associated protein 2A (SKP2A), an E3 ligase SCF complex constituting F-box protein, was also reported as a new auxin receptor, involved in cell division [94]. Furthermore, the lack of direct measurements of auxin–protein interactions has confounded the structure–function relationship analysis, and an overall common structural feature that includes all of the auxins has not been unambiguously established [95]. These interacting proteins are receptors, transporters, and metabolic enzymes involved in biosynthesis,

TABLE 7.2.2 Common auxins

Trivial name	Class	Abbreviation	IUPAC name	Structure	Molecular formula	Molecular weight (g/mol)
Indole-3-acetic acid	Indole	IAA	2-(1H-indol-3-yl)acetic acid	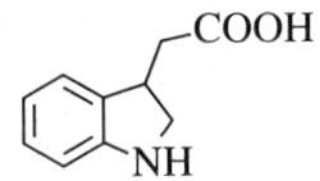	$C_{10}H_9NO_2$	175.19
Indole-3-butyric acid	Indole	IBA	4-(1H-indol-3-yl)butanoic acid	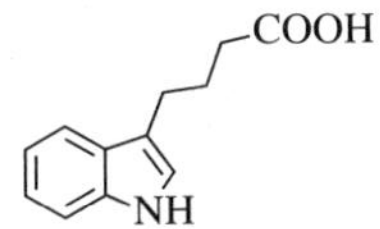	$C_{12}H_{13}NO_2$	203.24
4-Chloroindole-3-acetic acid	Indole	4-Cl-IAA	2-(4-chloro-1H-indol-3-yl) acetic acid	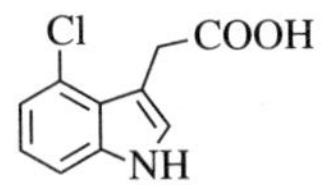	$C_{10}H_8ClNO_2$	209.63
Indole-3-pyruvic acid	Indole	IPA	3-(1H-indol-3-yl)-2-oxopropanoic acid	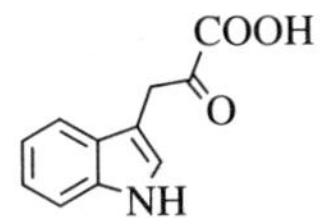	$C_{11}H_9NO_3$	203.20
Phenylacetic acid	Benzene	PAA	2-phenylacetic acid	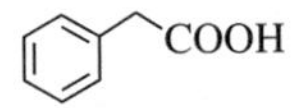	$C_8H_8O_2$	136.15
Naphthalene-1-acetic acid	Naphthalene	NAA	2-naphthalene-1-ylacetic acid	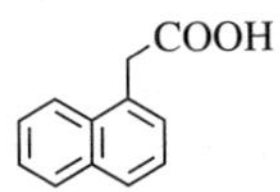	$C_{12}H_{10}O_2$	186.21
2,4-Dichlorophenoxyacetic acid	Phenol	2,4-D	2-(2,4-dichlorophenoxy) acetic acid	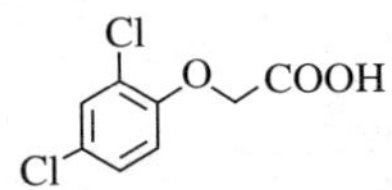	$C_8H_6Cl_2O_3$	221.04

degradation, and conjugation [96, 97]. Besides, most of the endogenous auxins are not found as a free and biologically active form, but as conjugates [98]. A plethora of conjugates are reported. Low-molecular-weight ester conjugates with sugars or amide conjugates with amino acids are involved in auxin storage, transport, or homeostasis (reviewed in Reference [99]). The function of high-molecular-weight conjugates with, for example, proteins or glycoproteins is not completely clear [100]. It gets even more complicated as different plant species have different conjugate profiles and distinct catabolism mechanisms [99]. Surprisingly, the set of free auxins isolated from plants is less complex, and only a few molecules have been chemically identified in extracts: IAA, indole-3-butyric acid (IBA), phenylacetic acid (PAA), and 4-Cl-IAA [101–103]. Interestingly, the synthetic auxins exert usually the strongest effects in tissue culture. Many of these synthetic auxins are used as herbicides and in particular useful against dicotyledonous weeds. A common effect of synthetic auxins is that they induce ectopic cell division, but most of the damage is presumably exerted by an increased production of ABA and ethylene ([104] and refs. therein). Therefore, some synthetic auxins, in particular 2,4-dichlorophenoxyacetic acid (2,4-D), are mostly used to induce the formation of callus.

Some of the synthetic auxin-like molecules turned out to show an inhibitory effect on auxin-mediated physiological responses. These compounds have been named antiauxins, but because of possible confusion with regard to the mode of action—several have been shown to inhibit auxin transport rather than to prevent the binding of auxin to its receptor—this term is no longer used in auxin research. The most commonly used auxin transport inhibitors are *N*-1-naphthylphthalamic acid (NPA), 2,3,5-triiodobenzoic acid (TIBA), and *p*-chlorophenoxyisobutyric acid (PCIB). The mode of action has not yet been fully clarified, but they seem to block cycling of auxin efflux carriers between the plasma membrane and endosomal compartments [105]. Moreover, NPA also inhibits an ABC subfamily B auxin efflux transporter [106, 107]. Recently, auxin transport has been shown to be selectively inhibited by alkoxy-auxins [108]. These compounds do not exert auxin signaling responses and therefore hold the premise to separate transport-dependent processes from signaling processes. Some other auxin antagonists preventing binding with the TIR1 receptor were found by synthesizing variants of IAA [109]. Interestingly, the variants with short side chains showed auxin activity. The use in tissue culture of molecules with auxin antagonistic action is limited to situations where one aims to reduce the activity of auxin. This can sometimes also be achieved by simply transferring the explant to auxin-free medium. Morphogenesis, somatic embryogenesis, and adventitious organogenesis often ensue when auxin is removed from the medium [110]. In a few examples, the application of auxin antagonists promotes or speeds up somatic embryogenesis [111], and in some cases, it inhibits this process [112]. These opposite effects have to be attributed either to the differential specificity with regard to the treated plant species or to a differential requirement during the consecutive developmental steps in embryogenesis and organogenesis. In the end, the application of a single auxin might be a too simplified approach to manipulate and redirect plant development with the purpose of clonal reproduction and rooting.

The search for new auxins is still going on. Through compound screens of large chemical libraries such as the DIVERSet library from ChemBridge Corporation, many more auxins or auxin antagonists have been identified [113–117]. Screening of compounds of the Korea Chemical Bank resulted in the identification of a new transport inhibitor [118]. The availability of new compounds may boost *in vitro* tissue culture research and help to realize higher reproduction rates and uniformity in root induction.

7.2.4 HORMONES IN TISSUE CULTURE APPLICATIONS

7.2.4.1 Induction of Callus

Most plant tissues and even single microspores and protoplasts react to auxin by ectopic cell divisions, which on solid medium lead to the formation of callus. By and large, synthetic auxins like 2,4-D and NAA are most frequently used for callus induction. The powerful response to 2,4-D may be attributed to its strong accumulation inside the cell [119] and the inability to be exported out of the plant cell [120]. Moreover, 2,4-D and other auxinic herbicides seem to persist for a long time with little apparent conversion that may also explain the sustained cell divisions [121]. Monocotyledonous plants are usually more resistant to 2,4-D than dicotyledonous plants, but 2,4,5-trichlorophenoxyacetic acid (2,4,5-T) and 3,6-dichloroanisic acid (dicamba) have been successfully applied to induce callus in grasses [122, 123]. Woody species also show a more restricted sensitivity to auxin with 4-amino-3,5,6-trichloropicolinic acid (picloram) as a potent callus-inducing auxin (e.g., Reference [124]). During the initial phase of callus induction, the presence of cytokinin may be required to sustain growth. After a series of subcultures, the dependency for cytokinin may be lost, a phenomenon that is known as cytokinin habituation [125]. Cytokinin habituation has been shown to correlate with an increase in expression of cytokinin signal transduction genes [126] and has been shown to be installed upon a single application of the diphenylurea compound DPU in *Phaseolus lunatus* [37]. Surprisingly, the cytokinin TDZ can also induce callus in a variety of plant culture systems [34], probably because, besides binding the cytokinin receptor in *Arabidopsis* [16, 41], it also regulates auxin signaling, presumably by influencing polar auxin transport [127].

Callus comes in various shapes, colors, and textures, indicating that auxin-mediated dedifferentiation of the explant cells is not complete and that some level of differentiation is maintained in the presence of auxin. The same explant may generate several types of callus that have different morphogenetic regeneration capacities. Most studies show that the tissue, incubation conditions, light, and temperature are critical parameters determining the regenerative capacity of callus, suggesting that environmental factors are more important than the auxin applied [128]. The growth of callus has been associated with increased genetic and epigenetic changes that could also be a source for variation in callus types [129]. The prolonged cultivation of callus usually leads to selection with the fastest-growing cells taking over the slow-growing ones, generating a more uniform growth pattern. 2,4-Dichlorophenoxyacetic acid has been reported to inflict cytogenetic damage in plants (e.g., Reference [130]) and in mosquito [131]. Because polyploid cells tend to pass quicker through the cell cycle, 2,4-D has been used to generate polyploid cultures [132]. This technique is still exploited to produce diploid regenerants from haploid microspores and ovules [133]. The novel auxin-like compounds have not been systematically analyzed with respect to genetic stability. Alternative compounds that do not induce somaclonal variation would be very valuable for tissue culture applications [134]. Irvine et al. [135] tested 79 compounds for their callus-inducing capacity on sugarcane and selected picloram as an alternative for 2,4-D, but in later studies, it seemed also to induce genetic variability [136].

7.2.4.2 Adventitious Rooting

Adventitious roots arise from non-root tissue, in contrast to lateral roots that originate from the pericycle, an organogenesis-competent cell layer surrounding the central cylinder of the root [137]. In natural conditions, adventitious roots emerge upon wounding or from preformed primordia that, for example, activate after a flood. In tissue culture, cut shoots are obviously wounded, and this can suffice to stimulate adventitious rooting, but generally an auxin, mostly IAA or IBA, is supplied to promote root organogenesis. Usually, the latter has a higher root-inducing capacity because it is more stable and can be metabolized to IAA [138], but the response is also species dependent [139]. Adventitious roots arise directly from a differentiated cell or indirectly from callus tissue. Callus-derived roots do not provide a vascular connection and therefore are inefficient in transport of nutrients. For this reason, classical auxins, and not callus-inducing auxins like 2,4-D, are used for micropropagation. Unfortunately, the root induction capacity of auxin is not universal, and some species are difficult to root. The mechanisms underlying recalcitrance to root is not fully resolved, but cytokinin accumulation during the reproduction phase of micropropagation is in many cases causing a poor root induction in subsequent stages. Some cytokinins like BA are metabolized to stable N-glucosides, which accumulate in the shoot base where they block rooting [140]. Molecular studies have shown that adventitious rooting is a heritable quantitative genetic trait [141–151] involving multiple genes (Reference [139] and refs. therein) and, therefore, recalcitrance for rooting may be difficult to overcome.

Nevertheless, exploring new compounds can in some cases resolve the problem. For example, wounding stimulates in addition to auxin accumulation also the biosynthesis of phenolic compounds, some of which have positive or negative effects on rooting. The formation of polyphenolic compounds is a by-product from tissue culture and wounding and occurs frequently in woody species where it has been shown to inhibit root formation. Some of the phenolics, like flavonoids, stimulate adventitious rooting, presumably by influencing auxin transport [152]. Other compounds may prevent degradation of auxins by inhibiting decarboxylation [153, 154]. De Klerk et al. [154] tested different phenolic compounds in combination with a suboptimal IAA concentration and found that ferulic acid, a methylated *ortho*-diphenol, had the strongest effect on rooting. For other small molecules like polyamines and ethylene, there are contradicting reports on their root induction capacities [139].

7.2.4.3 Optimization of Micropropagation

In vitro shoots can be derived directly from shoot primordia, via *de novo* shoot morphogenesis or via somatic embryogenesis. Each method requires an appropriate combination of plant growth regulators, which mainly consists of a combination of cytokinin and auxin. The most popular way of micropropagation is via shoot culture, because this method is easy and rapid and does not involve a callus step and hence has a low risk of generating somaclonal variation [155]. The shoots are transplanted on a medium with cytokinins, often combined with a smaller amount of auxin, in order to inhibit apical dominance and to stimulate bud outgrowth [128]. Apical dominance can also be disrupted by manually tipping the shoots or applying an antiauxin [134]. The rate of propagation of a specific species may vary depending on the growth regulator applied. For example, 2-isopentenyladenine (2-iP) is very effective for shoot proliferation in plant species from the *Ericaceae* family [156], while TDZ is highly effective for woody plants [157] or for certain legumes [158].

However, optimal shoot multiplication conditions are sometimes disadvantageous for other processes, such as general fitness or subsequent rooting. Therefore, topolins are in some cases preferred for multiplication, as a number of reports indicate that they have less negative side effects. Werbrouck et al. [159] showed that *meta*-topolin (mT) combined a good axillary shoot production with root formation in micropropagated *Spathiphyllum*. The main metabolite of mT was its *O*-glucoside, which was degraded easily during acclimatization and hence showed no negative effect on rooting during acclimatization. Salvi et al. [160] studied the effect of various cytokinins for *in vitro* propagation of *Curcuma longa*. *meta*-Topolin caused a higher shoot multiplication rate and greener and more robust shoots compared to BA. Bairu et al. [161] reported that mT was the preferred cytokinin both in terms of multiplication rate and rooting of *Aloë polyphylla*. It completely controlled the hyperhydricity problem in this species. Also application of the non-active or less active nucleosides, precursors of the free cytokinin bases, seems to be beneficial in some cases. For example, Baroja-Fernández et al. [162] reported how a low dose of the riboside of mT could stimulate plantlet growth and survival of a weakly growing potato cultivar. Bogaert et al. [163] used a leaf-variegated petunia for micropropagation experiments and found that the use of *meta*-methoxytopolin riboside (MemTR) resulted in the production of a considerable number of new variegated shoots, with only a small number of offtype, fully green or albino shoots, while BA induced many non-variegated shoots. These results suggested that in petunia, MemTR stimulates axillary shoot production in a way that maintains the histogenic integrity of the shoot meristem, whereas BA-derived primordia are mainly adventitious in origin and are composed of cells with a single genotype. Bogaert et al. [163] also demonstrated that MemTR had a much better anti-senescing effect in micropropagated roses than BA, mT, MemT, FmT, or FmTR. Bairu et al. [164] compared the effect of mT, mTR, MemT, and MemTR to BA, on the micropropagation of the banana cultivars 'Williams' and 'Grand Naine'. Superior multiplication rates were recorded for mTR treatments. Plants were categorized as normal or abnormal based upon morphological appearance, and an abnormality index was calculated. Plants multiplied on BA showed the highest abnormality index. On an equimolar concentration of MemTR, the plantlets showed the lowest abnormality index. Hence, topolins or ribosides are valuable compounds to test in the optimization of multiplication protocols. However, the use of these compounds does not guarantee a minimalization of the disadvantageous side effects of cytokinin treatments. Indeed, several studies reported BA or other cytokinins as the least deleterious [165].

If apical dominance cannot be broken, nodes can be used for propagation as each node contains a bud. In this case, it is often unnecessary to add hormones to the medium as shoots elongate easily on hormone-free medium [128]. Alternatively, explants can be propagated through *de novo* or adventitious shoot organogenesis. Adventitious shoot formation has the advantage that it can occur in any type of explant in which a subpopulation of cells dedifferentiates and starts to divide, and hence, it can also be used for the initiation of a culture [128]. The pool of dividing cells forms either a meristemoid mass of cells that directly generates a new shoot or an undifferentiated callus. Not every callus is the same. Some variations in appearance are usually observed as well as variations in regeneration capacity. Organs that originate from callus are identified as indirect and typically show an increased risk in somaclonal variation. Somaclonal variants show inheritable phenotypic changes compared to the mother plant from which the explant is derived and are normally unwanted during mass clonal production of elite plant selections. The frequency at which somaclonal variation occurs depends on the plant species propagated and on the cultivation protocol. In general, indirect shoot organogenesis involves two steps: first the explant is incubated on callus-inducing medium (CIM), usually containing a synthetic auxin, to initiate cell divisions [166], and second, it is incubated on shoot-inducing medium (SIM), usually containing a high cytokinin dose. Similar to direct shoot organogenesis, a subpopulation of cells, here within the callus, becomes competent and develops into shoot progenitor cells. Thus, callus is not a homogeneous mass of cells, and cell–cell interactions are likely contributing to organ initiation [167, 168]. Besides the hormones in the media, many other factors influence the capacity for shoot regeneration. What is generally described as the "quality" of an explant determines whether regeneration is successful. These quality parameters are the age, type, size and position of the explant, period of incubation, and other conditioning factors [169]. Developing a protocol for shoot regeneration is therefore an empirical undertaking.

Although there are some indicators for the acquirement of shoot competence [170, 171], it is still unknown which are the triggers for cells to become competent for shoot regeneration. Interestingly, competence of callus is not determined toward either root or shoot formation. The final identity of the organ emerging from regenerating callus depends on the composition of the medium and can be reversed by changing the auxin-to-cytokinin ratio. Hence, cells acquire organogenesis competence, rather than shoot-forming competence [170, 172], which corresponds with the finding that the initially formed shoot primordium in *Arabidopsis* root explants actually resembles a lateral root meristem [168, 173]. The auxin in the CIM induces the G1-to-S transition of the cell cycle [174] and is critically important for the formation of a lateral root primordium [175]. On the contrary, cytokinin blocks the lateral root development [176] and converses the initial root primordium into a shoot primordium [173]. Depending on the plant species, different cytokinins may induce more efficiently shoot formation. For example, zeatin and 2-iP are better than BA for inducing shoots from *Arabidopsis* root explants, while BA is more efficient for cotyledon explants [177]. Sometimes, phenylurea derivatives give better results than the classical cytokinins [178] or can induce shoot regeneration where classical cytokinins fail (e.g., Reference [158]).

7.2.4.4 Somatic Embryogenesis

The emergence of bipolar structures that resemble somatic embryos was first discovered in carrot cultures [179]. On virtually every type of carrot tissue, it is possible to induce callus that has regenerative capacity. Before the somatic embryos appear, the carrot callus produces cell clumps, proembryogenic masses (PEM), that start to develop when the 2,4-D levels drop or the PEM are cultivated in the absence of 2,4-D. The callus cells that are committed to develop embryos, called embryonic cells, accumulate starch and form a small cluster that only proceeds further to embryos when the auxin is no longer inhibitory. Therefore, the formation of somatic embryos has to be seen as a spontaneous process and not as induced by auxin [180].

Herbaceous plants show a fair chance of producing embryonic callus, but woody species show more recalcitrance. There are major genotype or cultivar differences, and the choice of explant can be of paramount importance for obtaining embryogenic callus [181]. It usually helps to start with juvenile explants, in particular immature embryos, as a suitable source for generating embryos. In analogy to adventitious shoot or root formation, indirect—via an intermediary callus phase—or direct somatic embryos are distinguished. However, direct somatic embryogenesis is rare and does usually not allow mass propagation. A callus, on the contrary, produces numerous small embryoids, which is required for commercial production. Because callus usually cannot be cultured in the absence of an auxin source, it is difficult to dissect the role of auxin in cell proliferation and the induction of embryogenesis. 2,4-Dichlorophenoxyacetic acid is the most commonly used auxin to stimulate callus growth and hence appears frequently in somatic embryogenesis protocols. Antiauxins have also been shown to be effective in inducing somatic embryos in a few cases [134]. In some species, the addition of cytokinins promotes callus growth and consequently is also part of some embryogenesis induction protocols. Moreover, to induce somatic embryogenesis, TDZ can substitute for both auxin and cytokinin [34]. In view of the economic potential of somatic embryogenesis, researchers have tested many kinds of additives, including other hormones and chemicals as well as "nutritive" chemicals [182]. However, so far, a true inducer of somatic embryogenesis has not been reported. Embryogenic callus is usually very compact and consists of small cytoplasmic dense cells. When embryogenic cultures are established, the callus is transferred to hormone-free medium, upon which the embryogenic cells develop into mature somatic embryos. ABA and media with a reduced osmotic potential are often used to arrest embryonic growth and to install dormancy for easier germination. The somatic embryos finally develop into plants in the absence of any growth regulator [183].

7.2.5 CONCLUDING REMARKS

In vitro tissue culture is, besides its use in basic research, of great value for propagating plants. The application of plant growth regulators is essential in many protocols, and since the discovery of the first phytohormone, a lot of different plant hormones have been tested. In the last decade, however, the screening of compounds for tissue culture purposes is rare, and despite the diversity of plant growth regulators already available, most published tissue culture methods use a limited selection. The discovery of new plant hormones such as strigolactone and new chemicals with hormone-like action like pyrabactin supports the idea that the screenings done in the past have not exhausted the realm of useful plant growth regulators. It is up to the tissue culturist to recognize this potential and further explore the chemical world possibly through compound screening and automation of tissue culture methods and by including new model plant species and tissue culture methods.

REFERENCES

1. Went, F. W. (1926). On growth accelerating substances in the coleoptile of *Avena sativa*. *Proceedings of the Section of Sciences, Koninklijke Nederlandse Akademie van Wetenschappen*, *30*, 10–19.
2. Went, F. W. (1928). Wuchsstoff und wachstum. *Recueil des Travaux Botaniques Neerlandais*, *25*, 1–116.
3. Haberlandt, G. (1913). Zur physiologie der zellteilung. *Sitzungsberichte der Königlich Preussischen Akademie der Wissenschaften*, *I*, 318–345.
4. Miller, C. O., et al. (1955). Kinetin, a cell division factor from deoxyribonucleic acid. *Journal of American Chemical Society*, *77*, 1392.
5. Barciszewski, J., et al. (1996). Evidence for the presence of kinetin in DNA and cell extracts. *FEBS Letters*, *393*, 197–200.
6. Raman, N., Elumalai, S. (1996). Presence of cytokinin in the root nodules of *Casuarina equisetifolia*. *Indian Journal of Experimental Biology*, *34*, 577–579.
7. Miller, C. O. (1961). Kinetin and related compounds in plant growth. *Annual Review of Plant Physiology*, *12*, 395–408.
8. Skoog, F., Miller, C. O. (1957). Chemical regulation of growth and organ formation in plant tissue cultured in vitro. *Symposia of the Society for Experimental Biology*, *11*, 118–131.
9. Park, S.-Y., et al. (2009). Abscisic acid inhibits type 2C protein phosphatases via the PYR/PYL family of START proteins. *Science*, *324*, 1068–1071.
10. Umehara, M., et al. (2008). Inhibition of shoot branching by new terpenoid plant hormones. *Nature*, *455*, 195–200.

11. Mok, M. C., *Cytokinins and plant development – an overview*. In: Mok, D. W. S., Mok, M. C., editors. Cytokinins: Chemistry, activity and function, CRC Press, Boca Raton, 1994, 155–166.
12. Horgan, R., et al. (1975). A new cytokinin from *Populus x robusta*. *Phytochemistry*, *14*, 1005–1008.
13. Ge, L., et al. (2004). Analysis of some cytokinins in coconut (*Cocos nucifera* L.) water by micellar electrokinetic capillary chromatography after solid-phase extraction. *Journal of Chromatography A*, *1048*, 119–126.
14. Ge, L., et al. (2005). Analysis of positional isomers of hydroxylated aromatic cytokinins by micellar electrokinetic chromatography. *Electrophoresis*, *26*, 1768–1777.
15. Sakakibara, H. (2006). Cytokinins: Activity, biosynthesis, and translocation. *Annual Review of Plant Biology*, *57*, 431–449.
16. Spíchal, L., et al. (2004). Two cytokinin receptors of *Arabidopsis thaliana*, CRE1/AHK4 and AHK3, differ in their ligand specificity in a bacterial assay. *Plant and Cell Physiology*, *45*, 1299–1305.
17. Hou, B., et al. (2004). *N*-glucosylation of cytokinins by glycosyltransferases of *Arabidopsis thaliana*. *Journal of Biological Chemistry*, *279*, 47822–47832.
18. Dixon, S. C., et al. (1989). Zeatin glycosylation enzymes in Phaseolus: Isolation of *O*-glucosyltransferase from *P. Lunatus* and comparison to *O*-xylosyltransferase from *P. Vulgaris*. *Plant Physiology*, *90*, 1316–1321.
19. Shantz, E. M., Steward, F. C. (1955). The identification of compound A from coconut milk as 1,3-diphenylurea. *Journal of American Chemical Society*, *74*, 6133–6135.
20. Ricci, A., Bertoletti, C. (2009). Urea derivatives on the move: Cytokinin-like activity and adventitious rooting enhancement depend on chemical structure. *Plant Biology*, *11*, 262–272.
21. Bruce, M. I., Zwar, J. A. (1966). Cytokinin activity of some substituted areas and thioureas. *Proceedings of the Royal Society Series B: Biological Sciences*, *165*, 245–265.
22. Bruce, M. I., Zwar, J. A., Kefford, N. P. (1965). Chemical structure and plant kinin activity – the activity of urea and thiourea derivatives. *Life Sciences*, *4*, 461–466.
23. Okamoto, T., et al. (1981). 4-Pyridylureas are surprisingly potent cytokinins – the structure–activity relationship. *Chemical & Pharmaceutical Bulletin*, *29*, 3748–3750.
24. Takahashi, S., et al. (1978). Cytokinin activity of *N*-phenyl-*N'*-(4-pyridyl)urea derivatives. *Phytochemistry*, *17*, 1201–1207.
25. Arndt, F., Rusch, R., Stilfried, H. V. (1976). SN 49537, a new cotton defoliant. *Plant Physiology*, *57*, 99.
26. Mok, M. C., et al. (1982). Cytokinin activity of *N*-phenyl-*N'*-1,2,3-thiadiazol-5-ylurea (thidiazuron). *Phytochemistry*, *21*, 1509–1511.
27. Ricci, A., et al. (2004). Effect of Cl-substitution on rooting-or cytokinin-like activity of diphenylurea derivatives. *Journal of Plant Growth Regulation*, *23*, 261–268.
28. Ricci, A., et al. (2001). Cytokinin-like activity of *N'*-substituted *N*-phenylureas. *Plant Growth Regulation*, *34*, 167–172.
29. Yonova, P. A., Stoilkova, G. M. (2004). Synthesis and biological activity of urea and thiourea derivatives from 2-aminoheterocyclic compounds. *Journal of Plant Growth Regulation*, *23*, 280–291.
30. Song, X. J., Tan, X. H., Wang, Y. G. (2007). Synthesis and biological activity of some novel *N*-aryl-*N'*-[5-(pyrid-4-yl)-1,3,4-thiadiazol-2-yl]ureas. *Phosphorus Sulfur and Silicon and the Related Elements*, *182*, 1907–1913.

31. Ku, Y. G., et al. (2010). Affecting *Asparagus officinalis* shoot and root growth characteristics with CPPU foliar sprays and soil drench. *Korean Journal of Horticultural Science & Technology, 28,* 167–171.

32. Suttle, J. C. (2008). Effects of synthetic phenylurea and nitroguanidine cytokinins on dormancy break and sprout growth in *Russet Burbank* minitubers. *American Journal of Potato Research, 85,* 121–128.

33. Kapchina-Toteva, V., van Telgen, H. J., Yakimova, E. (2000). Role of phenylurea cytokinin CPPU in apical dominance release in in vitro cultured *Rosa hybrida* L. *Journal of Plant Growth Regulation, 19,* 232–237.

34. Murthy, B. N. S., Murch, S. J., Saxena, P. K. (1998). Thidiazuron: A potent regulator of *in vitro* plant morphogenesis. *In Vitro Cellular & Developmental Biology – Plant, 34,* 267–275.

35. Carra, A., et al. (2006). Diphenylurea derivatives induce somatic embryogenesis in citrus. *Plant Cell Tissue and Organ Culture, 87,* 41–48.

36. Ricci, A., et al. (2006). Diheteroarylurea derivatives as adventitious rooting adjuvants in mung bean shoots and M26 apple rootstock. *Plant Growth Regulation, 50,* 201–209.

37. Mok, M. C., et al. (1979). Induction of cytokinin autonomy by *N,N′*-diphenylurea in tissue cultures of *Phaseolus lunatus* L. *Proceedings of the National Academy of Sciences of the United States of America, 76,* 3880–3884.

38. Chatfield, J. M., Armstrong, D. J. (1986). Regulation of cytokinin oxidase activity in callus tissues of *Phaseolus vulgaris* L. cv Great Northern. *Plant Physiology, 80,* 493–499.

39. Laloue, M., Fox, J. E. (1989). Cytokinin oxidase from wheat: Partial purification and general properties. *Plant Physiology, 90,* 899–906.

40. Kopečný, D., et al. (2010). Phenyl- and benzylurea cytokinins as competitive inhibitors of cytokinin oxidase/dehydrogenase: A structural study. *Biochimie, 92,* 1052–1062.

41. Yamada, H., et al. (2001). The Arabidopsis AHK4 histidine kinase is a cytokinin-binding receptor that transduces cytokinin signals across the membrane. *Plant and Cell Physiology, 42,* 1017–1023.

42. Hothorn, M., Dabi, T., Chory, J. (2011). Structural basis for cytokinin recognition by *Arabidopsis thaliana* histidine kinase 4. *Nature Chemical Biology, 7,* 766–768.

43. Skoog, F., et al. (1967). Cytokinins – structure/activity relationships. *Phytochemistry, 6,* 1169–1192.

44. Doležal, K., et al. (2007). Preparation, biological activity and endogenous occurrence of N^6-benzyladenosines. *Bioorganic & Medicinal Chemistry, 15,* 3737–3747.

45. Doležal, K., et al. (2006). Preparation and biological activity of 6-benzylaminopurine derivatives in plants and human cancer cells. *Bioorganic & Medicinal Chemistry, 14,* 875–884.

46. Holub, J., et al. (1998). Biological activity of cytokinins derived from *ortho-* and *meta-*hydroxybenzyladenine. *Plant Growth Regulation, 26,* 109–115.

47. Szüčová, L., et al. (2009). Synthesis, characterization and biological activity of ring-substituted 6-benzylamino-9-tetrahydropyran-2-yl and 9-tetrahydrofuran-2-ylpurine derivatives. *Bioorganic & Medicinal Chemistry, 17,* 1938–1947.

48. Zatloukal, M., et al. (2008). Novel potent inhibitors of *A. thaliana* cytokinin oxidase/dehydrogenase. *Bioorganic & Medicinal Chemistry, 16,* 9268–9275.

49. Marival-Hodebar, L., et al. (1999). Fluorinated isopentenyladenines: Synthesis and cytokinin activity. *Israel Journal of Chemistry*, *39*, 155–161.
50. Haidoune, M., et al. (1998). Synthesis and cytokinin activity of new zeatin derivatives. *Journal of Agricultural and Food Chemistry*, *46*, 1577–1582.
51. Suttle, J. C., Mornet, R. (2005). Mechanism-based irreversible inhibitors of cytokinin dehydrogenase. *Journal of Plant Physiology*, *162*, 1189–1196.
52. Tao, G. Q., et al. (1991). Inhibitors of cytokinin metabolism III. The inhibition of cytokinin *N*-glucosylation in radish cotyledons. *Journal of Plant Growth Regulation*, *10*, 179–185.
53. Spíchal, L., et al. (2009). The purine derivative PI-55 blocks cytokinin action via receptor inhibition. *FEBS Journal*, *276*, 244–253.
54. Nisler, J., et al. (2010). Cytokinin receptor antagonists derived from 6-benzylaminopurine. *Phytochemistry*, *71*, 823–830.
55. Arata, Y., et al. (2010). The phenylquinazoline compound S-4893 is a non-competitive cytokinin antagonist that targets Arabidopsis cytokinin receptor CRE1 and promotes root growth in Arabidopsis and rice. *Plant and Cell Physiology*, *51*, 2047–2059.
56. Iwamura, H., et al. (1975). Synthesis and anticytokinin activity of 4-substituted-7-(β-D-ribofuranosyl)-pyrrolo[2,3-D]pyrimidines. *Phytochemistry*, *14*, 2317–2321.
57. Skoog, F., et al. (1975). Anticytokinin activity of substituted pyrrolo[2,3-D]pyrimidines. *Proceedings of the National Academy of Sciences of the United States of America*, *72*, 3508–3512.
58. Iwamura, H., et al. (1974). Anticytokinin activity of 4-furfurylamino-7-(β-D-ribofuranosyl) pyrrolo[2,3-D] pyrimidine. *Biochemical and Biophysical Research Communications*, *57*, 412–416.
59. Skoog, F., et al. (1973). Cytokinin antagonists – synthesis and physiological effects of 7-substituted 3-methylpyrazolo[4,3-D]pyrimidines. *Phytochemistry*, *12*, 25–37.
60. Hecht, S. M., et al. (1971). Cytokinins – development of a potent antagonist. *Proceedings of the National Academy of Sciences of the United States of America*, *68*, 2608–2610.
61. Spíchal, L., et al. (2007). Classical anticytokinins do not interact with cytokinin receptors but inhibit cyclin-dependent kinases. *Journal of Biological Chemistry*, *282*, 14356–14363.
62. Veselý, J., et al. (1994). Inhibition of cyclin-dependent kinases by purine analogs. *European Journal of Biochemistry*, *224*, 771–786.
63. Sharma, P. S., Sharma, R., Tyagi, R. (2008). Inhibitors of cyclin dependent kinases: Useful targets for cancer treatment. *Current Cancer Drug Targets*, *8*, 53–75.
64. DeAzevedo, W. F., et al. (1997). Inhibition of cyclin-dependent kinases by purine analogues – crystal structure of human cdk2 complexed with roscovitine. *European Journal of Biochemistry*, *243*, 518–526.
65. Havlíček, L., et al. (1997). Cytokinin-derived cyclin-dependent kinase inhibitors: Synthesis and cdc2 inhibitory activity of olomoucine and related compounds. *Journal of Medicinal Chemistry*, *40*, 408–412.
66. Kryštof, V., et al. (2002). Synthesis and biological activity of olomoucine II. *Bioorganic & Medicinal Chemistry Letters*, *12*, 3283–3286.
67. Murai, N., et al. (1980). Relationships between cytokinin production, presence of plasmids, and fasciation caused by strains of *Corynebacterium fascians*. *Proceedings of the*

National Academy of Sciences of the United States of America: Biological Sciences, *77*, 619–623.

68. Armstrong, D. J., et al. (1976). Cytokinins in *Corynebacterium fascians* cultures – isolation and identification of 6-(4-hydroxy-3-methyl-cis-2-butenylamino)-2-methylthiopurine. *Plant Physiology*, *58*, 749–752.
69. Vereecke, D., et al. (2000). The *Rhodococcus fascians*–plant interaction: Morphological traits and biotechnological applications. *Planta*, *210*, 241–251.
70. Joshi, M. V., Loria, R. (2007). *Streptomyces turgidiscabies* possesses a functional cytokinin biosynthetic pathway and produces leafy galls. *Molecular Plant–Microbe Interactions*, *20*, 751–758.
71. Omer, Z. S., et al. (2004). 5′-Deoxyisopentenyladenosine and other cytokinins in culture filtrates of the bacterium *Pantoea agglomerans*. *Physiologia Plantarum*, *121*, 439–447.
72. Evidente, A., et al. (1989). 2′-Deoxyzeatin riboside and other cytokinins in culture filtrates of *Pseudomonas amygdali*. *Phytochemistry*, *28*, 2603–2607.
73. Surico, G., et al. (1985). A cytokinin from the culture filtrate of *Pseudomonas syringae pv. savastanoi*. *Phytochemistry*, *24*, 1499–1502.
74. Sassa, T., et al. (1998). Structural confirmation of cotylenin A, a novel fusicoccane-diterpene glycoside with potent plant growth-regulating activity from *Cladosporium fungus sp. 501-7W*. *Bioscience Biotechnology and Biochemistry*, *62*, 1815–1818.
75. Sassa, T., Tojyo, T., Munakata, K. (1970). Isolation of a new plant growth substance with cytokinin-like activity. *Nature*, *227*, 379.
76. Yamamoto-Yamaguchi, Y., et al. (2001). Induction of the monocytic differentiation of myeloid leukaemia cells by cotylenin A, a plant growth regulator. *British Journal of Haematology*, *112*, 697–705.
77. Vanneste, S., Friml, J. (2009). Auxin: A trigger for change in plant development. *Cell*, *136*, 1005–1016.
78. Hagen, G., Guilfoyle, T. (2002). Auxin-responsive gene expression: Genes, promoters and regulatory factors. *Plant Molecular Biology*, *49*, 373–385.
79. Debergh, P. C., Read, P. E., *Micropropagation*. In: Debergh, P. C., Zimmerman, R. H., editors. Micropropagation: Technology and application, Kluwer Academic Publishers, Dordrecht, 1991, 1–14.
80. Koepfli, J. B., Thimann, K. V., Went, F. W. (1938). Phytohormones: Structure and physiological activity. I. *Journal of Biological Chemistry*, *122*, 763–780.
81. Thimann, K. V. (1958). Auxin activity of some indole derivatives. *Plant Physiology*, *33*, 311–321.
82. Robison, M. M., Robison, B. L. (1956). 7-Azaindole. 3. Syntheses of 7-aza analogs of some biologically significant indole derivatives. *Journal of the American Chemical Society*, *78*, 1247–1251.
83. Hamilton, R. H., Kivilaan, A., McManus, J. M. (1960). Biological activity of tetrazole analogues of indole-3-acetic acid and 2,4-dichlorophenoxyacetic acid. *Plant Physiology*, *35*, 136–140.
84. Tan, X., et al. (2007). Mechanism of auxin perception by the TIR1 ubiquitin ligase. *Nature*, *446*, 640–645.
85. Dharmasiri, N., Dharmasiri, S., Estelle, M. (2005). The F-box protein TIR1 is an auxin receptor. *Nature*, *435*, 441–445.

86. Kepinski, S., Leyser, O. (2005). The *Arabidopsis* F-box protein TIR1 is an auxin receptor. *Nature*, *435*, 446–451.
87. Dharmasiri, N., et al. (2005). Plant development is regulated by a family of auxin receptor F box proteins. *Developmental Cell*, *9*, 109–119.
88. Greenham, K., et al. (2011). The AFB4 auxin receptor is a negative regulator of auxin signaling in seedlings. *Current Biology*, *21*, 520–525.
89. Calderon Villalobos, L. I. A., et al. (2012). A combinatorial TIR1/AFB-Aux/IAA co-receptor system for differential sensing of auxin. *Nature Chemical Biology*, *8*, 477–485.
90. Braun, N., et al. (2008). Conditional repression of AUXIN BINDING PROTEIN1 reveals that it coordinates cell division and cell expansion during postembryonic shoot development in *Arabidopsis* and tobacco. *The Plant Cell*, *20*, 2746–2762.
91. Robert, S., et al. (2010). ABP1 mediates auxin inhibition of clathrin-dependent endocytosis in *Arabidopsis*. *Cell*, *143*, 111–121.
92. Xu, T., et al. (2010). Cell surface- and Rho GTPase-based auxin signaling controls cellular interdigitation in *Arabidopsis*. *Cell*, *143*, 99–110.
93. Yamagami, M., et al. (2004). Two distinct signaling pathways participate in auxin-induced swelling of pea epidermal protoplasts. *Plant Physiology*, *134*, 735–747.
94. Jurado, S., et al. (2010). The *Arabidopsis* cell cycle F-box protein SKP2A binds to auxin. *The Plant Cell*, *22*, 3891–3904.
95. Ferro, N., et al. (2010). Route to novel auxin: Auxin chemical space toward biological correlation carriers. *Chemical Reviews*, *110*, 4690–4708.
96. Chapman, E. J., Estelle, M. (2009). Mechanism of auxin-regulated gene expression in plants. *Annual Review of Genetics*, *43*, 265–285.
97. Del Bianco, M., Kepinski, S. (2011). Context, specificity, and self-organization in auxin response. *Cold Spring Harbor Perspectives in Biology*, *3*, a001578.
98. Tam, Y. Y., Epstein, E., Normanly, J. (2000). Characterization of auxin conjugates in Arabidopsis. Low steady-state levels of indole-9-acetyl-aspartate indole-3-acetyl-glutamate, and indole-3-acetyl-glucose. *Plant Physiology*, *123*, 589–595.
99. Bajguz, A., Piotrowska, A. (2009). Conjugates of auxin and cytokinin. *Phytochemistry*, *70*, 957–969.
100. Ludwig-Müller, J. (2011). Auxin conjugates: Their role for plant development and in the evolution of land plants. *Journal of Experimental Botany*, *62*, 1757–1773.
101. Wightman, F., Lighty, D. L. (1982). Identification of phenylacetic acid as a natural auxin in the shoots of higher plants. *Physiologia Plantarum*, *55*, 17–24.
102. Epstein, E., Ludwig-Müller, J. (1993). Indole-3-butyric acid in plants – occurrence, synthesis, metabolism and transport. *Physiologia Plantarum*, *88*, 382–389.
103. Marumo, S., et al. (1968). Isolation of a novel auxin methyl 4-chloroindoleacetate from immature seeds of *Pisum sativum*. *Agricultural and Biological Chemistry*, *32*, 117.
104. Grossmann, K. (2009). Auxin herbicides: Current status of mechanism and mode of action. *Pest Management Science*, *66*, 113–120.
105. Geldner, N., et al. (2001). Auxin transport inhibitors block PIN1 cycling and vesicle trafficking. *Nature*, *413*, 425–428.
106. Nagashima, A., Uehara, Y., Sakai, T. (2008). The ABC subfamily B auxin transporter AtABCB19 is involved in the inhibitory effects of *N*-1-naphthyphthalamic acid on the

phototropic and gravitropic responses of *Arabidopsis* hypocotyls. *Plant and Cell Physiology*, *49*, 1250–1255.

107. Geisler, M., et al. (2005). Cellular efflux of auxin catalyzed by the *Arabidopsis* MDR/PGP transporter AtPGP1. *The Plant Journal*, *44*, 179–194.
108. Tsuda, E., et al. (2011). Alkoxy-auxins are selective inhibitors of auxin transport mediated by PIN, ABCB, and AUX1 transporters. *Journal of Biological Chemistry*, *286*, 2354–2364.
109. Hayashi, K., et al. (2008). Small-molecule agonists and antagonists of F-box protein–substrate interactions in auxin perception and signaling. *Proceedings of the National Academy of Sciences of the United States of America*, *105*, 5632–5637.
110. Tran Thanh Van, K. M. (1981). Control of morphogenesis in in vitro cultures. *Annual Review of Plant Physiology*, *32*, 291–311.
111. Newcomb, W., Wetherel, D. F. (1970). Effects of 2,4,6-trichlorophenoxyacetic acid on embryogenesis in wild carrot tissue cultures. *Botanical Gazette*, *131*, 242–245.
112. Fujimura, T., Komamine, A. (1979). Synchronization of somatic embryogenesis in a carrot cell-suspension culture. *Plant Physiology*, *64*, 162–164.
113. Christian, M., et al. (2008). Identification of auxins by a chemical genomics approach. *Journal of Experimental Botany*, *59*, 2757–2767.
114. Savaldi-Goldstein, S., et al. (2008). New auxin analogs with growth-promoting effects in intact plants reveal a chemical strategy to improve hormone delivery. *Proceedings of the National Academy of Sciences of the United States of America*, *105*, 15190–15195.
115. Armstrong, J. I., et al. (2004). Identification of inhibitors of auxin transcriptional activation by means of chemical genetics in *Arabidopsis*. *Proceedings of the National Academy of Sciences of the United States of America*, *101*, 14978–14983.
116. Sungur, C., et al. (2007). The small molecule 2-furylacrylic acid inhibits auxin-mediated responses in *Arabidopsis thaliana*. *Plant and Cell Physiology*, *48*, 1693–1701.
117. Grozinger, C. M., et al. (2001). Identification of a class of small molecule inhibitors of the sirtuin family of NAD-dependent deacetylases by phenotypic screening. *Journal of Biological Chemistry*, *276*, 38837–38843.
118. Kim, J.-Y., et al. (2010). Identification of an ABCB/P-glycoprotein-specific inhibitor of auxin transport by chemical genomics. *Journal of Biological Chemistry*, *285*, 23309–23317.
119. Meijer, E. A., de Vries, S. C., Mordhorst, A. P. (1999). Co-culture with *Daucus carota* somatic embryos reveals high 2,4-D uptake and release rates of *Arabidopsis thaliana* cultured cells. *Plant Cell Reports*, *18*, 656–663.
120. Delbarre, A., et al. (1996). Comparison of mechanisms controlling uptake and accumulation of 2,4-dichlorophenoxy acetic acid, naphthalene-1-acetic acid, and indole-3-acetic acid in suspension-cultured tobacco cells. *Planta*, *198*, 532–541.
121. Grossmann, K. (2005). What it takes to get a herbicide's mode of action. Physionomics, a classical approach in a new complexion. *Pest Management Science*, *61*, 423–431.
122. Gray, D. J., Conger, B. V. (1985). Influence of dicamba and casein hydrolysate on somatic embryo number and culture quality in cell-suspensions of *Dactylis glomerata* (*Gramineae*). *Plant Cell Tissue and Organ Culture*, *4*, 123–133.
123. Heyser, J. W., et al. (1983). High-frequency, long-term regeneration of rice from callus-culture. *Plant Science Letters*, *29*, 175–182.

124. Beyl, C. A., Sharma, G. C. (1983). Picloram induced somatic embryogenesis in *Gasteria* and *Haworthia*. *Plant Cell Tissue and Organ Culture*, *2*, 123–132.
125. Meins, F. (1989). Habituation – heritable variation in the requirement of cultured plant-cells for hormones. *Annual Review of Genetics*, *23*, 395–408.
126. Pischke, M. S., et al. (2006). A transcriptome-based characterization of habituation in plant tissue culture. *Plant Physiology*, *140*, 1255–1278.
127. Suttle, J. C. (1988). Disruption of the polar auxin transport-system in cotton seedlings following treatment with the defoliant thidiazuron. *Plant Physiology*, *86*, 241–245.
128. George, E. F., Debergh, P. C., *Micropropagation: Uses and methods*. In: George, E. F., Hall, M. A., De Klerk, G. J., editors. Plant propagation by tissue culture, Springer, Dordrecht, 2008, 29–64.
129. Smulders, M. J. M., de Klerk, G. J. (2011). Epigenetics in plant tissue culture. *Plant Growth Regulation*, *63*, 137–146.
130. Bayliss, M. W. (1977). Effects of 2,4-D on growth and mitosis in suspension cultures of *Daucus carota*. *Plant Science Letters*, *8*, 99–103.
131. Ali, M. N., Ahmad, W. (1996). Genotoxicity of 2,4-dichlorophenoxyacetic acid (2,4-D) on filarial vector, *Culex pipiens fatigans*. *Journal of Environmental Biology*, *17*, 331–337.
132. Kubalakova, M., Dolezel, J., Lebeda, A. (1996). Ploidy instability of embryogenic cucumber (*Cucumis sativus* L.) callus culture. *Biologia Plantarum*, *38*, 475–480.
133. Kasha, K. J., et al. (2006). Haploid production and chromosome doubling. *Acta Horticulturae*, *725*, 817–827.
134. Machakova, I., Zazimalova, E., George, E. F., *Plant growth regulators I: Introduction: auxins, their analogues and inhibitors*. In: George, E. F., Hall, M. A., De Klerk, G. J., editors. Plant propagation by tissue culture, Springer, Dordrecht, 2008, 175–204.
135. Irvine, J. E., Fitch, M., Moore, P. H. (1983). The induction of callus in sugarcane tissue-cultures by selected chemicals. *Plant Cell Tissue and Organ Culture*, *2*, 141–149.
136. Soniya, E. V., Banerjee, N. S., Das, M. R. (2001). Genetic analysis of somaclonal variation among callus-derived plants of tomato. *Current Science*, *80*, 1213–1215.
137. Casimiro, I., et al. (2001). Auxin transport promotes Arabidopsis lateral root initiation. *The Plant Cell*, *13*, 843–852.
138. Bartel, B., LeClere, S., Magidin, M., Zolman, B. K. (2001). Inputs to the active indole-3-acetic acid pool: *De novo* synthesis, conjugate hydrolysis, and indole-3-butyric acid β-oxidation. *Journal of Plant Growth Regulation*, *20*, 198–216.
139. Li, S. W., et al. (2009). Mediators, genes and signaling in adventitious rooting. *Botanical Review*, *75*, 230–247.
140. Werbrouck, S. P. O., et al. (1995). The metabolism of benzyladenine in *Spathiphyllum floribundum* 'Schott Petite' in relation to acclimatization problems. *Plant Cell Reports*, *14*, 662–665.
141. King, J. J., Stimart, D. P. (1998). Genetic analysis of variation for auxin-induced adventitious root formation among eighteen ecotypes of *Arabidopsis thaliana L. heynh*. *Journal of Heredity*, *89*, 481–487.
142. Marques, C. M., et al. (1999). Genetic dissection of vegetative propagation traits in *Eucalyptus tereticornis* and *E. globulus*. *Theoretical and Applied Genetics*, *99*, 936–946.

143. Han, K. H., Bradshaw, H. D., Gordon, M. P. (1994). Adventitious root and shoot regeneration *in vitro* is under major gene control in an F2 family of hybrid poplar (*Populus trichocarpa x deltoides*). *Forest Genetics*, *1*, 139–146.

144. Shepherd, M., et al. (2008). Mapping species differences for adventitious rooting in a *Corymbia torelliana x Corymbia citriodora* subspecies variegata hybrid. *Tree Genetics & Genomes*, *4*, 715–725.

145. Shepherd, M., et al. (2006). Congruence in QTL for adventitious rooting in *Pinus elliottii x Pinus caribaea* hybrids resolves between and within-species effects. *Molecular Breeding*, *18*, 11–28.

146. Ochoa, I. E., Blair, M. W., Lynch, J. P. (2006). QTL analysis of adventitious root formation in common bean under contrasting phosphorus availability. *Crop Science*, *46*, 1609–1621.

147. Horii, H., et al. (2006). Quantitative trait loci for adventitious and lateral roots in rice. *Plant Breeding*, *125*, 198–200.

148. Mano, Y., et al. (2005). QTL mapping of adventitious root formation under flooding conditions in tropical maize (*Zea mays* L.) seedlings. *Breeding Science*, *55*, 343–347.

149. Mano, Y., et al. (2005). Identification of QTL controlling adventitious root formation during flooding conditions in teosinte (*Zea mays* ssp. *huehuetenangensis*) seedlings. *Euphytica*, *142*, 33–42.

150. Zheng, B. S., et al. (2003). Mapping QTLs and candidate genes for rice root traits under different water-supply conditions and comparative analysis across three populations. *Theoretical and Applied Genetics*, *107*, 1505–1515.

151. Price, A. H., Tomos, A. D. (1997). Genetic dissection of root growth in rice (*Oryza sativa* L.). II. Mapping quantitative trait loci using molecular markers. *Theoretical and Applied Genetics*, *95*, 143–152.

152. Murphy, A., Peer, W. A., Taiz, L. (2000). Regulation of auxin transport by aminopeptidases and endogenous flavonoids. *Planta*, *211*, 315–324.

153. Wilson, P. J., van Staden, J. (1990). Rhizocaline, rooting cofactors, and the concept of promoters and inhibitors of adventitious rooting – a review. *Annals of Botany*, *66*, 479–490.

154. De Klerk, G. J., et al. (2011). Effects of phenolic compounds on adventitious root formation and oxidative decarboxylation of applied indoleacetic acid in Malus 'Jork 9.' *Plant Growth Regulation*, *63*, 175–185.

155. Piccioni, E., et al. (1997). Estimating alfalfa somaclonal variation in axillary branching propagation and indirect somatic embryogenesis by RAPD fingerprinting. *International Journal of Plant Sciences*, *158*, 556–562.

156. Norton, M. E., Norton, C. R. (1985). In vitro propagation of *Ericaceae*: A comparison of the activity of the cytokinins N^6-benzyladenine and N^6-isopentenyladenine in shoot proliferation. *Scientia Horticulturae*, *27*, 335–340.

157. Huetteman, C. A., Preece, J. E. (1993). Thidiazuron – a potent cytokinin for woody plant-tissue culture. *Plant Cell Tissue and Organ Culture*, *33*, 105–119.

158. Malik, K. A., Saxena, P. K. (1992). Thidiazuron induces high-frequency shoot regeneration in intact seedlings of pea (*Pisum sativum*), chickpea (*Cicer arietinum*) and lentil (*Lens culinaris*). *Australian Journal of Plant Physiology*, *19*, 731–740.

159. Werbrouck, S. P. O., et al. (1996). Meta-topolin, an alternative to benzyladenine in tissue culture? *Physiologia Plantarum*, *98*, 291–297.

160. Salvi, N. D., George, L., Eapen, S. (2002). Micropropagation and field evaluation of micropropagated plants of turmeric. *Plant Cell, Tissue and Organ Culture*, *68*, 143–151.

161. Bairu, M. W., et al. (2007). Optimizing the micropropagation protocol for the endangered *Aloe polyphylla*: Can meta-topolin and its derivatives serve as replacement for benzyladenine and zeatin? *Plant Cell Tissue and Organ Culture*, *90*, 15–23.

162. Baroja-Fernández, E., et al. (2002). Aromatic cytokinins in micropropagated potato plants. *Plant Physiology and Biochemistry*, *40*, 217–224.

163. Bogaert, I., et al. (2006). New aromatic cytokinins can make the difference. *Acta Horticulturae*, *725*, 265.

164. Bairu, M. W., et al. (2008). The role of topolins in micropropagation and somaclonal variation of banana cultivars 'Williams' and 'Grand Naine' (*Musa* spp. AAA). *Plant Cell Tissue and Organ Culture*, *95*, 373–379.

165. Aremu, A. O., et al. (2012). Topolins: A panacea to plant tissue culture challenges? *Plant Cell Tissue and Organ Culture*, *108*, 1–16.

166. Cary, A. J., Che, P., Howell, S. H. (2002). Developmental events and shoot apical meristem gene expression patterns during shoot development in *Arabidopsis thaliana*. *The Plant Journal*, *32*, 867–877.

167. Gordon, S. P., et al. (2007). Pattern formation during de novo assembly of the *Arabidopsis* shoot meristem. *Development*, *134*, 3539–3548.

168. Sugimoto, K., Jiao, Y. L., Meyerowitz, E. M. (2010). *Arabidopsis* regeneration from multiple tissues occurs via a root development pathway. *Developmental Cell*, *18*, 463–471.

169. Gahan, P. B., George, E. F., *Adventitious regeneration*. In: George, E. F., Hall, M. A., De Klerk, G. J., editors. Plant propagation by tissue culture, Springer, Dordrecht, 2008, 355–402.

170. Motte, H., et al. (2011). *CUC2* as an early marker for regeneration competence in *Arabidopsis* root explants. *Journal of Plant Physiology*, *168*, 1598–1601.

171. Ochatt, S. J., et al. (2010). Competence versus recalcitrance for *in vitro* regeneration. *Notulae Botanicae Horti Agrobotanici Cluj-Napoca*, *38*, 102–108.

172. Pernisová, M., et al. (2009). Cytokinins modulate auxin-induced organogenesis in plants via regulation of the auxin efflux. *Proceedings of the National Academy of Sciences of the United States of America*, *106*, 3609–3614.

173. Atta, R., et al. (2008). Pluripotency of Arabidopsis xylem pericycle underlies shoot regeneration from root and hypocotyl explants grown *in vitro*. *The Plant Journal*, *57*, 626–644.

174. Dubrovsky, et al. (2008). Auxin acts as a local morphogenetic trigger to specify lateral root founder cells. *Proceedings of the National Academy of Sciences of the United States of America*, *105*, 8790–8794.

175. Himanen, K., et al. (2002). Auxin-mediated cell cycle activation during early lateral root initiation. *The Plant Cell*, *14*, 2339–2351.

176. Laplaze, L., et al. (2007). Cytokinins act directly on lateral root founder cells to inhibit root initiation. *The Plant Cell*, *19*, 3889–3900.

177. Zhao, Q. H., Fisher, R., Auer, C. (2002). Developmental phases and *STM* expression during *Arabidopsis* shoot organogenesis. *Plant Growth Regulation*, *37*, 223–231.

178. Read, P. E., Fellman, C. D., Zimmerman, E. (1986). New plant growth regulators for cuttings and for tissue culture. *Proceedings of the International Plant Propagator's Society*, *35*, 78–84.

179. Wetherell, D. F., Halperin, W. (1963). Embryos derived from callus tissue cultures of wild carrot. *Nature*, *200*, 1336–1337.

180. Thorpe, T. A., *In vitro embryogenesis in plants*, Kluwer, Dordrecht, 1995, 5–14.

181. Lo Schiavo, F., *Early events in embryogenesis*. In: Bajaj, Y. P. S., editor. *Biotechnology in agriculture and forestry.* Somatic embryogenesis and synthetic seed I., Springer-Verlag, Berlin, 1995, 20–29.

182. Jimenez, V. M. (2005). Involvement of plant hormones and plant growth regulators on *in vitro* somatic embryogenesis. *Plant Growth Regulation*, *47*, 91–110.

183. Von Arnold, S., *Somatic embryogenesis*. In: George, E. F., Hall, M. A., De Klerk, G. J., editors. Plant propagation by tissue culture, Springer, Dordrecht, 2008, 335–354.

INDEX

Note: Page numbers in *italics* refer to Figures; those in **bold** to Tables.

Plant Chemical Biology, First Edition. Edited by Dominique Audenaert and Paul Overvoorde.